Numerical Integration

NATO ASI Series

Advanced Science Institutes Series

A Series presenting the results of activities sponsored by the NATO Science Committee, which aims at the dissemination of advanced scientific and technological knowledge, with a view to strengthening links between scientific communities.

The Series is published by an international board of publishers in conjunction with the NATO Scientific Affairs Division

A Life Sciences	Plenum Publishing Corporation
B Physics	London and New York
C Mathematical and Physical Sciences	Kluwer Academic Publishers
	Dordrecht, Boston and London
D Behavioural and Social Sciences	
E Applied Sciences	
F Computer and Systems Sciences	Springer-Verlag
G Ecological Sciences	Berlin, Heidelberg, New York, London,
H Cell Biology	Paris and Tokyo
I Global Environmental Change	

NATO-PCO-DATA BASE

The electronic index to the NATO ASI Series provides full bibliographical references (with keywords and/or abstracts) to more than 30000 contributions from international scientists published in all sections of the NATO ASI Series.
Access to the NATO-PCO-DATA BASE is possible in two ways:

– via online FILE 128 (NATO-PCO-DATA BASE) hosted by ESRIN,
Via Galileo Galilei, I-00044 Frascati, Italy.

– via CD-ROM "NATO-PCO-DATA BASE" with user-friendly retrieval software in English, French and German (© WTV GmbH and DATAWARE Technologies Inc. 1989).

The CD-ROM can be ordered through any member of the Board of Publishers or through NATO-PCO, Overijse, Belgium.

Series C: Mathematical and Physical Sciences - Vol. 357

Numerical Integration
Recent Developments, Software and Applications

edited by

Terje O. Espelid
Department of Informatics,
University of Bergen, Bergen, Norway

and

Alan Genz
School of Electrical Engineering and Computer Science,
Washington State University, Pullman, WA, U.S.A.

SPRINGER-SCIENCE+BUSINESS MEDIA, B.V.

Proceedings of the NATO Advanced Research Workshop on
Numerical Integration: Recent Developments, Software and Applications
Bergen, Norway
June 17–21, 1991

ISBN 978-94-010-5169-9   ISBN 978-94-011-2646-5 (eBook)

DOI 10.1007/978-94-011-2646-5

Printed on acid-free paper

Dedicated to

JAMES N. LYNESS

on the Occasion of his Sixtieth Birthday

October 16, 1992

TABLE OF CONTENTS

Preface   xi

Numerical Integration Rules

Ronald Cools
   A Survey of Methods for Constructing Cubature Formulae   1

Karin Gatermann
   Linear Representations of Finite Groups and the Ideal Theoretical
   Construction of G-Invariant Cubature Formulas   25

Hans J. Schmid and H. Berens
   On the Number of Nodes of Odd Degree Cubature Formulae for
   Integrals with Jacobi Weights on a Simplex   37

Klaus-Jürgen Förster
   On Quadrature Formulae near Gaussian Quadrature   45

Ian Sloan
   Numerical Integration in High Dimensions - the Lattice Rule Approach   55

Harald Niederreiter
   Existence Theorems for Efficient Lattice Rules   71

Bernard Bialecki
   SINC Quadratures for Cauchy Principal Value Integrals   81

Philip Rabinowitz and William E. Smith
   Interpolatory Product Integration in the Presence of
   Singularities: L_p Theory   93

David B. Hunter
   The Numerical Evaluation of Definite Integrals Affected by
   Singularities Near the Interval of Integration   111

Nikolaos I. Ioakimidis
   Application of Computer Algebra Software to the Derivation of
   Numerical Integration Rules for Singular and Hypersingular Integrals   121

Numerical Integration Error Analysis

Walter Gautschi
   Remainder Estimates for Analytic Functions   133

Helmut Brass
    Error Bounds Based on Approximation Theory   147

Knut Petras
    One Sided L_1-Approximation and Bounds for Peano Kernels   165

Ricolindo Cariño, Ian Robinson and Elise De Doncker
    An Algebraic Study of the Levin Transformation in
    Numerical Integration   175

Numerical Integration Applications

Günther Hämmerlin
    Developments in Solving Integral Equations Numerically   187

Christoph Schwab and Wolfgang L. Wendland
    Numerical Integration of Singular and Hypersingular Integrals
    in Boundary Element Methods   203

James N. Lyness
    On Handling Singularities in Finite Elements   219

Ken Hayami
    A Robust Numerical Integration Method for 3-D Boundary Element
    Analysis and its Error Analysis using Complex Function Theory   235

Jarle Berntsen
    On the Numerical Calculation of Multidimensional Integrals
    Appearing in the Theory of Underwater Acoustics   249

Alan Genz
    Statistics Applications of Subregion Adaptive Multiple
    Numerical Integration   267

Frank Stenger, Brian Keyes, Mike O'Reilly and Ken Parker
    The Sinc Indefinite Integration and Initial Value Problems   281

Numerical Integration Algorithms and Software

Patrick Keast
    Software for Integration over Triangles and General Simplices   283

Ricolindo Cariño, Ian Robinson and Elise De Doncker
    An Algorithm for Automatic Integration of Certain Singular
    Functions over a Triangle   295

Ronald Cools and Ann Haegemans
    CUBPACK: Progress Report   305

Elise de Doncker and John Kapenga
    Parallel Cubature on Loosely Coupled Systems   317

Marc Beckers and Ann Haegemans
    Transformation of Integrands for Lattice Rules   329

Terje O. Espelid
    DQAINT: An Algorithm for Adaptive Quadrature over
    a Collection of Finite Intervals   341

Christoph Schwab
    A Note on Variable Knot, Variable Order Composite Quadrature
    for Integrands with Power Singularities   343

Avram Sidi
    Computation of Oscillatory Infinite Integrals by Extrapolation
    Methods   349

Appendix: Final Program   353

List of Participants   357

List of Contributors   361

Index   365

PREFACE

This volume contains refereed papers and extended abstracts of papers presented at the NATO Advanced Research Workshop entitled 'Numerical Integration: Recent Developments, Software and Applications', held at the University of Bergen, Bergen, Norway, June 17-21, 1991. The Workshop was attended by thirty-eight scientists. A total of eight NATO countries were represented. Eleven invited lectures and twenty-three contributed lectures were presented, of which twenty-five appear in full in this volume, together with three extended abstracts and one note.

The main focus of the workshop was to survey recent progress in the theory of methods for the calculation of integrals and show how the theoretical results have been used in software development and in practical applications. The papers in this volume fall into four broad categories: numerical integration rules, numerical integration error analysis, numerical integration applications and numerical integration algorithms and software.

It is five years since the last workshop of this nature was held, at Dalhousie University in Halifax, Canada, in 1986. Recent theoretical developments have mostly occurred in the area of integration rule construction. For polynomial integrating rules, invariant theory and ideal theory have been used to provide lower bounds on the numbers of points for different types of multidimensional rules, and to help in structuring the nonlinear systems which must be solved to determine the points and weights for the rules. Many new optimal or near optimal rules have been found for a variety of integration regions using these techniques. The review paper by Cools in this volume summarizes much of this work. For high dimensional multiple integrals, the theory of lattice rules has recently been extensively expanded. Research work in this area has been led by Lyness, Niederreiter and Sloan, with new results which deal with classification of lattice rules, lattice rules of rank greater than one, existence of efficient lattice rules and improved error bounds for lattice rules. Theoretical work has also continued in the area of error bounds, and error analysis using asymptotic expansions; review papers by Brass and Gautschi in this volume summarize some of the progress in this area. Recent work led by Rabinowitz has studied different types of singular integrands, including product integrands. Continuing work led by Stenger has shown how Sinc methods may be used effectively for wider classes of problems. The work by Sidi has lead to a better understanding about what types of extrapolation methods are suitable for infinite range oscillatory integrands.

Effective and efficient software is now available for a variety of integration problems. The most important multidimensional regions are hypercubes and simplices. Better algorithms based on globally adaptive subdivision strategies can now be used for one and higher dimensional integrals, with new implementations by Berntsen, Cools, Espelid, Genz, and Haegemans that include recent work by Berntsen and Espelid on more reliable and robust error estimation methods.

A major area of practical application for the theoretical work in numerical integration has occurred in the area of integral equations and the related boundary element method. Recent work has focussed on the development of efficient methods for treating the different types of kernels that arise in applications. These include weakly singular, Cauchy singular and hyper-singular kernels. Theoretical error results have been useful in selecting effective transformation and extrapolation methods for these problems. Papers in this volume by Hämmerlin, Hayami, Lyness, Schwab and Wendland summarize much of this work.

James N. Lyness has made extensive contributions to both the theoretical and practical aspects of numerical integration over the last thirty years. This work includes rule construction, error analysis, applications and software development for single and multidimensional integrals. His work has had a major influence on most of the people who have worked on numerical integration problems over the last twenty years. Since many of these people attended the Workshop, the editors decided to dedicate these Proceedings to James, who celebrates his sixtieth birthday on October 16, 1992.

We wish to acknowledge the financial support received from the NATO Science Committee. We are also grateful to the University of Bergen Department of Informatics, the University of Bergen, Faculty of Science, the City of Bergen and the Norwegian Marshall Fund for additional financial support. We also wish to thank Marit Nordvik for assistance with running the Workshop and preparation of this volume, and Tor Bastiansen for help with the financial management of the Workshop. In addition we would like to thank James Lyness and Pat Keast for help with the initial organization of the Workshop, and Tor Sørevik for help with the final organization and running of the Workshop. Finally, our thanks go to all contributors and participants who made the Workshop a success.

Terje O. Espelid
Bergen

Alan Genz
Pullman

A SURVEY OF
METHODS FOR CONSTRUCTING CUBATURE FORMULAE

RONALD COOLS
Department of Computer Science
Katholieke Universiteit Leuven
Celestijnenlaan 200A
B-3001 Heverlee, Belgium
Email: ronald@cs.kuleuven.ac.be

ABSTRACT. In this survey we distinguish two approaches to the problem of constructing cubature formulae: the invariant theoretical and the ideal theoretical approach. Both approaches are described theoretically for arbitrary dimensions. Methods for constructing cubature formulae are described for 2-dimensional regions.

1 Introduction to the problem

An integral I is a linear functional

$$I[f] := \int_{\Omega} w(x)f(x)dx$$

where the region $\Omega \subset \mathbb{R}^n$. We will always assume that $w(x) \geq 0$, $\forall x \in \mathbb{R}^n$, i.e. I is a positive functional. It is often desirable to approximate $I[f]$ by a weighted sum of function values

$$I[f] \simeq Q[f] := \sum_{i=1}^{N} w_i f(y^{(i)}) \qquad (1)$$

with $w_i \in \mathbb{R}$ and $y^{(i)} \in \mathbb{R}^n$. If $n = 1$ then Q is called a *quadrature formula*. If $n \geq 2$ then Q is called a *cubature formula*. The choice of the knots $y^{(i)}$ and weights w_i is independent of the function f. They are chosen so that the formula gives a good approximation for some class of functions.

One usually prefers cubature formulae where the knots are inside the region Ω because it is not always known in advance whether it is possible to evaluate the integrand in a point outside Ω. According to [37] a *good quadrature or cubature formula* has all knots $y^{(i)}$ inside the region Ω and all weights w_i positive.

The vector space of all polynomials in n variables is denoted by $\mathcal{P}^n$. The subspace of all polynomials from $\mathcal{P}^n$ of degree at most d is denoted by $\mathcal{P}^n_d$.

1

T. O. Espelid and A. Genz (eds.), Numerical Integration, 1–24.

There are several criteria to specify and classify cubature formulae. We will only use the following.

Definition 1.1 *A cubature formula Q for an integral I has degree d if*

$$Q[f] = I[f], \forall f \in \mathcal{P}_d^n \tag{2}$$

and

$$\exists g \in \mathcal{P}_{d+1}^n \text{ such that } Q[g] \neq I[g].$$

Because $\mathcal{P}_d^n$ is a vector space, the conditions (2) are equivalent with

$$Q[f_j] = I[f_j], j = 1, 2, \ldots, \dim \mathcal{P}_d^n \tag{3}$$

where the f_j form a basis for $\mathcal{P}_d^n$. If the basis and the number of knots N are fixed, then conditions (3) form a system of nonlinear equations

$$\sum_{i=1}^{N} w_i f_j(y^{(i)}) = I[f_j], j = 1, 2, \ldots, \dim \mathcal{P}_d^n. \tag{4}$$

Each equation in (4) is a polynomial equation. Each knot introduces $n + 1$ unknowns: the weight and the n coordinates of the knot. A cubature formula of degree (at least) d with N knots is thus determined by a system of $\dim \mathcal{P}_d^n$ nonlinear equations in $N(n + 1)$ unknowns.

Definition 1.2 *If the weights of a cubature formula of degree d are uniquely determined by the knots, the cubature formula is called an interpolatory cubature formula.*

In contrast with one dimension, in two and more dimensions the variety of regions is unlimited. Following the notation of Stroud [45] we mention the following standard regions in this paper:

C_n : The n-dimensional cube $[-1, 1]^n$ with weight function $w(x) = 1$.

C_2^α : The square $[-1, 1]^2$ with weight function $w(x_1, x_2) = (1 - x_1^2)^\alpha (1 - x_2^2)^\alpha$.

S_2 : The unit circle with weight function $w(x) = 1$.

$E_2^{r^2}$: The plane with weight function $w(x_1, x_2) = \exp(-x_1^2 - x_2^2)$.

T_2 : The unit triangle with weight function $w(x) = 1$.

2 The first 100 years

In this section we will give a very short (incomplete) history of the construction of cubature formulae. The book of Stroud [45] contains an overview of the results on multiple integration that were known in 1971. The book of Davis and Rabinowitz [17] contains one chapter with a short overview of results on multiple integration that were known in 1984.

The first cubature formulae were constructed by J.C. Maxwell in 1877 [31]. He constructed a formula of degree 7 with 13 knots for the square C_2 and a cubature formula of degree

7 with 27 knots for the cube C_3 by solving the system of nonlinear equations (4). His formula for the square has one negative weight and the formula for the cube has several knots outside the region. Maxwell wrote the following about this: "... we meet with the curious result, that in certain cases the solution indicates that we are to employ values of the function some of which correspond to values of the variables outside the limits of integration. ... This, of course, renders the method useless in determining the integral from the measured values of the [integrand] as when we wish to determine the weight of a brick from the specific gravities of samples taken from 27 selected places in the brick, for we are directed by the method to take some of the samples from places outside the brick." The first two cubature formulae were not good.

The zeros of one-dimensional orthogonal polynomials can be used as knots of quadrature formulae with all knots inside the interval and all weights positive. This theory does not appear to carry over to several dimensions. However, as in the one-dimensional case, there is a close connection between multivariate orthogonal polynomials and cubature formulae.

P. Appell seems to be the first, in 1890, to investigate the relation between orthogonal polynomials in two variables and cubature formulae for two-dimensional integrals [1]. The results of J. Radon, published in 1948 [38], are much better known. He constructed a cubature formula of degree 5 with 7 knots for the square C_2, the circle S_2 and the triangle T_2 using the common zeros of 3 orthogonal polynomials of degree 3.

During the sixties, A. Stroud and I.P. Mysovskikh studied the relation between orthogonal polynomials and cubature formulae for n-dimensional regions.

In the late sixties, the researchers working at the construction of cubature formulae were split in two groups.

The first group directly attacked the system of nonlinear equations (4). They used consistency conditions to write down a system of nonlinear equations that could be expected to have a solution. This approach requires arbitrary starting values and some luck. This approach originates from P. Rabinowitz and N. Richter [37] and is further developed by F. Mantel, P. Keast and J. Lyness. See e.g. [27, 28, 30].

The second group used the relation between orthogonal polynomials and cubature formulae. A. Stroud, R. Franke, R. Piessens and A. Haegemans used this approach. See e.g. [19, 24, 25, 45]. The theory of the relation between orthogonal polynomials and cubature formulae was deepened in 1973 by H.M. Möller with the introduction of polynomial ideals [33].

Both approaches have their merits and are now still being used. In this paper the reader is introduced to both the invariant theoretical and the ideal theoretical approach. Although the theory deals with arbitrary dimensions, the presentation of applications is restricted to two dimensions.

3 Construction of cubature formulae using invariant theory

3.1 Introduction to invariant theory

In this section we describe the part of invariant theory needed to understand the sequel. Let $\mathcal{G}$ be a group of linear transformations.

Definition 3.1 *The $\mathcal{G}$-orbit of a point $y \in \mathbb{R}^n$ is the set of points $\{g(y)|\forall g \in \mathcal{G}\}$.*

Definition 3.2 *A set $\Omega \subset \mathbb{R}^n$ is said to be invariant with respect to a group $\mathcal{G}$ if $g(\Omega) = \Omega$ for all $g \in \mathcal{G}$. An invariant polynomial of $\mathcal{G}$ is a polynomial ϕ which is left unchanged by every transformation in $\mathcal{G}$.*

The vector space of all invariant polynomials of $\mathcal{G}$ is denoted by $\mathcal{P}^n(\mathcal{G})$ and the subspace of $\mathcal{P}^n(\mathcal{G})$ with only the polynomials of degree $\leq d$ is denoted by $\mathcal{P}^n_d(\mathcal{G})$.

Definition 3.3 *Let $\phi_1, \phi_2, \ldots, \phi_l$ be invariant polynomials of $\mathcal{G}$. $\phi_1, \phi_2, \ldots, \phi_l$ form an integrity basis for the invariant polynomials of $\mathcal{G}$ $\Leftrightarrow$ any invariant polynomial of $\mathcal{G}$ is a polynomial in $\phi_1, \phi_2, \ldots, \phi_l$.*
Each polynomial ϕ_i is called a basic invariant polynomial of $\mathcal{G}$.

Because the degree of a polynomial is left unchanged by a linear transformation of the variables, one can restrict the search of basic invariant polynomials to homogeneous polynomials. If the number of basic invariant polynomials $l > n$ then there exist polynomials equations, called syzygies, relating $\phi_1, \phi_2, \ldots, \phi_l$.

Some properties are summarized in the following theorems.

Theorem 3.1 *[26] There always exists a finite integrity basis for the invariant polynomials of a finite group $\mathcal{G}$.*

Theorem 3.2 *[5, 43] Let $\mathcal{G}$ be a finite group acting on the n-dimensional vector space $\mathbb{R}^n$. $\mathcal{G}$ is a finite reflection group if and only if the invariant polynomials of $\mathcal{G}$ have an integrity basis consisting of n homogeneous polynomials which are algebraically independent.*

Let C_i be the number of linearly independent homogeneous invariant polynomials of degree i of a group $\mathcal{G}$. A useful formula for the C_i is given by the following.

Theorem 3.3 *[32] Let $\omega_1(g), \omega_2(g), \ldots, \omega_n(g)$ be the eigenvalues of $g \in \mathcal{G}$. Then*

$$\sum_{i=0}^{\infty} C_i t^i = \frac{1}{|\mathcal{G}|} \sum_{g \in \mathcal{G}} \frac{1}{(1 - \omega_1(g)t)(1 - \omega_2(g)t) \ldots (1 - \omega_n(g)t)}.$$

The power series in Theorem 3.3 is known as the Molien-series of $\mathcal{G}$. It is important because

$$\dim \mathcal{P}^n_d(\mathcal{G}) = \sum_{i=0}^{d} C_i.$$

3.2 Relation between invariant theory and cubature formulae

If a region Ω is invariant with respect to some group $\mathcal{G}$ and if the weight function $w(x)$ also is invariant with respect to $\mathcal{G}$, then it seems reasonable to look for cubature formulae where the set of knots is a union of $\mathcal{G}$-orbits.

Definition 3.4 *A cubature formula is said to be invariant with respect to a group $\mathcal{G}$ if the region Ω and the weight function $w(x)$ are invariant with respect to $\mathcal{G}$, and if the set of knots is a union of $\mathcal{G}$-orbits. The knots of one and the same orbit have the same weight.*

The usefulness of invariant theory, in the context of constructing cubature formulae, is highlighted by the following result due to Sobolev :

Theorem 3.4 (Sobolev's theorem) *[44] Let the formula (1) be invariant with respect to $\mathcal{G}$. The cubature formula Q has degree d if*

$$I[f] = Q[f], \forall f \in \mathcal{P}_d^n(\mathcal{G})$$

and

$$\exists g \in \mathcal{P}_{d+1}^n \text{ such that } I[g] \neq Q[g].$$

The number of equations and unknowns in (4) can be reduced by imposing some additional structure on the formula. Sobolev's theorem suggests that we look for invariant cubature formulae, that is, solutions of the equations

$$Q[\phi_i] = I[\phi_i], i = 1, 2, \ldots, \dim \mathcal{P}_d^n(\mathcal{G}), \tag{5}$$

where the ϕ_i form a basis for $\mathcal{P}_d^n(\mathcal{G})$.

The idea of demanding that a cubature formula has the same symmetries as the given integral, is as old as the construction of cubature formulae itself. Indeed, when Maxwell constructed cubature formulae for the square and the cube [31] he considered only cubature formulae that are invariant with respect to the groups of symmetries of these regions. Such formulae are called "fully symmetric".

3.3 Consistency conditions

In this section fully symmetric quadrature and cubature formulae for one and two-dimensional integrals will be considered.

Each orbit in an invariant cubature formula introduces a number of unknowns in the nonlinear equations and gives a number of knots in the cubature formula. The different types of orbits are described in Tables 3.1 and 3.2.

type	generator	number of unknowns	number of points in an orbit	unknowns
0	(0)	1	1	weight
1	(a)	2	2	a, weight

Table 3.1 : different types of Z_2-orbits

type	generator	number of unknowns	number of points in an orbit	unknowns
0	(0,0)	1	1	weight
1	(a,0)	2	4	a, weight
2	(a,a)	2	4	a, weight
3	(a,b)	3	8	a, b, weight

Table 3.2 : different types of $\mathcal{D}_4$-orbits

Let K_i be the number of orbits of type i in an invariant quadrature or cubature formula. One does not expect a solution of a system of nonlinear equations if there are more equations than unknowns. Based on this, Rabinowitz and Richter [37] introduced the notion of consistency conditions. A consistency condition is an inequality for the K_i that must be satisfied in order to obtain a system of nonlinear equations where the number of unknowns is greater than or equal to the number of equations in each subsystem. Cubature formulae which do not satisfy the consistency conditions are called "fortuitous" and are supposed to be rare.

If $n = 1$, $\mathcal{G} = Z_2$ and the homogeneous basic invariant polynomial is $\phi := x^2$. The consistency conditions are very familiar in this case.

Theorem 3.5 *The consistency conditions for one-dimensional symmetric quadrature formulae of degree $2k - 1$ are :*

$$
\begin{aligned}
K_0 + \; 2K_1 \; &\geq k \\
2K_1 \; &\geq k - 1 \\
K_0 \; &\leq 1
\end{aligned}
$$

A quadrature formula that satisfies these conditions has $N = K_0 + 2K_1$ knots.

If $n = 2$ then $\mathcal{G} = \mathcal{D}_4$ and one can choose $\phi_1 := x^2 + y^2$ and $\phi_2 = x^2 y^2$ as basic invariant polynomials in x and y. The consistency conditions are now less obvious.

Theorem 3.6 *[30] The consistency conditions for two-dimensional fully symmetric cubature formulae of degree $2k - 1$ are :*

$$
\begin{aligned}
3K_3 \; &\geq \; A_1(k) - k \\
2K_2 + 3K_3 \; &\geq \; A_1(k) \\
2K_1 + 3K_3 \; &\geq \; A_1(k) \\
K_0 + 2K_1 + 2K_2 + 3K_3 \; &\geq \; A_1(k) + k \\
K_0 \; &\leq \; 1
\end{aligned}
$$

$$
A_1(k) = \begin{cases} \frac{(k-1)^2}{4} & \text{if } k \text{ is odd} \\ \frac{k}{2}\left(\frac{k}{2} - 1\right) & \text{if } k \text{ is even} \end{cases}
$$

A cubature formula that satisfies these conditions has $N = K_0 + 4K_1 + 4K_2 + 8K_3$ knots.

The construction of a cubature formula with the lowest possible number of knots, requires two steps :

1. First, solve the integer programming problem:

$$\text{minimize } N(K_i : i = 0, 1, \ldots)$$

   where the integers K_i satisfy the consistency conditions.

2. Solve the system of nonlinear equations (5). If no solution of the nonlinear equations is found, then another (non optimal) solution of the consistency conditions must be tried.

Consistency conditions can be derived for every structure and dimension. They can help to set up a system of nonlinear equations where in each subsystem the number of unknowns is larger than or equal to the number of equations.

We want to emphasize that consistency conditions are not sufficient and not necessary conditions. Even if a system of equations has more unknowns than equations, it might not have a real solution. Furthermore, fortuitous cubature formulae are known. E.g. in [16] fortuitous formulae of arbitrary degree were constructed for C_2^α.

The classical theory of moments for polynomials can be used in combination with consistency conditions as a tool to construct good cubature formulae [18]. One can also derive sufficient conditions to demand that the weights are positive [13].

3.4   Groups of rotations

There is no reason why a cubature formula should have the same structure as the integral. (What should a formula for a circle look like ?) Cubature formulae that are invariant with respect to a subgroup of the symmetry group of the integral were already investigated by Radon [38]. Humans seem to have a preference for certain symmetries. Symmetry with respect to the axes is studied regularly (e.g. [24, 25]) but symmetry with respect to the diagonals has been used only recently [16]. The symmetry groups are nevertheless isomorphic. Rotational symmetries turned up by surprise in [34].

In this section we want to construct cubature formulae for integrals that are invariant with respect to $\Re_4$: the subgroup of orientation preserving transformations from the dihedral group $\mathcal{D}_4$.

The system of nonlinear equations that determines the cubature formulae, is described in the following theorem. The basic invariant polynomials are written using polar coordinates.

Theorem 3.7 *[11] Let $\phi_1 := r^2$, $\phi_2 := r^4 \cos(4\theta)$ and $\phi_3 := r^4 \sin(4\theta)$. An invariant cubature formula with respect to $\Re_4$ of degree $d := 2k - 1$, is a solution of the system of nonlinear equations*

$$Q[\phi_1^s \phi_2^t \phi_3^v] = I[\phi_1^s \phi_2^t \phi_3^v]$$

with $2s + 4t + 4v \leq d, s, t \in \mathbb{N}$, $v \in \{0, 1\}$.
The number of equations $= \dim \mathcal{P}_d^2(\Re_4) = \lfloor \frac{k^2+1}{2} \rfloor$.

The different types of orbits for $\mathfrak{R}_4$ are described in Table 3.3.

type	generator	number of unknowns	number of points in a $\mathfrak{R}_4$-orbit	unknowns
0	$(0,0)$	1	1	weight
1	$(a,0)$	2	4	a, weight
2	(a,a)	2	4	a, weight
3	(a,b)	3	4	a, b, weight

Table 3.3 : different types of $\mathfrak{R}_4$-orbits.

Let K_i be the number of $\mathfrak{R}_4$-orbits of type i in an invariant cubature formula. One of the consistency conditions is that the number of unknowns in the system of nonlinear equations is greater than or equal to the number of equations:

$$K_0 + 2(K_1 + K_2) + 3K_3 \geq \dim \mathcal{P}_d^2(\mathfrak{R}_4). \tag{6}$$

The number of knots in the cubature formula is

$$N = K_0 + 4K_1 + 4K_2 + 4K_3.$$

Mantel stated [29] that a cubature formula of degree 11 with the structure $[0,0,0,6]$ exists for fully symmetric regions, and cubature formulae with structure $[0,1,0,5]$ or $[0,0,1,5]$ do not. These last 2 structures violate consistency condition (6). But, suppose a cubature formula with structure $[0,0,0,6]$ is known for a fully symmetric region that is also circular symmetric, e.g. S_2, $E_2^{r^2}$. Then this cubature formula remains a cubature formula of the same degree if all points are rotated around the origin by an arbitrary angle. So, one can always rotate a cubature formula with structure $[0,0,0,6]$ such that a formula with structure $[0,1,0,5]$ or $[0,0,1,5]$ is obtained. The consistency conditions seems to be inconsistent !

We can now give two comments on the use of structures:

- If the region is circular symmetric, a rotated cubature formula remains a solution of the system of nonlinear equations. This freedom results in the following modified consistency condition:

$$K_0 + 2(K_1 + K_2) + 3K_3 \geq \dim \mathcal{P}_d^2(\mathfrak{R}_4) - 1. \tag{7}$$

If the region is not circular symmetric, then condition (6) is all right. So, if one constructs formulae which are invariant with respect to a subgroup of the symmetry group of a region, one may not forget this.

- If one studies $\mathfrak{R}_4$-invariant regions, then orbits of type 1 and 2 are artificial and only make sense if one wants to reduce the degree of freedom.

The search for $\mathfrak{R}_4$-invariant cubature formulae for $\mathcal{D}_4$-invariant regions has produced interesting results. The cubature formulae of degree 9 and 11 for C_2 with the lowest possible number of knots (resp. 17 and 24) are $\mathfrak{R}_4$-invariant [11, 34]. The formula with the lowest known number of knots of degree 13 for C_2 has 33 knots and is also $\mathfrak{R}_4$-invariant [11].

3.5   Cubature formula for circular symmetric regions

In the previous section we were confronted with the special nature of circular symmetric regions. For these regions $I[r^i \cos(j\theta)] = 0, \forall i, j \in \mathbb{N}_0$. In this section this special nature will be exploited.

In this section a *basic Q_M cubature rule operator* will be denoted by

$$Q_M(r, \alpha)[f] := \frac{1}{M} \sum_{j=0}^{M-1} f(r, \alpha + \frac{2\pi j}{M})$$

with f a function of the polar coordinates (r, α), $r = 0$ or $\cos(2M\alpha) = 1$. Then $Q_M(r, \alpha)[f]$ requires 1 or M function values.

In [7, 9, 23], cubature formulae of the form

$$Q[f] := \sum_{i=1}^{mmax} \sum_{t=1}^{a_i} w_{it} Q_{M_i}(r_{it}, \alpha_i)[f] + w_0 f(0, .) \tag{8}$$

were considered. Here, a_i is the number of M_i-gons, each M_i-gon is fully determined by one vertex (r_{it}, α_i), and w_{it} is a weight. $mmax$ and the a_i s depend on the degree of the cubature formula.

Only cubature formulae with knots that are the vertices of M_m-gons are considered, with

$$M_1 := A.2, \alpha_1 := 0$$

$$M_m := A.2^{m-1}, \alpha_m := \frac{\pi}{M_m}, m = 2, 3, \ldots, mmax \tag{9}$$

$$for \ A \in \mathbb{N} \setminus \{0, 1\}.$$

The formulae (8) are invariant with respect to $\mathcal{D}_{M_1}$. According to Sobolev's theorem, such a cubature formula has degree d if it is exact for a basis of $\mathcal{P}_d^2(\mathcal{D}_{M_1})$. Because of the special structure, the number of polynomials for which the cubature formula must be exact can be further reduced. Here we only want to sketch the advantages of this type of cubature formulae. For details we refer to [6, 7, 9].

A first advantage of the special structure (9) is that the number of nonlinear equations is reduced. E.g.: A $\mathcal{D}_4$-invariant cubature formula of degree 35 is determined by a system of 90 equations (5). The formula with the special structure (9) is a solution of a system of 60 equations.

A second advantage of the special structure is that this large system of nonlinear equations is split into several smaller systems which can be solved sequentially. E.g.: For the formula of degree 35 mentioned before, the largest system contains 18 equations.

A third advantage is that each of these systems of nonlinear equations can be solved easily, because they have the same form as the systems that determine a quadrature formula.

The problem is thus reduced from solving one large system of nonlinear equations to solving more, but smaller systems of nonlinear equations sequentially. A program has been written that computes automatically a cubature formula of a given degree d with given structure parameter A for a region for which the moments are known [7]. The number of knots is a function of the degree of precision and A. An optimal choice for A can easily be made.

The construction of cubature formulae with a special structure became easy not only because of this structure and the properties of the region, but also because of the choice of a basis for $\mathcal{P}^2(\mathcal{D}_{M_1})$.

3.6 Only $\mathcal{D}_4$-invariant regions

In this section a basis for $\mathcal{P}^2(\mathcal{D}_4)$ that makes the construction of a cubature formula for a $\mathcal{D}_4$-invariant region easier, will be constructed. What is special about this basis is that it exists for $\mathcal{D}_4$-invariant regions that are *not* circular symmetric.

In [47], one searched for a cubature formula of degree 19 with 68 knots of the form:

$$Q[f] := \sum_{i=1}^{4} w_i Q_i^{(1)}[f] + \sum_{i=5}^{8} w_i Q_i^{(2)}[f] + \sum_{i=9}^{13} w_i Q_i^{(3)}[f],$$

where

$$
\begin{aligned}
Q_i^{(1)}[f] &:= f(x_i, y_i) + f(-x_i, y_i) + f(x_i, -y_i) + f(-x_i, -y_i) + \\
&\quad\ f(y_i, x_i) + f(-y_i, x_i) + f(y_i, -x_i) + f(-y_i, -x_i) \\
Q_i^{(2)}[f] &:= f(x_i, 0) + f(-x_i, 0) + f(0, x_i) + f(0, -x_i) \\
Q_i^{(3)}[f] &:= f(x_i, x_i) + f(-x_i, x_i) + f(x_i, -x_i) + f(-x_i, -x_i),
\end{aligned}
$$

for a $\mathcal{D}_4$-invariant region. Such a formula, with structure parameters $[0, 4, 5, 4]$, satisfies the consistency conditions of Theorem 3.6.

Let $s(x, y) := x^2 + y^2$ and $u(x, y) := x^2 y^2$. Then, every $\mathcal{D}_4$-invariant polynomial f is a polynomial in s and u:

$$f(x, y) = g(s(x, y), u(x, y)).$$

Consequently

$$Q[f] = \sum_{i=1}^{4} 8 w_i g(s_i, u_i) + \sum_{i=5}^{8} 4 w_i g(s_i, 0) + \sum_{i=9}^{13} 4 w_i g(s_i, \frac{s_i^2}{4}),$$

where

$$
\begin{aligned}
s_i &:= s(x_i, y_i), & u_i &:= u(x_i, y_i), & i &= 1, \ldots, 4 \\
s_i &:= s(x_i, 0), & & & i &= 5, \ldots, 8 \\
s_i &:= s(x_i, x_i), & & & i &= 9, \ldots, 13.
\end{aligned}
$$

As $\dim \mathcal{P}_{19}^2(\mathcal{D}_4) = 30$, the formula is determined by a system of 30 polynomial equations in the 30 unknowns:

$$
\begin{aligned}
w_i, s_i & \qquad i = 1, \ldots, 13 \\
u_i, & \qquad i = 1, \ldots, 4.
\end{aligned}
$$

A straightforward choice for the basis is $\{s^i u^j : 0 \leq 2i + 4j \leq 19, i, j \in \mathbb{N}\}$ but we want a basis that makes the construction really easy.

Let $v(s, u) = s^2 - 4u$ and $t(s, u) = u(s^2 - 4u)$. Let the polynomials

$$
\begin{aligned}
r_1(s, u) &:= s^3 + a_1 s^2 + b_1 u + c_1 s + d_1 \\
r_2(s, u) &:= su + a_2 s^2 + b_2 u + c_2 s + d_2 \\
r_3(s) &:= s^4 + a_3 s^3 + b_3 s^2 + c_3 s + d_3 \\
r_4(s) &:= s^5 + a_4 s^4 + b_4 s^3 + c_4 s^2 + d_4 s + h_5
\end{aligned}
$$

be such that they vanish is some of the knots:

$$
\begin{aligned}
r_1(s_i, u_i) &= 0, & i &= 1, \ldots, 4 \\
r_2(s_i, u_i) &= 0, & i &= 1, \ldots, 4 \\
r_3(s_i) &= 0, & i &= 5, \ldots, 8 \\
r_4(s_i) &= 0, & i &= 9, \ldots, 13.
\end{aligned}
$$

One can prove that if $b_1 \neq 0$, another basis for $\mathcal{P}^2_{19}(\mathcal{D}_4)$ is given by

$$
\begin{array}{llllll}
(c1): & t\, r_1, & t\, r_1\, s, & t\, r_1\, s^2, & t\, r_1\, u, & \\
(c2): & t\, r_2, & t\, r_2\, s, & t\, r_2\, s^2, & t\, r_2\, u, & \\
(c3): & t, & t\, s, & t\, u, & t\, s^2, & \\
(c4): & v\, r_3, & v\, r_3\, s, & v\, r_3\, s^2, & v\, r_3\, s^3, & \\
(c5): & v, & v\, s, & v\, s^2, & v\, s^3, & \\
(c6): & r_4, & r_4\, s, & r_4\, s^2, & r_4\, s^3, & r_4\, s^4, \\
(c7): & 1, & s, & s^2, & s^3, & s^4.
\end{array}
$$

The equation

$$
\sum_{i=1}^{4} 8 w_i g(s_i, u_i) + \sum_{i=5}^{8} 4 w_i g(s_i, 0) + \sum_{i=9}^{13} 4 w_i g\left(s_i, \frac{s_i^2}{4}\right) = I[g(s(x, y), u(x, y))], \qquad (10)
$$

has to be satisfied for all polynomials $g(s, u)$ belonging to this basis.

If in (10) g is replaced by the basis elements in (c1), the left hand side of (10) vanishes and a linear system determining a_1, b_1, c_1, d_1 is obtained. Similarly, if g is replaced by the basis elements in (c2), a linear system determining a_2, b_2, c_2, d_2 is obtained. Then $(s_1, u_1), (s_2, u_2), (s_3, u_3), (s_4, u_4)$ are the solutions of the system

$$
\begin{cases}
r_1(s, u) &= 0 \\
r_2(s, u) &= 0.
\end{cases}
$$

This is the most difficult step in the construction procedure ! If g is replaced by the basis elements in (c3), a linear system determining w_1, w_2, w_3, w_4 is obtained. If g is replaced by the basis elements in (c4), a linear system determining a_3, b_3, c_3, d_3 is obtained. Then, s_5, s_6, s_7, s_8 are the zeros of $r_3(s)$. If g is replaced by the basis elements in (c5), a linear system determining w_5, w_6, w_7, w_8 is obtained. If g is replaced by the basis elements in (c6), a linear system determining a_4, b_4, c_4, d_4, h_4 is obtained. Then, $s_9, s_{10}, s_{11}, s_{12}, s_{13}$ are the zeroes of $r_4(s)$. Finally, if g is replaced by the basis elements in (c7), a linear system determining $w_9, w_{10}, w_{11}, w_{12}, w_{13}$ is obtained.

The necessary computations are easy. In [47] the systems have been solved using computer algebra for C_2. The cubature formula has 68 real nodes, among which 8 lie outside the integration region. Note that $b_1 = 0$ if the integral I is $\mathcal{D}_8$-invariant. So, the procedure cannot be used for circular symmetric regions.

3.7 Evaluation

Invariant theory and consistency conditions are very useful for constructing a system of nonlinear equations that determines a cubature formula with a particular structure. In general, the system of nonlinear equations is still too large to solve completely with currently available tools. One can try to find a basis such that the equations are easy to solve. Typically the equations are then of the same form as the equations that determine a quadrature formula. This was illustrated in the previous sections for 2-dimensional regions but it is also true for higher dimensions (see [2]).

4 Construction of cubature formulae using ideal theory

4.1 Orthogonal polynomials

One can define orthogonality analogous to the one-dimensional case.

Definition 4.1 *A polynomial $f \in \mathcal{P}^n$ is I-orthogonal to a polynomial $g \in \mathcal{P}^n$ if $I[fg] = 0$.*

Definition 4.2 *A polynomial $f \in \mathcal{P}^n$ of degree d for which $I[fg] = 0$, $\forall g \in \mathcal{P}^n_{d-1}$ is called an orthogonal polynomial for the integral I.*

Definition 4.3 *A polynomial $f \in \mathcal{P}^n$ is d-orthogonal for the integral I if $I[fg] = 0$ for all g that satisfy $fg \in \mathcal{P}^n_d$.*

For a given integral I there exist $\dim \mathcal{P}^n_d - \dim \mathcal{P}^n_{d-1}$ unique orthogonal polynomials of degree d of the form

$$P^{a_1, a_2, \ldots, a_n} := x_1^{a_1} x_2^{a_2} \ldots x_n^{a_n} + S \tag{11}$$

with $a_i \in \mathbb{N}, \sum_{i=1}^n a_i = d$ and $S \in \mathcal{P}^n_{d-1}$. The polynomials (11) are called *basic orthogonal polynomials*.

4.2 Introduction to ideal theory

In this section we describe the part of ideal theory needed to understand the sequel.

Definition 4.4 *A polynomial ideal $\mathfrak{A}$ is a subset of the ring of polynomials in n variables $\mathcal{P}^n$ such that if $f_1, f_2 \in \mathfrak{A}$ and $g_1, g_2 \in \mathcal{P}^n$, then $f_1 g_1 + f_2 g_2 \in \mathfrak{A}$.*

Definition 4.5 *The zero set of an ideal $\mathfrak{A}$ is*

$$NG(\mathfrak{A}) := \{y \in \mathbb{C}^n | \forall f \in \mathfrak{A} : f(y) = 0\}.$$

One can define several types of bases for an ideal $\mathfrak{A}$, but if $NG(\mathfrak{A})$ is a finite set of points, a basis for the ideal $\mathfrak{A}$ consists of at least n polynomials. The most important basis from our point of view is the H-basis.

Definition 4.6 *Let $\mathfrak{A}$ be a polynomial ideal. $\{f_1, f_2, \ldots, f_s\} \subset \mathfrak{A}$ is an H-basis for $\mathfrak{A}$ if for all $f \in \mathfrak{A}$ polynomials $g_1, g_2, \ldots, g_s$ exist such that*

$$f = \sum_{i=1}^{s} f_i g_i \text{ and } \deg(f_i g_i) \leq \deg(f), i = 1, 2, \ldots, s.$$

Theorem 4.1 *[33] For any polynomial ideal an H-basis exists.*

An important function of a polynomial ideal is the Hilbert function [26].

Definition 4.7 *The Hilbert function $\mathcal{H}$ is defined as*

$$
\begin{aligned}
\mathcal{H}(k; \mathfrak{A}) &:= \dim \mathcal{P}_k^n - \dim(\mathfrak{A} \cap \mathcal{P}_k^n), k \in \mathbb{N} \\
&:= 0, -k \in \mathbb{N}_0.
\end{aligned}
$$

When talking about the zero set of a polynomial ideal, the Hilbert function turns out to be very useful. It is obvious that the Hilbert function is an increasing function : $\mathcal{H}(k + 1; \mathfrak{A}) \geq \mathcal{H}(k; \mathfrak{A}), \forall k \in \mathbb{N}$.

Theorem 4.2 *[22] If $\mathcal{H}(k; \mathfrak{A}) = \mathcal{H}(K; \mathfrak{A})$ for all $k \geq K$ holds for a sufficiently large K, the polynomials in $\mathfrak{A}$ have exactly $\mathcal{H}(K; \mathfrak{A})$ common zeros, if these are counted with multiplicities.*

4.3 Relation between ideal theory and cubature formulae

In this section we will explain why polynomial ideals are useful for the construction of cubature formulae. The relation between polynomial ideals, Hilbert function and cubature formulae is given by the following theorem, which is a generalization of theorems of Stroud [45].

Theorem 4.3 *[33] Let I be an integral over an n-dimensional region and $d \in \mathbb{N}$. Let $\{y^{(1)}, y^{(2)}, \ldots, y^{(N)}\} \subset \mathbb{C}^n$ and*

$$\mathfrak{A} := \{f \in \mathcal{P}^n | f(y^{(i)}) = 0, i = 1, 2, \ldots, N\}.$$

Then the following statements are equivalent.

1) $f \in \mathfrak{A} \cap \mathcal{P}_d^n \Rightarrow I[f] = 0.$

2) *There exists a cubature formula Q (1) such that $I[f] = Q[f], \forall f \in \mathcal{P}_d^n$, with at most $\mathcal{H}(d; \mathfrak{A})$ weights different from zero.*

As an application we will prove the well-known bound for the number of knots in a cubature formula.

Theorem 4.4 *If an interpolatory cubature formula of degree d for an integral over a n-dimensional region has N knots, then*

$$\binom{n + \lfloor d/2 \rfloor}{n} \leq N \leq \binom{n + d}{n}$$

Proof :

a) No polynomial of degree $\leq \lfloor \frac{d}{2} \rfloor$ can be zero at all knots of the cubature formula. Thus $\mathfrak{A} \cap \mathcal{P}_{\lfloor d/2 \rfloor}^n = \{0\}$. Therefore, according to Theorem 4.3,

$$N = \mathcal{H}(d; \mathfrak{A}) \geq \mathcal{H}\left(\left\lfloor \frac{d}{2} \right\rfloor; \mathfrak{A}\right) = \dim \mathcal{P}_{\lfloor d/2 \rfloor}^n = \binom{n + \lfloor d/2 \rfloor}{n}.$$

b) From the definition of Hilbert function, it follows immediately that

$$N = \mathcal{H}(d; \mathfrak{A}) \leq \dim \mathcal{P}_d^n = \binom{n + d}{d}.$$

$\square$

H-bases are important because of the following.

Theorem 4.5 *[33] If $\{f_1, f_2, \ldots, f_s\}$ is an H-basis of a polynomial ideal $\mathfrak{A}$ and if the set of common zeros of $f_1, f_2, \ldots, f_s$ is finite and nonempty, then the following statements are equivalent :*

1) *There is a cubature formula of degree d for the integral I which has as knots the common zeros of $f_1, f_2, \ldots, f_s$. These zeros may be multiple, leading to the use of function derivatives in the cubature formula.*

2) f_i is d-orthogonal for I, $i = 1, 2, \ldots, s$.

An immediate consequence of Theorem 4.5 is given in the following corollary.

Corollary 4.1 *Every polynomial of degree $\tau \leq d$ that is zero at all knots of a cubature formula of degree d, is orthogonal to all polynomials of degree $\leq d - \tau$.*

Combining Max Noether's Theorem [35] and Theorem 4.5 gives the following.

Corollary 4.2 *If f_i are polynomials of degree τ_i, $i = 1, 2, \ldots, n$, with no common zero at infinity, I an integral, $d \in \mathbb{N}$ and $N = \prod_{i=1}^{n} \tau_i$, then the following statements are equivalent :*

1) f_i *is d-orthogonal for I, $i = 1, 2, \ldots, n$.*

2) *The set of common zeros of $f_1, f_2, \ldots, f_n$ can be used as knots in a cubature formula of degree d. These zeros may be multiple, leading to the use of function derivatives in the cubature formula.*

If a cubature formula is constructed using Corollary 4.2 then it is always possible that some knots have a weight equal to zero. In this case the Hilbert function can be found without any difficulty.

Corollary 4.3 *[6] Let $f_1, f_2, \ldots, f_n \in \mathcal{P}^n$ have only finite common zeros, $\tau_i := \deg(f_i)$ and $\mathfrak{A}$ the ideal generated by $\{f_1, f_2, \ldots, f_n\}$. Let f_i be d-orthogonal for I, $i=1,2, \ldots, n$. If $k := \min\{\tau_i | i = 1, 2, \ldots, n\}$ then the Hilbert function is*

$$\mathcal{H}(d; \mathfrak{A}) = H(d, n) - \sum_{i=1}^{n} H(d - \tau_i, n)$$

for $d \leq 2k - 1$, with

$$H(t, n) \quad := \quad \dim \mathcal{P}_t^n (t \geq 0)$$
$$:= \quad 0 (t < 0).$$

Example : Let $L_k(x)$ be the Legendre polynomial of degree k in the variable x. The zeros of $L_k(x)$ are the knots of the Gauss-Legendre quadrature formula of degree $2k - 1$ for C_1.

Let $f_i := L_k(x_i)$, $i = 1, 2, \ldots, n$, $n \in \mathbb{N}_0$. According to Corollary 4.2, the k^n common zeros of $f_1, f_2, \ldots, f_n$ can be used as knots in a cubature formula of degree $2k - 1$ for C_n. All these points are used in the Gauss-product formula.

From Corollary 4.3 we know that at most

$$\mathcal{H}(2k - 1; (f_1, f_2, \ldots, f_n)) = \begin{pmatrix} 2k - 1 + n \\ n \end{pmatrix} - n \begin{pmatrix} k - 1 + n \\ n \end{pmatrix}$$

weights are different from zero if these points are used in an interpolatory cubature formula. For $n = 3$ the number of nonzero weights is bounded by

$$\mathcal{H}(2k - 1; (f_1, f_2, f_3)) = \frac{5k^3 + 3k^2 - 2k}{6}$$

and for $k \geq 3$ this is lower than k^3. Thus, the k^3 common zeros of f_1, f_2 and f_3 do not uniquely determine the weights of a cubature formula of degree $2k - 1$. Hence the Gauss-product formula is not always an interpolatory cubature formula.

Theorems 4.3 and 4.5 do not guarantee that the knots of the cubature formula are real and that the weights of the cubature formula are positive. H.J. Schmid found a way to study cubature formulae with real knots and positive weights. First we introduce the concept of fundamental set.

Definition 4.8 *A set of polynomials $S \subset \mathcal{P}^n$ is fundamental of degree d whenever* $(\dim \mathcal{P}_d^n - \dim \mathcal{P}_{d-1}^n)$ *linearly independent polynomials of the form*

$$x_1^{a_1} x_2^{a_2} \ldots x_n^{a_n} + S_j \ \text{with} \ S_j \in \mathcal{P}_{d-1}^n, j = 1, 2, \ldots, (\dim \mathcal{P}_d^n - \dim \mathcal{P}_{d-1}^n) \ \text{and} \ \sum_{i=1}^{n} a_i = d$$

are in span(S).

Theorem 4.6 *[40] Let $\{R_1, R_2, \ldots, R_t\} \subset \mathcal{P}_{d+1}^n$ be a set of independent d-orthogonal polynomials that is fundamental of degree $d + 1$. Let $\mathfrak{A} := (R_1, R_2, \ldots, R_t)$ and $V := \text{span}\{R_1, R_2, \ldots, R_t\}$. Let $N + t = \dim \mathcal{P}_{d+1}^n$ and U an arbitrary but fixed vector space such that $\mathcal{P}_{d+1}^n = V \oplus U$. Then the following statements are equivalent :*

1) *There exists an interpolatory cubature formula of degree d*

$$Q[f] := \sum_{i=1}^{N} w_i f(y^{(i)}), y^{(i)} \in \mathbb{R}^n, w_i > 0$$

   with $\{y^{(1)}, y^{(2)}, \ldots, y^{(N)}\} \subset NG(\mathfrak{A})$.

2) *$\mathfrak{A}$ and U are characterized by :*

   (a) *$\mathfrak{A} \cap U = (0)$*

   (b) *$I[f^2 - R^+] > 0$ for all $f \in U$, where $R^+ \in \mathfrak{A}$ is chosen such that $f^2 - R^+ \in \mathcal{P}_d^n$.*

3) *$\mathfrak{A}$ is a real ideal and $|NG(\mathfrak{A}) \cap \mathbb{R}^n| = N$. The knots of the cubature formula are the elements of $NG(\mathfrak{A}) \cap \mathbb{R}^n$.*

A starting point in Theorem 4.6 is that the ideal $\mathfrak{A}$ is fundamental of degree $d + 1$. In general, $\mathfrak{A}$ will be fundamental of degree $l, l + 1, \ldots$ where $\lfloor d/2 \rfloor + 1 \leq l \leq d + 1$. Let m be such that $\mathfrak{A}$ is fundamental of degree m but is not fundamental of degree $m - 1$. One can try to determine a set of polynomials of degree m that is a basis of an ideal that satisfies the conditions of Theorem 4.6. This idea was first suggested by Morrow and Patterson [36] for 2-dimensional regions. It was further developed by Schmid [40, 42].

4.4 On lower bounds for the number of knots

Ideal theory can be used to obtain lower bounds for the number of knots in a (pair of) cubature formula(e). Lower bounds are important. Any method to construct cubature formulae depends (implicitly or explicitly) on some lower bound. If a lower bound is known, then usually a method to construct cubature formulae that attain this bound, is known. Good lower bounds can learn us that some cubature formulae cannot exist, although their structure satisfies the consistency conditions.

It is well known that the number of knots N in a cubature formula of degree $2k - 1$ satisfies $N \geq \dim \mathcal{P}_{k-1}^n$ (Theorem 4.4). Any higher lower bound must take into account some information about the region and weight function.

For a central symmetric integral (i.e. where region and weight function remain invariant after reflection through the center) H.M. Möller found a better lower bound.

Theorem 4.7 *[33]*
Let γ := *the dimension of the space of even polynomials of* $\mathcal{P}_k^n$ *and*
     δ := *the dimension of the space of odd polynomials of* $\mathcal{P}_k^n$.
The minimum number of knots in a cubature formula of degree $2k - 1$ *for a central symmetric region* $\Omega \subset \mathbb{R}^n, n \geq 1$, *is*

$$N_{hmm} = 2\delta \qquad \text{if } k \text{ is even} \quad \text{and}$$
$$N_{hmm} = 2\gamma - 1 \quad \text{if } k \text{ is odd.}$$

Furthermore, cubature formulae with N_{hmm} *knots are central symmetric and have all weights positive.*

For $n = 1$, Theorem 4.4 and Theorem 4.7 give the same lower bound. For $n = 2$

$$N_{hmm} = \frac{k(k+1)}{2} + \left\lfloor \frac{k}{2} \right\rfloor = \dim \mathcal{P}_{k-1}^2 + \left\lfloor \frac{k}{2} \right\rfloor. \tag{12}$$

This is also the lower found if the region Ω is a triangle [3, 34].

Together with the lower bound (12), Möller obtained a method for constructing cubature formulae with N_{hmm} knots. Morrow and Patterson [36] also developed a method for constructing cubature formulae with N_{hmm} knots for two-dimensional fully symmetric product regions. But soon it became clear that cubature formulae with N_{hmm} knots do not always exist (see [36, 46]).

Definition 4.9 *A cubature formula* Q_1 *is partially embedded in a cubature formula* Q_2 *if* Q_1 *and* Q_2 *have some knots in common.*
A cubature formula Q_1 *is (totally) embedded in a cubature formula* Q_2 *if all knots of* Q_1 *are used by* Q_2.

Definition 4.10 *A pair of cubature formulae is said to be a good pair if it has the following properties:*

1) *If a knot is used by both Q_1 and Q_2, then the corresponding weights in the two cubature formulae should be different.*

2) *All knots have to be inside the region and the weights of the highest degree formula have to be positive.*

This definition of *good pair* was first introduced in [4].

Theorem 4.8 *[15] The minimum number of knots in a good pair of quadrature or cubature formulae of degree $2k - 1$ and $2k + m, m \in \mathbb{N}$ is*

$$\binom{k-1+n}{n} + \binom{k+n}{n}, n \in \mathbb{N}_0.$$

In the one-dimensional case the lower bound of Theorem 4.8 is attained for two different values of m, if the Gauss-Kronrod formula exists. In the multi-dimensional case the lower bound of Theorem 4.8 can only be attained for $m \leq 1$.

4.5 The T-method for constructing cubature formulae

In this section we will briefly describe a method developed by H.J. Schmid [39].

Consider a 2-dimensional central symmetric region. Following the idea mentioned in section 4.3, Schmid considered the case where the ideal $\mathfrak{A}$ associated with a cubature formula of degree $2k - 1$, is fundamental of degree $k + 1$. All cubature formulae with N_{hmm} and $N_{hmm} + 1$ knots are included in this case [6].

Let $R_0, R_1, \ldots, R_{k+1}$ be linearly independent polynomials of degree $k + 1$ in 2 variables, x and y. According to Corollary 4.1, these polynomials are orthogonal to all polynomials of degree $k - 2$ if they become zero in the knots of a cubature formula of degree $2k - 1$. Thus, the R_is can be written as

$$R_i := P^{k+1-i,i} + \sum_{j=0}^{k} \beta_{ij} P^{k-j,j} + \sum_{j=0}^{k-1} \gamma_{ij} P^{k-1-j,j}, i = 0, 1, \ldots, k+1 \tag{13}$$

where the $P^{a,b}$ are basic orthogonal polynomials (11). The β_{ij} and γ_{ij} are parameters which have to be determined such that the R_is belong to an ideal $\mathfrak{A}$ that satisfies the conditions of Theorem 4.6. Because the integral is central symmetric, the basic orthogonal polynomials have a special form and the β_{ij} vanish.

The T-method is based on the following observations :

a) Let $T_i := yR_i - xR_{i+1}, i = 0, 1, \ldots, k$. Then T_i is a polynomial of degree k and T_i has to be orthogonal to all polynomials of degree $k - 1$.

b) The polynomials $xT_i, yT_i, i = 0, 1, \ldots, k$ are of degree $k + 1$ and they belong to $\mathfrak{A}$. Thus $xT_i, yT_i \in span\{R_0, R_1, \ldots, R_{k+1}\}$.

Both conditions lead to necessary conditions for the γ_{ij} : linear and quadratic equations in the γ_{ij} s. For details we refer to [6, 39, 42]. From Theorem 4.6 (2b) Schmid obtained a set of inequalities for the γ_{ij}. These inequalities together with the linear and quadratic equations give necessary and sufficient conditions for the γ_{ij} so that all conditions of Theorem 4.6 are satisfied. The T-method has been used by Schmid [41] to construct cubature formulae of degree ≤ 9 for C_2^α. We used it to construct cubature formulae of degree ≤ 13 for $C_2, S_2, E_2^{r^2}$ without finding new ones. The only new results we obtained, are that a cubature formula of degree 13 with 31 knots does not exist for S_2 and $E_2^{r^2}$. Cools and Schmid [16] used the method to construct cubature formulae with N_{hmm} knots of arbitrary degree for $C_2^{0.5}$ and $C_2^{-0.5}$. A remarkable observation is that these formulae are invariant under reflection through the diagonals.

4.5.1   The S-method for constructing cubature formulae

The S-method was suggested by us in an attempt to find a method that is less dependent on Möller's lower bound than the T-method. If the T-method is used to construct symmetric cubature formulae for a 2-dimensional symmetric integral then $\gamma_{ij} = 0$ if $i + j$ is odd in the polynomials R_i (13). The polynomials R_i can be divided in two sets : $A := \{R_i | i$ is even$\}$ and $B := \{R_i | i$ is odd$\}$. Instead of demanding that $(A \cup B) \subset \mathfrak{A}$, as in the T-method, we demand that $A \subset \mathfrak{A}$ or $B \subset \mathfrak{A}$. We assign $C := A$ and $q := 0$ if we want to investigate the case $A \subset \mathfrak{A}$. We assign $C := B$ and $q := 1$ if we want to investigate the case $B \subset \mathfrak{A}$. The S-method is based on the following observations :

a) Let $S_i := y^2 R_i - x^2 R_{i+2}, i = q, q + 2, \ldots, k - 1$. Then S_i is a polynomial of degree $k + 1$ and S_i must be orthogonal to all polynomials of degree $k - 2$.

b) Because S_i has degree $k + 1$, $S_i \in span(C)$.

Both conditions lead to necessary conditions for the γ_{ij} : linear and quadratic equations in the γ_{ij} s. Later, we proved necessary and sufficient conditions for our method. These are also linear and quadratic equations.

The S-method has been used to construct cubature formulae of degree 13 with 36, 35 and 34 knots for several standard regions and of degree 17 for C_2. For details we refer to [6, 10, 12].

4.6   Pairs of embedded cubature formulae

The construction of pairs or sequences of embedded cubature formulae has also been done using ideal theory. The two most important strategies for constructing such sequences are

1) *bottom up*, better known as optimal addition of knots. This is based on the same idea as Gauss-Kronrod-Patterson sequences.

2) *top down*, i.e. starting from a high degree formula, lower degree embedded interpolatory formulae are constructed using a subset of the given points.

We will now only mention some results for the construction of 2-dimensional cubature formulae that illustrate these 2 strategies.

Theorem 4.9 *[8] If a central symmetric cubature formula of degree $2k - 1$ for a central symmetric integral has N_{hmm} or $N_{hmm} + 1$ knots, there exist two orthogonal polynomials of degree $k + 1$ that become zero at these knots. Furthermore, the set of all common zeros of these polynomials can be used as knots of a cubature formula of degree $2k + 1$, if all these common zeros are finite.*

If Theorem 4.9 is used to construct a pair of cubature formulae of degrees $2k - 1$ and $2k + 1$ the total number of function evaluations is $(k + 1)^2$. This is equal to our lower bound (Theorem 4.8).

Theorem 4.10 *[6, 14] Assume T_1 and T_2 have the following properties.*

1) *T_1 and T_2 are orthogonal polynomials of degree k for an integral I.*

2) *T_1 and T_2 have exactly k^2 distinct finite common zeros.*

Then there exists a subset of the k^2 common zeros of T_1 and T_2 with

$$\begin{array}{ll} k^2 - \frac{(t-1)(t-2)}{2} & if \quad t \leq k \\ \frac{(2k-t+2)(2k-t+1)}{2} & if \quad t > k \end{array}$$

elements which can be used as knots in a cubature formula of degree $2k - t$ for the integral I.

For $t = 1$ and $t = 3$ a pair of embedded cubature formulae with minimal number of points is obtained.

4.7   Evaluation

Ideal theory is a powerful tool for theoretical investigations of cubature formulae. The most complex concepts of ideal theory have only been used to develop construction methods and to prove theorems about cubature formulae. The reader probably noticed that we do not need the most sophisticated part of ideal theory to construct formulae. This is one of the beautiful aspects of ideal theory. Ideal theory does not distinguish between fortuitous and non-fortuitous formulae. The construction methods we described require the solution of systems of linear and quadratic equations. These systems are in general smaller than the systems that determine the formulae.

5 Conclusion

Both the invariant theoretical and the ideal theoretical approach have their merits and intrinsic problems. It is therefore worthwhile to combine the two approaches.

At the moment we see two ways to do this. First, one can study the ideal of invariant polynomials associated with an invariant cubature formula (e.g. [48]). Second, and more promising, is the combination of representation theory of finite groups and ideal theory as introduced in [20, 21].

A lot of questions are waiting for an answer in this area. A lot of rain will fall in Bergen before constructing cubature formulae for three- and higher-dimensional regions becomes as easy (?) as constructing cubature formulae for two-dimensional regions.

References

[1] P. Appell. Sur une classe de polynome à deux variables et le calcul approche des integrales double. *Ann. Fac. Sci. Univ. Toulouse*, 4:H1–H20, 1890.

[2] M. Beckers and A. Haegemans. Construction of three-dimensional invariant cubature formulae. Report TW 85, Dept. of Computer Science, K.U. Leuven, 1986.

[3] H. Berens and H.J. Schmid. On the number of nodes of odd degree cubature formulae for integrals with Jacobi weights on a simplex. This volume, 1992.

[4] J. Berntsen and T.O. Espelid. On the construction of higher degree three-dimensional embedded integration rules. *SIAM J. Numer. Anal.*, 25:222–234, 1988.

[5] C. Chevalley. Invariants of finite groups generated by reflections. *Amer. J. Math.*, 77:778–782, 1955.

[6] R. Cools. *The construction of cubature formulae using invariant theory and ideal theory.* PhD thesis, Katholieke Universteit Leuven, 1989.

[7] R. Cools and A. Haegemans. Automatic computation of knots and weights of cubature formulae for circular symmetric planar regions. Report TW 77, Dept. of Computer Science, K.U. Leuven, 1986.

[8] R. Cools and A. Haegemans. Optimal addition of knots to cubature formulae for planar regions. *Numer. Math.*, 49:269–274, 1986.

[9] R. Cools and A. Haegemans. Automatic computation of knots and weights of cubature formulae for circular symmetric planar regions. *J. Comput. Appl. Math.*, 20:153–158, 1987.

[10] R. Cools and A. Haegemans. Construction of fully symmetric cubature formulae of degree $4k - 3$ for fully symmetric planar regions. *J. Comput. Appl. Math.*, 17:173–180, 1987.

[11] R. Cools and A. Haegemans. Another step forward in searching for cubature formulae with a minimal number of knots for the square. *Computing*, 40:139–146, 1988.

[12] R. Cools and A. Haegemans. Construction of symmetric cubature formulae with the number of knots (almost) equal to Möller's lower bound. In H. Brass and G. Hämmerlin, editors, *Numerical Integration III*, pages 25–36. Birkhäuser Verlag, 1988.

[13] R. Cools and A. Haegemans. Why do so many cubature formulae have so many positive weights ? *BIT*, 28:792–802, 1988.

[14] R. Cools and A. Haegemans. On the construction of multi-dimensional embedded cubature formulae. *Numer. Math.*, 55:735–745, 1989.

[15] R. Cools and A. Haegemans. A lower bound for the number of function evaluations in an error estimate for numerical integration. *Constr. Approx.*, 6:353–361, 1990.

[16] R. Cools and H.J. Schmid. Minimal cubature formulae of degree $2k - 1$ for two classical functionals. *Computing*, 43:141–157, 1989.

[17] P.J. Davis and P. Rabinowitz. *Methods of numerical integration.* Academic Press, 1984.

[18] T.O. Espelid. On the construction of good fully symmetric integration rules. *SIAM J. Numer. Anal.*, 24:855–881, 1987.

[19] R. Franke. Orthogonal polynomials and approximate multiple integration. *Siam J. Numer. Anal.*, 8:757—765, 1971.

[20] K. Gatermann. Gruppentheoretische Konstruction von symmetrischen Kubaturformeln. Technical Report TR 90-1, Konrad-Zuse-zentrum für Informationstechnik Berlin, 1990.

[21] K. Gatermann. Linear representations of finite groups and the idealtheoretical construction of G-invariant cubature formulas. This volume, 1992.

[22] W. Gröbner. *Moderne Algebraische Geometrie.* Springer Verlag, Wien, 1949.

[23] A. Haegemans. Circularly symmetrical integration formulas for two-dimensional circularly symmetrical regions. *BIT*, 16:52–59, 1976.

[24] A. Haegemans and R. Piessens. Construction of cubature formulas of degree eleven for symmetric planar regions, using orthogonal polynomials. *Numer. Math.*, 25:139–148, 1976.

[25] A. Haegemans and R. Piessens. Construction of cubature formulas of degree seven and nine symmetric planar regions, using orthogonal polynomials. *SIAM J. Numer. Anal.*, 14:492–508, 1977.

[26] D. Hilbert. Über die Theorie der algebraischen Formen. *Math. Ann.*, 36:473–534, 1890.

[27] P. Keast and J.N. Lyness. On the structure of fully symmetric multidimensional quadrature rules. *SIAM. J. Numer. Anal.*, 16:11–29, 1979.

[28] J.N. Lyness and D. Jespersen. Moderate degree symmetric quadrature rules for the triangle. *J. Inst. Maths Applics.*, 15:19–32, 1975.

[29] F. Mantel. Non-fortuitous, non-product, non-fully symmetric cubature structures (abstract). In P. Keast and G. Fairweather, editors, *Numerical Integration*, page 205. Reidel Publishing Company, 1987.

[30] F. Mantel and P. Rabinowitz. The application of integer programming to the computation of fully symmetric integration formulas in two and three dimensions. *SIAM J. Numer. Anal.*, 14:391–425, 1977.

[31] J.C. Maxwell. On approximate multiple integration between limits of summation. *Proc. Cambridge Philos. Soc.*, 3:39–47, 1877.

[32] T. Molien. Über die Invarianten der linearen Substitutiongruppen. *Berliner Sitzungsberichte*, pages 1152–1156, 1898.

[33] H.M. Möller. *Polynomideale und Kubaturformeln*. PhD thesis, Universität Dortmund, 1973.

[34] H.M. Möller. Kubaturformeln mit minimaler Knotenzahl. *Numer. Math.*, 25:185–200, 1976.

[35] H.M. Möller. On the construction of cubature formulae with few nodes using Gröbner bases. In P. Keast and G. Fairweather, editors, *Numerical Integration*, pages 177–192. Reidel Publ. Comp., 1987.

[36] C.R. Morrow and T.N.L. Patterson. Construction of algebraic cubature rules using polynomial ideal theory. *SIAM J. Numer. Anal.*, 15:953–976, 1978.

[37] P. Rabinowitz and N. Richter. Perfectly symmetric two-dimensional integration formulas with minimal number of points. *Math. Comp.*, 23:765–799, 1969.

[38] J. Radon. Zur mechanischen Kubatur. *Monatsh. Math.*, 52:286–300, 1948.

[39] H.J. Schmid. Construction of cubature formulae using real ideals. In W. Schempp and K. Zeller, editors, *Multivariate approximation theory*, pages 359–377, Stuttgart, 1979. Birkhäuser Verlag.

[40] H.J. Schmid. Interpolatorische Kubaturformeln und reelle Ideale. *Math. Z.*, 170:267–282, 1980.

24

[41] H.J. Schmid. Interpolatorische Kubaturformeln. *Dissertationes Mathematicae*, CCXX, 1983.

[42] H.J. Schmid. Minimal cubature formulae and matrix equations. Unpublished, 1991.

[43] G.C. Shephard and J.A. Todd. Finite unitary reflection groups. *Canad. J. Math.*, 6:274–304, 1954.

[44] S.L. Sobolev. Cubature formulas on the sphere invariant under finite groups of rotations. *Soviet Math.*, 3:1307–1310, 1962.

[45] A.H. Stroud. *Approximate calculation of multiple integrals.* Prentice Hall, 1971.

[46] P. Verlinden and R. Cools. Minimal cubature formulae of degree $4k + 1$ for integrals with circular symmetry. Report TW 137, Dept. of Computer Science, K.U. Leuven, 1990.

[47] P. Verlinden and R. Cools. A cubature formula of degree 19 with 68 nodes for integration over the square. Report TW 149, Dept. of Computer Science, K.U. Leuven, 1991.

[48] P. Verlinden, R. Cools, D. Roose, and A. Haegemans. The construction of cubature formulae for a family of integrals: a bifurcation problem. *Computing*, 40:337–346, 1988.

LINEAR REPRESENTATIONS OF FINITE GROUPS AND THE IDEAL THEORETICAL CONSTRUCTION OF G–INVARIANT CUBATURE FORMULAS

KARIN GATERMANN
Konrad-Zuse-Zentrum Berlin
Heilbronner Str. 10
D–1000 Berlin 31
Germany
Email: Gatermann@sc.ZIB-Berlin.de

ABSTRACT. Under some conditions the common zeros of some sufficiently orthogonal polynomials may be chosen as the nodes of a cubature formula (Möller). Once the nodes are known the weights are determined by the solution of a system of linear equations. It is shown how in this approach the theory of linear representations of finite groups exploites the symmetry which is an additionally required property of cubature formulas for regions such as the triangle, square or the tetrahedron. By means of this theory reasonable parameter dependent orthogonal polynomials are chosen and equations for these parameters are determined from some necessary conditions, the existence of the so-called syzygies.

1. Introduction

The numerical evaluation of integrals

$$I(f) := \iint\limits_{\Omega} w(x)f(x)\,dx, \quad \Omega \subseteq \mathbb{R}^N, \; w, f \in C(\mathbb{R}^N), \; w \geq 0,$$

is approximately done by a cubature formula $Q(f) = \sum_{i=1}^{n} A_i f(y_i)$.

The integral is assumed to be G–invariant, i.e.

$$I(f(tx)) = I(f(x)), \quad \forall t \in G, \quad f \in C(\mathbb{R}^N),$$

where G is a finite group of affine linear transformations $t : \mathbb{R}^N \longrightarrow \mathbb{R}^N$. We are concerned with the computation of the nodes $y_i \in \mathbb{R}^N$ and weights $A_i \in \mathbb{R}$ of a G–invariant cubature formula, i.e.

$$Q(f(t(x))) = Q(f(x)), \quad \forall t \in G, \quad f \in C(\mathbb{R}^N).$$

A G–invariant formula has a G–invariant set of nodes and $A_i = A_j$ if a $t \in G$ exists with $t(y_i) = y_j$.

A cubature formula is said to have degree d if $I_{|\mathbb{P}_d} = Q_{|\mathbb{P}_d}$, but $I_{|\mathbb{P}_{d+1}} \neq Q_{|\mathbb{P}_{d+1}}$, where $\mathbb{P}_d \subset \mathbb{R}[x]$ is the vector space of all polynomials p with $\deg(p) \leq d$.

25

T. O. Espelid and A. Genz (eds.), Numerical Integration, 25–35.
© *1992 Kluwer Academic Publishers.*

In a generalization of the one-dimensional case ($N = 1$) in the construction procedure of a cubature formula of degree d for a given integral and given degree d the common zeros of some polynomials $p_j, j = 1, \ldots, m$ under certain conditions may be chosen to be the nodes (MÖLLER [7], [8], STROUD [13]). One of these conditions is that the polynomials are a special ideal basis of the ideal generated by these polynomials, which is equivalent to the existence of some syzygies $\sum_{j=1}^{m} q_j p_j = 0$, $q_j \in \mathbb{R}[x]$. Finally, the weights are determined from the solution of a system of linear equations.

On these lines one tries to start with some p_j depending on parameters which are determined by some necessary conditions, the existence of syzygies. Exploiting additionally the symmetry, the T-method (SCHMID [10]) and the S-method (COOLS, HAEGEMANS [1]) for example are based on this idea.

In this paper it is explained that the theory of linear representations (see SERRE [11] or STIEFEL, FÄSSLER [12]) simplifies this way of construction of G-invariant cubature formulas twice: It brings clarity and transparency to the possible choices of polynomials p_j and to the necessary conditions, the existence of the syzygies.

As examples formulas of degree 5 for the tetrahedron T_3 and a sequence of formulas for the triangle T_2 are computed using the Computer Algebra System REDUCE.

2. Cubature formulas and ideals

The integral I is assumed to be *positive*, i.e.

$$I(f^2) > 0, \quad \forall f \in C(\Omega) - \{0\}.$$

Because of this property, I gives an inner product. Let p vanish in the nodes y_i of a cubature formula of degree d. Since

$$I(fp) = Q(fp) = 0$$

for all $f \in \mathbb{R}[x]$ with $fp \in \mathbb{P}_d$, polynomials p with the property

$$I(fp) = 0, \quad \forall f \in \mathbb{R}[x] \quad \text{with} \quad (fp) \in \mathbb{P}_d$$

are defined to be *d-orthogonal* [7]. Let K_s denote the vector space of all polynomials of degree s (and 0) which are $(2s - 1)$-orthogonal.

Lemma 2.1 : *Each d-orthogonal p of degree $s \leq d$ has a representation*

$$p = p_s + \sum_{k=d-s+1}^{s-1} p_k \quad \text{with} \quad p_s \in K_s, p_s \neq 0, \ p_k \in K_k.$$

The set of polynomials $p \in \mathbb{R}[x]$ vanishing in all nodes y_i of a cubature formula Q is an ideal $\mathcal{A}$, i.e.

$$p_1, p_2 \in \mathcal{A} \quad \Rightarrow \quad f_1 p_1 + f_2 p_2 \in \mathcal{A}, \quad \forall f_1, f_2 \in \mathbb{R}[x].$$

There is a one-to-one-correspondence between the ideals consisting of polynomials vanishing (may be multiple) in some points and the ideals $\mathcal{A} = (p_1, \ldots, p_m)$ generated by some polynomials $p_1, \ldots, p_m \in \mathbb{R}[x]$.

A vector space basis of a direct complement M_k of $\mathbb{P}_{k-1}$ in $\mathbb{P}_k$ is called a *fundamental system* [10]. An ideal $\mathcal{A}$ has finitely many common zeros, iff there is a fundamental system in $\mathcal{A}$ of some degree K.

If we want the common zeros of some polynomials to be suitable as nodes of a cubature formula of degree d, the polynomials have to be d–orthogonal. This is not sufficient because for example $p := f_1 p_1 + f_2 p_2 \in \mathbb{P}_d$, $f_1 p_1 \notin \mathbb{P}_d$ and $I(p) \neq 0$ is possible even if p_1 and p_2 are d–orthogonal.

Definition 2.2 *(Möller [8]): The polynomials $p_1, \ldots, p_m$ are an H-basis of the ideal $\mathcal{A} = (p_1, \ldots, p_m)$ if for all $p \in \mathcal{A}$ polynomials $f_j, j = 1, \ldots, m$ exist with*

$$p = \sum_{j=1}^{m} f_j p_j \quad and \quad deg(f_j p_j) \leq deg(p).$$

The following theorem plays a central role in the computation of cubature formulas.

Theorem 2.3 *(Möller [7]): Let $\mathcal{A} = (p_1, \ldots, p_m) \subset \mathbb{R}[x]$.*

i.) $p_1, \ldots, p_m$ are d–orthogonal.

ii.) $p_1, \ldots, p_m$ are an H-basis of $\mathcal{A}$.

iii.) The common zeros $y_i, i = 1, \ldots, n$ of $p_1, \ldots, p_m$ are real and simple.

If conditions i.)-iii.) hold, then a cubature formula Q with nodes y_i of degree $\geq d$ exists.

Theorem 2.3 suggests the following construction procedure for a cubature formula for a given degree d. Choose some parameter dependent, d–orthogonal polynomials p_j, such that they are linearly independent. Moreover the leading homogenous parts have to be linearly independent and there should be a fundamental system in the generated ideal. When the parameters are determined from necessary conditions for the property of H-basis, then the common zeros of p_j are computed. If they are real and simple, the weights are given by the solution of a system of linear equations.

For a characterization of H-basis we introduce, for two vector spaces $M, E \subset \mathbb{R}[x]$, the notation $M \odot E$ for the space spanned by all products $f \cdot p, f \in M, p \in E$. Let $M_k = \mathbb{P}_k = \{0\}$ for $k < 0$. For given $p_1, \ldots, p_m$ let $E_l, l = s, \ldots, s + \nu$ denote the spaces spanned by those p_j which have degree exactly l. Define

$$W_k := \bigoplus_{l=s}^{s+\nu} (\mathbb{P}_{k-l} \odot E_l), \quad T_k := \bigoplus_{l=s}^{s+\nu} (M_{k-l} \odot E_l).$$

Proposition 2.4 : *Let* $p_1, \ldots, p_m \in \mathbb{R}[x]$ *and* E_l, W_k, T_k *as above. The following conditions are equivalent.*

i.) $p_1, \ldots, p_m$ *form an H-basis of* $\mathcal{A} = (p_1, \ldots, p_m)$.

ii.) $W_k \cap \mathbb{P}_{k-1} \subseteq W_{k-1}, \quad \forall k \in \mathbb{N}$.

iii.) $T_k \cap \mathbb{P}_{k-1} \subseteq W_{k-1}, \quad \forall k \in \mathbb{N}$.

Enlarging k gradually from $s+1$ to some $s+t$, the restricted condition iii.) gives for one k equations for the parameters in the following way. Compute a basis of $S_{k-1} := T_k \cap \mathbb{P}_{k-1}$ and from this compute a basis Q_λ of a direct complement V_{k-1} of W_{k-1} in $W_{k-1} + S_{k-1}$. This computation of the *critical polynomials* Q_λ may be done with a Computer Algebra System like REDUCE by checking the leading terms of the polynomials. If the p_j form an H-basis, then $V_{k-1} = \{0\}$. Thus demanding $Q_\lambda = 0$ gives equations for the parameters by checking coefficients.

$Q_\lambda = 0$ coresponds to a nontrivial representation of zero $\sum_{j=1}^m g_j p_j = 0$, where the tuple $g \in \mathbb{R}[x]^m$ is called *syzygie*. As the space of syzygies is a module M over $\mathbb{R}[x]$ ($f_1 g + f_2 \tilde{g} \in M, \forall g, \tilde{g} \in M, f_1, f_2 \in \mathbb{R}[x]$), it has a finite module basis consisting of syzygies with low degree of g_j. To investigate this module basis is in general a difficult question because it depends on the number of p_j, their degrees, and on p_j itself. But first this implies that it suffices to show condition iii.) for some k only. Second it reflects the fact that enlarging k to $k+1$ gradually and demanding $x_i Q_\lambda = 0$ gives no supplementary equations. In other words V_k may be chosen such that $M_1 \odot V_{k-1} \subseteq V_k$. This has to be considered if the number of equations for the parameters is determined in advance without actual computations. The number of equations from demanding $Q_\lambda = 0$ for one critical polynomial equals the dimension of the smallest vector space which contains Q_λ as an element for all parameter values, for example Q_λ is a combination of d–orthogonal polynomials.

3. Representation Theory

Of special interest are the integrals T_2 and T_3 with weight function $w \equiv 1$ and regions

$$\Omega = \{(x_1, x_2) \in \mathbb{R}^2 \mid x_1, x_2 \geq 0, x_1 + x_2 \leq 1\},$$
$$\Omega = \{(x_1, x_2, x_3) \in \mathbb{R}^3 \mid x_1, x_2, x_3 \geq 0, x_1 + x_2 + x_3 \leq 1\},$$

respectively, where the notations agree with those in STROUD [13]. The integral T_2 is invariant with respect to the group $D_3 = \{id, r, r^2, s, sr, sr^2\}$, where

$$r : \mathbb{R}^2 \longrightarrow \mathbb{R}^2, \quad (x_1, x_2) \longrightarrow (x_2, 1 - x_1 - x_2) \quad \text{and}$$
$$s : \mathbb{R}^2 \longrightarrow \mathbb{R}^2, \quad (x_1, x_2) \longrightarrow (x_2, x_1)$$

correspond to a rotation and a reflection, respectively. The tetrahedron T_3 is invariant with respect to the group S_4 of all permutations of $x_1, x_2, x_3, 1 - x_1 - x_2 - x_3$. Cubature formulas for these integrals are demanded to be invariant as well.

Because in Section 4 the theory of linear representations is applied to the computation of G–invariant cubature formulas, the definitions and theorems are collected restricting ourselves to real or complex representations of finite groups such as D_3 and S_4.

Definition 3.5 (STIEFEL, FÄSSLER[12]): *Let G be a finite group.*

  i.) A linear representation ϑ of G in a (real or complex) vector space V is a mapping
  $\vartheta : G \longrightarrow \mathrm{Aut}(V)$ from G in the group $\mathrm{Aut}(V)$ of automorphisms of V satisfying

$$\vartheta(s)\vartheta(t) = \vartheta(st), \quad \forall s,t \in G.$$

  Each $\vartheta(t)$ has a representation as a matrix, which is denoted by $\vartheta(t)$ as well. The dimension of ϑ is the dimension of V as a vector space. The mapping

$$\psi : G \longrightarrow \mathbb{K}, \quad \psi(t) = trace(\vartheta(t)), \quad \mathbb{K} = \mathbb{C} \text{ or } \mathbb{R}$$

  is called the character *of ϑ.*

  ii.) Two representations ϑ_1, ϑ_2 of G in V_1 and V_2 are isomorphic if there is a linear isomorphism $\phi : V_1 \to V_2$ which transfers ϑ_1 in ϑ_2, i.e.

$$\phi \circ \vartheta_1(t) = \vartheta_2(t) \circ \phi, \quad \forall t \in G.$$

  iii.) A linear subspace $W \subset V$ is called G–invariant with respect to $\vartheta : G \to \mathrm{Aut}(V)$ if

$$\vartheta(t)(w) \in W, \quad \forall t \in G, \ w \in W.$$

  The restriction of ϑ to a G–invariant subspace $W \subset V$

$$\vartheta^W : G \longrightarrow \mathrm{Aut}(V), \quad \vartheta^W(t) := \vartheta(t)_{|W}$$

  is a subrepresentation *of ϑ.*

  iv.) The tensorproduct $\vartheta^1 \times \vartheta^2 : G \longrightarrow Aut(V_1 \times V_2)$ of two linear representations ϑ^1, ϑ^2 of G in V_1 and V_2 is given by

$$(\vartheta^1 \times \vartheta^2)(t) : V_1 \times V_2 \longrightarrow V_1 \times V_2 , \quad (\vartheta^1 \times \vartheta^2)(t)(v_1 \otimes v_2) = \vartheta^1(t)(v_1) \otimes \vartheta^2(t)(v_2),$$

  where $V_1 \times V_2$ is the tensorproduct of the vector spaces V_1, V_2.

Example 3.6 : *For the groups $G = D_3, S_4$ above a real linear representation is given by*

$$\vartheta : G \longrightarrow Aut(\mathbb{R}[x]), \quad \vartheta(t) : \mathbb{R}[x] \to \mathbb{R}[x], \quad p(x) \longrightarrow p(t(x)).$$

Then G–invariant subspaces are $\mathbb{P}_l$ or K_k, the vector space of orthogonal polynomials of degree k with respect to the G–invariant integral. Since the construction of cubature formulas involves the discussion of spaces such as $M \odot E$, we introduce the tensorproduct. The subrepresentation $\vartheta^{M \odot E}$ is isomorphic to $\vartheta^M \times \vartheta^E$, if $\dim(M \odot E) = \dim M \cdot \dim E$.

For every G–invariant subspace $W \subset V$ there is always a complementary G–invariant subspace $W' \subset V$, $V = W \oplus W'$, $\vartheta = \vartheta^W \oplus \vartheta^{W'}$. If the only G–invariant subspaces of V are the trivial ones $\{0\}$ and V, the representation ϑ and the underlying space V are called *irreducible*. We have the following canonical decomposition of V in G–invariant subspaces:

Theorem 3.7 ([12]): *For every finite group G there is a finite number h of nonisomorphic irreducible representations ϑ^i. For every linear representation $\vartheta : G \to \mathrm{Aut}(V)$ there is a unique so-called* canonical decomposition *of V*

$$V = \bigoplus_{i=1}^{h} V_i , \qquad (1)$$

such that the subrepresentations ϑ^{V_i}, $V_i \neq \{0\}$, have only irreducible subrepresentations isomorphic to ϑ^i.

The further decomposition of V into its irreducible subspaces is not unique in general. But the *multiplicity* c_i of occurence of an irreducible representation ϑ^i in ϑ is unique and we may write

$$\vartheta = \sum_{i=1}^{h} c_i \vartheta^i .$$

Analougsly, $\psi = \sum_{i=1}^{h} c_i \psi^i$ holds, where ψ, ψ^i are the characters corresponding to ϑ and the irreducible representations ϑ^i, respectively.

Considering first complex representations the computation of the canonical decomposition uses the characters ψ^i in the projections onto the isotypic components V_i

$$\Pi^i : V \to V_i, \quad \Pi^i(v) = \frac{n_i}{|G|} \sum_{t \in G} \vartheta(t)(v) \psi^i(t^{-1}) , \quad i = 1, \ldots, h, \qquad (2)$$

where n_i is the dimension of the corresponding irreducible representation ϑ^i.

If we let ϑ^1 be the *trivial representation* with $\psi^1(t) = 1$, $\forall t \in G$, then V_1 is the space of all G–*invariant* elements in V. In case $V = \mathbb{C}[x]$ the component V_1 is the ring of invariants.

The character corresponding to $\rho \times \vartheta$ is the product $\chi \cdot \psi$ of the characters χ and ψ corresponding to ρ and ϑ. Comparing the symmetry of $M \odot E$ for two G–invariant vector spaces $M, E \subseteq \mathbb{R}[x]$ in Example 3.6 with their tensorproduct the inequality $\psi^{M \odot E} \leq \psi^{M \otimes E} = \psi^M \cdot \psi^E$ holds, where $\psi^{M \odot E}, \psi^M, \psi^E$ are the corresponding characters. Here we shortly write $\psi \leq \chi$ for $c_i \leq b_i, i = 1, \ldots, h$ where $\psi = \sum c_i \psi^i, \chi = \sum b_i \psi^i$.

For irreducible representations with dimension $n_i > 1$ occurring with multiplicity $c_i > 1$, it is not sufficient to choose an arbitrary basis of the isotypic component V_i. A *symmetry adapted* basis

$$
\begin{matrix}
v_1^1 & \cdots & v_{n_i}^1 \\
\vdots & & \vdots \\
v_1^{c_i} & \cdots & v_{n_i}^{c_i}
\end{matrix}
\qquad (3)
$$

has to be constructed such that each row $v_1^k, \ldots, v_{n_i}^k$ spans a G–invariant space whose transformation by G is described by a given set of matrices $\vartheta^i(t) \in \mathbb{C}^{n_i, n_i}$ of the irreducible representation ϑ^i with elements $\vartheta^i(t)_{k,l}$. In [11], [12] the computation of such a symmetry adapted basis by means of projections is treated.

For $G = D_3$ and $G = S_4$ above the complex irreducible representations ϑ^i with (real) characters ψ^i are known [11]. In [4] tables with the products $\psi^i \psi^j$ represented as sums of the ψ^i are given. For these groups ϑ^i can be viewed as real irreducible representations and the projections are real mappings giving the real canonical decomposition. However, for other groups there may be a difference between complex and real irreducible representations.

4. Application of linear representations to the construction of invariant cubature formulas

In this section the construction of cubature formulas with Theorem 2.3 combined with the theory of linear representations is worked out. Let the integral I be G–invariant with respect to a group G of affine linear transformations $t : \mathbb{R}^N \to \mathbb{R}^N$ with real irreducible representations ϑ^i and corresponding characters $\psi^i, i = 1, \ldots, h$. We consider the linear representation $\vartheta(t)(p(x)) = p(t(x))$, $t \in G$, $p \in \mathbb{R}[x]$.

The ideal consisting of polynomials vanishing in the nodes of a G–invariant cubature formula is G–invariant because the set of zeros is G–invariant. If in addition the ideal $\mathcal{A}$ in Theorem 2.3 is G–invariant, then the set of common zeros is G–invariant and the weights may be chosen such that the cubature formula is G–invariant. Because every G–invariant vector space has a canonical decomposition, one restricts the approach of polynomials p_j to elements of the isotypic components of $\mathbb{R}[x]$ which is not a limitation of possible ideals.

Proposition 4.8 : *Let $\mathcal{A} = \sum_{i=1}^{h} \mathcal{A}^i$ be the real canonical decomposition of a G–invariant ideal $\mathcal{A} \subset \mathbb{R}[x]$. The ideal $\mathcal{A}$ has an H-basis*

$$p_j^i \in \mathcal{A}^i \subset \mathbb{R}[x], \quad j = 1, \ldots, m_i, i = 1, \ldots, h,$$

where the $p_j^i, j = 1, \ldots, m_i$ form a symmetry adapted basis, if $n_i > 1$.

Since every $\mathbb{P}_k \subset \mathbb{R}[x]$ has a G–invariant complement M_k in $\mathbb{P}_{k+1}$ with a canonical decomposition $M_k = \sum_{i=1}^{h} M_k^i$, the ring $\mathbb{R}[x]$ is structured by G:

$$\mathbb{R}[x] = \bigoplus_{k=0}^{\infty} \bigoplus_{i=1}^{h} M_k^i .$$

A special case are the isotypic components $M_k^i = K_k^i$ of K_k, the space of orthogonal polynomials of degree k. Obviously, $\mathcal{A}^i \subseteq \bigoplus_{k=0}^{\infty} K_k^i$.

Thus Proposition 4.8 together with Lemma 2.1 explains how to choose parameter dependent, d–orthogonal polynomials which for some parameter values hopefully form the H-basis of an ideal corresponding to a cubature formula of degree d:

$$p_j^i = p_{s_j}^i + \sum_{k=d-s_j+1}^{s_j-1} p_k^i, \quad j = 1, \ldots, m_i, \ i = 1, \ldots, h,$$

$p_{s_j}^i \in K_{s_j}^i - \{0\}, p_k^i \in K_k^i$. Here the polynomials $p_j^i, j = 1, \ldots, m_i$ form a symmetry adapted basis in case $n_i > 1$. In NOOIJEN, TE VELDE, BAERENDS [9] this approach was given, but they restricted this way of construction to two polynomials and to two-dimensional integrals.

The number of parameters depends on $\dim K_k^i$, the dimension n_i of ϑ^i, the degrees s_j, and the number m_i of polynomials p_j^i. Obviously, we have to choose $m_i \leq \dim K_s^i$ avoiding to violate the property of H-basis.

A preparatory step is the computation of a symmetry adapted basis of K_k^i by means of some projections like (2) applied to a basis of K_k.

Let $\mathcal{A}$ be the ideal generated by p_j^i and let $E_s, \ldots, E_{s+\nu}$ denote the G–invariant vector spaces spanned by the p_j^i of degree $l, l = s, \ldots, s + \nu$. In Section 1 the multiplicative structure of the ideal is used by investigation of W_k, T_k, S_k, V_k. Since $\mathbb{P}_l$ are G–invariant, M_l are chosen G–invariant. Consequently W_k, T_k, and S_{k-1} are G–invariant and have canonical decompositions

$$W_k = \sum_{i=1}^h W_k^i, \quad T_k = \sum_{i=1}^h T_k^i, \quad S_k = \sum_{i=1}^h S_k^i.$$

Before investigating the computation of critical polynomials in view of symmetry we discuss how to complete the approach with $E_{s+\nu+1}$. So let $E_s, \ldots, E_{s+\nu}$ be known and let W_k, T_k, S_{k-1} for $k = s + \nu + 1$ be computed. There is a G–invariant space R_k with $T_k = S_{k-1} \oplus R_k$. Obviously,

$$\psi^{T_k} = \psi^{S_{k-1}} + \psi^{R_k}, \qquad \psi^{R_k} \leq \psi^{M_k}.$$

In order of a minimal completition and of condition iii.) of Proposition 2.4, the approach E_k should satisfy $E_k \cap T_k = \{0\}$ and $\psi^{E_k} \leq \psi^{M_k} - \psi^{R_k}$. If the chances of an approach $E_{s+\nu+1}$ are investigated with the knowledge of $\psi^{E_l}, l = s+, \ldots, s+\nu$, but without computations and without knowledge of T_k, S_{k-1} and R_k, the character ψ^{E_k} may be chosen by the following considerations. The character of T_k satisfies $\psi^{T_k} \leq \chi^k$, where

$$\chi^k := \sum_{l=s}^{s+\nu} \psi^{M_{k-l}} \cdot \psi^{E_l}.$$

It is natural to assume R_k and ψ^{R_k} to be maximal. Especially, if $\chi^k \geq \psi^{M_k}$, a fundamental system of degree k may be assumed in $\mathcal{A}$. Based on the assumed ψ^{R_k} a character ψ^{E_k} may be chosen.

The following considerations are done for several k starting with $k = s+1$. If all $E_j, j \leq k$ are chosen, the next step in the algorithm is the computation of critical polynomials Q_λ, i.e. a direct complement V_{k-1} of W_{k-1} in $S_{k-1} + W_{k-1}$ is demanded to be $\{0\}$. Since V_{k-1} may be chosen G–invariant, we may concentrate on the isotypic components V_{k-1}^i, i.e. critical polynomials $Q_\lambda^i \in V_{k-1}^i$ have to be computed. For irreducible representations with dimension $n_i \geq 2$ it suffices to restrict to the first column of a symmetry adapted basis. As V_{k-1}^i is a direct complement of S_{k-1}^i in $S_{k-1}^i + W_{k-1}^i$, for the computation of a special Q_λ^i only some spaces M_j^i and polynomials p_j^i are involved which simplifies the computations

Table 1: Special cubature formulas for T_3 of degree 5 with 15 nodes

λ	References	Remark
$-\frac{17}{2}$	Grundmann,Möller [5]	member of a class of formulas
≈ 0.245	Grundmann,Möller [5]	$A_4 = 0$, 14 nodes
$\frac{9}{22}$	Stroud [13]	$T_3 : 5 - 1$
$\frac{29}{49}$	Keast [6]	$y_1 = \frac{1}{3}$
≈ 2.473		$A_4 = 0$, 14 nodes

a lot. Since $Q_\lambda^i \in V_{k-1}^i$ is a linear combination of d-orthogonal polynomials, the resulting number of equations is

$$\sum_{j=d-k+1}^{k-1} \frac{1}{n_i} \dim K_j^i \; - \; \frac{1}{n_i} \dim W_{k-1}^i \,.$$

The relevant symmetry types i are derived by comparision of $\psi^{S_{k-1}}$ and $\psi^{W_{k-1}}$. Enlarging k by 1 one has to observe that the space of possible syzygies spanned by $x_j Q_\lambda^i$ and the character $\psi^{M_1} \cdot \psi^{V_{k-1}}$ are already known.

If the p_j^i form an H-basis, then $W_{k-1} = \sum_{j=s}^{k-1} R_j$ and $\psi^{W_{k-1}} = \sum_{j=s}^{k-1} \psi^{R_j}$ are known. Without any actual computations of the Q_j^i he number of equations may be estimated just from the characters $\psi^{E_l}, l = s, \ldots, s+\nu$. For this estimation we have to enlarge k gradually and we have to assume the representations of the characters

$$\psi^{R_k} = \sum_{i=1}^h c_i^{R_k} \psi^i, \qquad \psi^{S_k} = \sum_{i=1}^h c_i^{S_k} \psi^i,$$

by comparison of χ^k and ψ^{M_k}. If the number of equations is not greater than the number of parameters the approach with polynomials p_j^i (or better the approach with $E_s, \ldots, E_{s+\nu}$) seems to be reasonable.

After computation of Q_λ^i and the solution of the resulting equations for the parameters the common zeros of p_j^i and the weights are computed, if the zeros are real and simple.

In [4] the $Ro3$–invariant cubature formula in [3] was verified with this algorithm.

Example 4.9 : *The algorithm above is used to construct S_4–invariant cubature formulas of degree 5. Using the notation in [11] S_4 has 5 irreducible representations with characters $\chi_0, \varepsilon, \theta, \psi, \varepsilon\psi$ of dimension $1, 1, 2, 3, 3$, respectively. Because*

$$\psi^{K_4} = 2 \cdot \chi_0 + 2 \cdot \theta + 2 \cdot \psi + \varepsilon\psi, \quad \psi^{K_3} = \chi_0 + 2 \cdot \psi + \varepsilon\psi,$$

it is reasonable to choose 6 polynomials of degree 3 such that $\psi^{E_3} = \psi + \varepsilon\psi$. Since $\psi^{M_1} = \psi$, $\psi^{M_4} = \psi^{K_4}$, and

$$\chi^4 = \psi^{M_1} \cdot \psi^{E_3} = \psi \cdot (\psi + \varepsilon\psi) = \chi_0 + \varepsilon + 2 \cdot \theta + 2\psi + 2\varepsilon\psi,$$

Table 2: D_3-invariant formula for T_2 of a sequence with 28 nodes of degree 9

x	y	$1 - x - y$	weight
$\frac{1}{3}$	$\frac{1}{3}$	$\frac{1}{3}$	$-6.97562388006918 \cdot 10^0$
$\frac{6+\sqrt{15}}{21}$	$\frac{6+\sqrt{15}}{21}$	$\frac{9-2\sqrt{15}}{21}$	$1.42504103834966 \cdot 10^{-2}$
$\frac{6-\sqrt{15}}{21}$	$\frac{6-\sqrt{15}}{21}$	$\frac{9+2\sqrt{15}}{21}$	$-9.50110616108759 \cdot 10^{-2}$
$\frac{1}{9}$	$\frac{1}{9}$	$\frac{7}{9}$	$1.22378153247155 \cdot 10^{-1}$
$\frac{4+\sqrt{15}}{9}$	$\frac{4-\sqrt{15}}{9}$	$\frac{1}{9}$	$8.15854354981127 \cdot 10^{-3}$
$\frac{2(5+\sqrt{15})}{55}$	$\frac{2(5+\sqrt{15})}{55}$	$\frac{35-4\sqrt{15}}{55}$	$2.37984719213637 \cdot 10^0$
$\frac{2(5-\sqrt{15})}{55}$	$\frac{2(5-\sqrt{15})}{55}$	$\frac{35+4\sqrt{15}}{55}$	$6.36757324486581 \cdot 10^{-3}$
$\frac{2(5+\sqrt{15})}{55}$	$\frac{2(5-\sqrt{15})}{55}$	$\frac{7}{11}$	$2.38626360941510 \cdot 10^{-2}$

$\psi^{R_4} = \chi_0 + 2\theta + \varepsilon\psi$ and $\psi^{S_3} = \varepsilon + \varepsilon\psi$ are assumed. Because $\psi^{E_3} = \psi + \varepsilon\psi$ the algorithm suggests to compute critical polynomials Q_λ, i.e. a basis of V_3 which has the character $\varepsilon + \varepsilon\psi$. But $\psi^{\mathbb{P}_3} - \psi^{E_3} = 3\chi_0 + \theta + 3\psi$ implies $V_3 = \{0\}$ in advance. Indeed the 15 common zeros

$$
\begin{array}{llll}
(y_1, & y_1, & y_1; & 1 - 3y_1)_{S_4} \\
(y_2, & y_2, & y_2; & 1 - 3y_2)_{S_4} \\
(y_3, & y_3, & 0.5 - y_3; & 0.5 - y_3)_{S_4} \\
(\frac{1}{4}, & \frac{1}{4}, & & \frac{1}{4})
\end{array}
\quad with \quad
\begin{array}{l}
y_1 = \frac{1}{8}\left(\frac{2\lambda-11}{2\lambda-7} + \sqrt{\frac{(-2\lambda+11)(2\lambda+21)}{7(2\lambda-7)^2}}\right) \\[8pt]
y_2 = \frac{1}{8}\left(\frac{2\lambda-11}{2\lambda-7} - \sqrt{\frac{(-2\lambda+11)(2\lambda+21)}{7(2\lambda-7)^2}}\right) \\[8pt]
y_3 = \frac{1}{4} + \frac{1}{2}\sqrt{\frac{2\lambda+3}{14(2\lambda+1)}} \quad ,
\end{array}
$$

for $\lambda \in (-\frac{21}{2}, -\frac{3}{2}) \cup (-\frac{1}{2}, \frac{11}{2}) \setminus \{\frac{7}{2}\}$ are the nodes of a continuum of S_4-invariant cubature formulas. Table 1 gives the values of λ for some known formulas and for a new 14-point-formula which unfortunately has points outside the region.

Example 4.10 : In [2] a D_3-invariant cubature formula for T_2 of degree 7 is given. Among its 16 nodes are the nodes of the formula $T_2 : 5 - 1$ (RADON, see [13]). There is a D_3-invariant cubature formula for T_2 of degree 9 with 28 nodes containing the 16 nodes of the formula above. This enlarges the pair of formulas to a sequence. The formula may be computed with a small modification of the outlined algorithm. D_3 has three irreducible representations with characters ψ^1, ψ^2, χ. It turns out that three 9-orthogonal polynomials p^2, p_1^3, p_2^3 of degree 6 are a good approach. Let $E_6 := \text{span}(p^2, p_1^3, p_2^3)$ and $\psi^{E_6} = \psi^2 + \chi$ where p_1^3, p_2^3 depend on one parameter. Additionally we restrict to $p^2, p_1^3, p_2^3 \in \mathcal{D}$, where $\mathcal{D}$ is the ideal of polynomials vanishing in the 16 nodes. Since

$$\chi^7 = \psi^{M_1} \cdot \psi^{E_6} = \chi(\psi^2 + \chi) = \psi^1 + \psi^2 + 2\chi,$$

and $\psi^{K_7} = \psi^1 + \psi^2 + 3\chi$ we assume $\psi^{R_7} = \chi^7, \psi^{S_6} = 0$. Since

$$\chi^8 = \psi^{M_2} \cdot \psi^{E_6} = (\psi^1 + \chi)(\psi^2 + \chi) = \psi^1 + 2\psi^2 + 3\chi,$$

we assume $\psi^{R_8} = \psi^1 + \psi^2 + 3\chi$ and $\psi^{S_7} = \psi^2$. Because $S_7 \subset \bigoplus_{j=2}^{7} K_j^2, \psi^{W_7} = \psi^1 + 2\psi^2 + 3\chi$, a necessary condition for p^2, p_1^3, p_2^3 forming an H-basis is the existence of a critical polynomial $Q^2 = 0$ of type 2. Because $Q^2 \in \mathbb{P}_5 \cap \mathcal{D}$, Q^2 is an element of a space of dimension one and thus exactly one equation for one parameter is obtained. The resulting cubature formula is given in Table 2.

References

[1] Cools, R. , Haegemans, A. (1987) ' Construction of fully symmetric cubature formulae of degree 4k-3 for fully symmetric planar regions', J. Comput. Appl. Math. 17, 173–180.

[2] Cools, R. , Haegemans, A. (1988) 'An embedded pair of cubature formulae of degree 5 and 7 for the triangle', BIT 28, 357–359.

[3] Gatermann, K. (1988) 'The construction of symmetric cubature formulas for the square and the triangle', Computing 40, 229–240.

[4] Gatermann, K. (1990) 'Gruppentheoretische Konstruktion von symmetrischen Kubaturformeln', Technical Report TR 90–1, Konrad–Zuse–Zentrum, Berlin.

[5] Grundmann, A. , Möller, H. M. (1978) 'Invariant integration formulas for the n-simplex by combinatorial methods', SIAM J. Numer. Anal. 15, 282–290.

[6] Keast, P. (1986) 'Moderate-degree tetrahedral quadrature formulas', Comput. Meths. Appl. Mech. Engng. 55, 339–348.

[7] Möller H. M. (1973) 'Polynomideale und Kubaturformeln', Thesis, Dortmund.

[8] Möller, H. M. (1987) 'On the construction of cubature formulae with few nodes using Groebner bases' in P. Keast, G. Fairweather (eds.) Numerical Integration Recent Developments, Software and Applications, D. Reidel Publishing Company, Dordrecht, 177–192.

[9] Nooijen, M. , te Velde, G. , Baerends, E. J. (1990) 'Symmetric numerical integration formulas for regular Polygons', SIAM J. Numer Anal. 27, 198–218.

[10] Schmid, H. J. (1980) 'Interpolatorische Kubaturformeln und reelle Ideale', Math. Z. 170, 267–282.

[11] Serre, J.-P. (1977) Linear Representations of Finite Groups, Springer, New York.

[12] Stiefel, E. , Fässler, A. (1979) Gruppentheoretische Methoden und ihre Anwendung', Teubner, Stuttgart.

[13] Stroud, A. H. (1971) Approximate Calculation of Multiple Integrals, Prentice-Hall, Englewood Cliffs, N. J.

ON THE NUMBER OF NODES OF ODD DEGREE CUBATURE FORMU- LAE FOR INTEGRALS WITH JACOBI WEIGHTS ON A SIMPLEX

H. BERENS and H. J. SCHMID
Mathematisches Institut
Bismarckstraße 1 1/2
D-8520 Erlangen

ABSTRACT. H.M. Möller improved the lower bound for the number of nodes in odd degree cubature formulae by an additional term. This term is the rank of a matrix depending on the moments of the integral considered. For the integrals in question the determination of this matrix and the computation of its rank is hard. However for constant weight function G.G. Rasputin succeeded in determining the rank. The approach presented here is based on the recursion formula for orthogonal polynomials. If the matrices involved in the recursion are known, it is not hard to factor Möller's matrix into a regular and a tri-diagonal matrix. The determination of the rank is easy. It will be derived for all integrals with Jacobi weight function.

1. Introduction

Let us denote by $\mathbb{P} = \mathbb{R}[x, y]$ the ring of polynomials of two variables with real coefficients, and let $\mathbb{P}_m$, $m = 0, 1, \ldots$, be the linear subspace spanned by

$$1, x, y, \ldots, x^m, x^{m-1}y, \ldots, xy^{m-1}, y^m.$$

We consider strictly positive linear functionals of the form

$$I : \mathbb{P}(\Omega) \to \mathbb{R} : f \mapsto I(f), \ \Omega \subseteq \mathbb{R}^2.$$

The orthogonal polynomials of degree k, $k = 0, 1, \ldots$, with respect to I will be denoted by

$$P_i^k = x^{k-i}y^i + Q_{ik}, \ Q_{ik} \in \mathbb{P}_{k-1}, \ i = 0, 1, \ldots, k;$$

the polynomials Q_{ik} are determined such that $I(P_i^k Q) = 0$ for all $Q \in \mathbb{P}_{k-1}$.

A cubature formula of degree m for I is a positive linear combination of point-evaluations

$$K : \mathbb{P} \to \mathbb{R} : f \mapsto K(f) = \sum_{i=0}^{N} C_i f(x_i, y_i), \ C_i > 0, (x_i, y_i) \in \Omega,$$

such that $K(f) = I(f)$ for all $f \in \mathbb{P}_m$ and that there is a least one $g \in \mathbb{P}_{m+1}$ satisfying $K(g) \neq I(g)$. Of special interest are cubature formulae with minimal N. As in the one-dimensional case, one obtains by the strict positivity

$$N \geq \dim \mathbb{P}_{[m/2]+1} = ([m/2] + 1)([m/2] + 2)/2.$$

T. O. Espelid and A. Genz (eds.), Numerical Integration, 37–44.

For odd m this bound will be attained for rather special integrals, while for classical integrals this bound is to weak. H. M. Möller improved the bound for odd degree as follows.

Theorem 1 ([15], [7]) If N is the number of nodes in a cubature formula of degree $m = 2k - 1$ then

$$N \geq \dim \mathbb{P}_{k-1} + \operatorname{rank} M_{k-1}^\star / 2,$$

where the entries of $M_{k-1}^\star \in \mathbb{R}^{k \times k}$ are defined by

$$m_{ij}^\star = I(P_{i-1}^k P_j^k - P_i^k P_{j-1}^k), \quad i,j = 1,2,\ldots,k.$$

To compute this lower bound one has to determine the matrix $M_{k-1}^\star$ and compute its rank. Since $M_{k-1}^\star$ is skew symmetric, its rank is even. However, the computation of $M_{k-1}^\star$ in general is difficult, in particular if the integral is not a tensor product of one-dimensional integrals. In the centrally symmetric case, where $I(x^{k-i} y^i) = 0$, $i = 0, 1, \ldots, k$, for all odd $k \in \mathbb{N}$, H. M. Möller proved

$$\operatorname{rank} M_{k-1}^\star = \begin{cases} k, & \text{if } k \text{ is even,} \\ k - 1, & \text{if } k \text{ is odd.} \end{cases}$$

without making use of the explicit form of $M_{k-1}^\star$.

For the Lebesgue integral over the simplex Möller computed the rank of $M_{k-1}^\star$ up to $k = 6$. For all $k \in \mathbb{N}$ G. G. Rasputin [19] derived the same behaviour of rank $M_{k-1}^\star$ as in the centrally symmetric case. We will generalize this result for all classical Jacobi weight functions by applying the recursion formula for the associated Jacobi polynomials.

In the following the domain of integration is $\Omega = \{(x,y) \in \mathbb{R}^2 : 0 \leq x, y, \ x + y \leq 1\}$. We define

$$I_{\alpha,\beta,\gamma} : \mathbb{P}(\Omega) \to \mathbb{R} : f \mapsto I_{\alpha,\beta,\gamma}(f)$$

by

$$I_{\alpha,\beta,\gamma}(f) = \int_0^1 \int_0^{1-x} f(x,y)(1 - x - y)^{\alpha-\beta-\gamma-2} x^\beta y^\gamma dy \, dx, \quad \alpha > \beta+\gamma+1, \beta > -1, \gamma > -1.$$

2.   Recursion formula for the Jacobi polynomials

There are several ways of generalizing the one-dimensional recursion formula in higher dimensions, e.g. [10], [11], [23]. We follow G. Renner's approach in [20]. To do so we need some special matrices and their elementary properties. The following notation will be used. The subindex k of a vector indicates that its dimension is $k + 1$, e.g. $P_k \in (\mathbb{P}_k)^{k+1}$ or

$a_k \in \mathbb{R}^{k+1}$. The subindex of a matrix A_k indicates either $A_k \in \mathbb{R}^{k+1 \times k+1}$ or $A_k \in \mathbb{R}^{k \times k+1}$. If matrices are transposed this holds vice versa. We need the following shift-matrices

$$F_k = \begin{pmatrix} & 0 & \\ \vdots & E_{k-1} & \\ & 0 & \end{pmatrix}, \quad L_k = \begin{pmatrix} & 0 \\ & E_{k-1} & \vdots \\ & 0 \end{pmatrix} \in \mathbb{R}^{k \times k+1},$$

where $E_{k-1} \in \mathbb{R}^{k \times k}$ is the identity. Multiplication of matrix A_k by $F_k^t(L_k^t)$ corresponds to erasing the first (last) column of A_k. Multiplication of $L_k(F_k)$ by A_k^t corresponds to erasing the last (first) row of A_k^t.

We introduce J_k to denote the counteridentity

$$J_k = \begin{pmatrix} 0 & 0 & \cdots & 0 & 1 \\ 0 & 0 & \cdots & 1 & 0 \\ \vdots & \vdots & & \vdots & \vdots \\ 0 & 1 & \cdots & 0 & 0 \\ 1 & 0 & \cdots & 0 & 0 \end{pmatrix} \in \mathbb{R}^{k+1 \times k+1}.$$

The following properties are evident,

$$F_k F_k^t = L_k L_k^t = E_{k-1}, \quad F_k L_{k+1} = L_k F_{k+1}, \quad J_k J_k = E_k, \quad F_k = J_{k-1} L_k J_k.$$

The orthogonal polynomials of degree k will be considered as polynomial vectors,

$$P_k = (P_0^k, P_1^k, \ldots, P_k^k)^t \in (\mathbb{P}_k)^{k+1}.$$

From this we obtain the symmetric moment-matrix

$$M_k = I(P_k P_k^t) \in \mathbb{R}^{k+1 \times k+1},$$

which is regular due to the strict positivity of I. We see that the matrix of Theorem 1 can be written as

$$M_{k-1}^\star = L_k M_k F_k^t - F_k M_k L_k^t.$$

In its essence this matrix goes back to J. Radon [18] who studied cubature formulae of degree 5 with 7 nodes. Radon showed that such formulae exist if and only if a matrix similar to $M_2^\star$ is of rank 2. Radon could not settle the problem whether there are integrals such that rank $M_2^\star = 0$; i.e., integrals admitting minimal formulae of degree 5 with 6 nodes. I.P. Mysovskikh [16], [17] constructed integrals of this type and drew attention to the rank condition. H. M. Möller [15] introduced $M_{k-1}^\star$ to prove Theorem 1. Considering our representation rank $M_{k-1}^\star = 0$ will be obtained if and only if M_k is a Hankel matrix.

The importance of $M_{k-1}^\star$ for cubature problems will be stressed by its tight connection with the matrices which appear in the recursion for the orthogonal polynomials. From this we obtain $M_{k-1}^\star = M_{k-1}^{\star\star} M_k$, where $M_{k-1}^{\star\star}$ is tri-diagonal for all classical integrals over the square, the circle and the triangle.

H. M. Möller [15] has described a method based on $M_{k-1}^\star$ to construct minimal cubature formulae of degree $2k - 1$ attaining the improved lower bound. In a different way H. J. Schmid [21] used $M_{k-1}^\star$ for the construction of formulae of degree $2k - 2$ and $2k - 1$. In both cases the factorization of $M_{k-1}^\star$ is useful.

Lemma 1 For $k = 1, 2, \ldots$ the following recursion holds

$$L_{k+1}P_{k+1} = xP_k - C_kP_k - D_k^tP_{k-1}, \quad F_{k+1}P_{k+1} = yP_k - \bar{C}_kP_k - \bar{D}_k^tP_{k-1},$$

where $C_k, \bar{C}_k \in \mathbb{R}^{k+1 \times k+1}$ and $D_k, \bar{D}_k \in \mathbb{R}^{k \times k+1}$. These matrices satisfy

$$C_kM_k = I(xP_kP_k^t), \qquad \bar{C}_kM_k = I(yP_kP_k^t),$$
$$D_k^tM_{k-1} = I(xP_kP_{k-1}^t) = M_kL_k^t, \qquad \bar{D}_k^tM_{k-1} = I(yP_kP_{k-1}^t) = M_kF_k^t$$

Proof.   The equations follow immediately from the definition and the orthogonality of the polynomial vectors P_i.   □

The conditions on D_k^t und $\bar{D}_k^t$ in the Lemma imply

$$M_{k-1}^\star = L_k\bar{D}_k^tM_{k-1} - F_kD_k^tM_{k-1} = (L_k\bar{D}_k^t - F_kD_k^t)M_{k-1}.$$

Introducing the matrix

$$M_{k-1}^{\star\star} = L_k\bar{D}_k^t - F_kD_k^t.$$

we have

$$\text{rank } M_{k-1}^\star = \text{rank } M_{k-1}^{\star\star}.$$

We will compute $M_{k-1}^{\star\star}$ for $I_{\alpha,\beta,\gamma}$ explicitly and determine its tri-diagonal form.

Lemma 2 The orthogonal polynomials with respect to $I_{\alpha,\beta,\gamma}$ satisfy the following recursion

$$P_i^{k+1} = xP_i^k - c_{i,i-1}P_{i-1}^k - c_{i,i}P_i^k - d_{i,i-2}P_{i-2}^{k-1} - d_{i,i-1}P_{i-1}^{k-1} - d_{i,i}P_i^{k-1},$$
$$i = 0, 1, \ldots, k, \quad d_{0,-2} = c_{0,-1} = d_{0,-1} = d_{1,-1} = 0,$$

where

$$c_{i,i-1}(\alpha, \beta, \gamma) = -\frac{2i(i+\gamma)}{(2k+\alpha-1)(2k+\alpha+1)},$$

$$c_{i,i}(\alpha, \beta, \gamma) = -\frac{2(k-i)(k-i+\beta) - (2(k-i)+\beta+1)(2k+\alpha-1)}{(2k+\alpha-1)(2k+\alpha+1)},$$

$$d_{i,i-2}(\alpha, \beta, \gamma) = \frac{i(i+\gamma)(i-1)(i+\gamma-1)}{(2k+\alpha-2)(2k+\alpha-1)^2(2k+\alpha)},$$

$$d_{i,i-1}(\alpha, \beta, \gamma) = i(i+\gamma)\frac{2(k-i)(k-i+\beta) - (2(k-i)+\beta+1)(2k+\alpha-2)}{(2k+\alpha-2)(2k+\alpha-1)^2(2k+\alpha)},$$

$$d_{i,i}(\alpha, \beta, \gamma) = \frac{(k-i)(k-i+\beta)(k+\alpha+i-1)(k+\alpha+i-\beta-1)}{(2k+\alpha-2)(2k+\alpha-1)^2(2k+\alpha)}.$$

Proof.   The orthogonal polynomials of degree k are of the form

$$P_i^k = \sum_{\nu=0}^{k-i}\sum_{\mu=0}^{i} a_{\nu,\mu}^{k,i} x^\nu y^\mu, \quad i = 0, 1, \ldots, k,$$

where

$$a_{\nu,\mu}^{k,i} = (-1)^{k-\nu-\mu} \frac{\Gamma(k-i+\beta+1)\Gamma(i+\gamma+1)}{\Gamma(\nu+\beta+1)\Gamma(\mu+\gamma+1)} \binom{k-i}{\nu} \binom{i}{\mu} \frac{\Gamma(k+\nu+\mu+\alpha)}{\Gamma(2k+\alpha)},$$

cf. [1], [23]. From the explicit form

$$\begin{aligned}
P_i^k = {}& x^{k-i}y^i - \frac{i(i+\gamma)}{2k+\alpha-1}x^{k-i}y^{i-1} - \frac{(k-i)(k-i+\beta)}{2k+\alpha-1}x^{k-i-1}y^i \\
&+ \frac{i(i-1)(i+\gamma)(i+\gamma-1)}{2(2k+\alpha-1)(2k+\alpha-2)}x^{k-i}y^{i-2} + \frac{i(i+\gamma)(k-i)(k-i+\beta)}{(2k+\alpha-1)(2k+\alpha-2)}x^{k-i-1}y^{i-1} \\
&+ \frac{(k-i)(k-i-1)(k-i+\beta)(k-i+\beta-1)}{2(2k+\alpha-1)(2k+\alpha-2)}x^{k-i-2}y^i + \quad \text{terms of degree } \leq k-3;
\end{aligned}$$

it is not difficult to compute the coefficients of the recursion. □

Lemma 3 The vector-valued form of the recursion reads

$$\begin{aligned}
L_{k+1}P_{k+1} &= xP_k - C_k(\alpha,\beta,\gamma)P_k - D_k^t(\alpha,\beta,\gamma)P_{k-1}, \\
F_{k+1}P_{k+1} &= yP_k - J_kC_k(\alpha,\gamma,\beta)J_kP_k - J_kD_k^t(\alpha,\gamma,\beta)J_{k-1}P_{k-1},
\end{aligned}$$

where the non-zero entries of $C_k(\alpha,\beta,\gamma) \in \mathbb{R}^{k+1 \times k+1}$ and $D_k(\alpha,\beta,\gamma) \in \mathbb{R}^{k \times k+1}$ are stated in the preceeding Lemma.

Proof. Replacing i by $k-i$ and x by y the first equation will be transformed into

$$L_{k+1}J_{k+1}P_{k+1} = J_kF_{k+1}P_{k+1} = yJ_kP_k - C_k(\alpha,\beta,\gamma)J_kP_k - D_k^t(\alpha,\beta,\gamma)J_{k-1}P_{k-1},$$

a recursion for $I(\alpha,\gamma,\beta)$. Multiplying this by J_k and interchanging β and γ in the matrices C_k and D_k^t the second recursion will be obtained. In particular the matrices in Lemma 1 are of the form

$$\bar{C}_k = J_kC_k(\alpha,\gamma,\beta)J_k, \quad \bar{D}_k^t = J_kD_k^tJ_{k-1}.$$

□

3. Lower bound

Setting

$$d_{i,j} = d_{i,j}(\alpha,\beta,\gamma) \quad \text{and} \quad \bar{d}_{i,j} = d_{k-i,k-1-j}(\alpha,\gamma,\beta),$$

the diagonal entries of $M_{k-1}^{\star\star}$ can be computed via

$$\begin{aligned}
m_{i,i-1}^{\star\star} &= \bar{d}_{i,i-1} - d_{i+1,i-1}, \quad i = 1,2\ldots,k-1, \\
m_{i,i+1}^{\star\star} &= \bar{d}_{i-1,i} - d_{i,i}, \quad i = 1,2,\ldots,k-1, \\
m_{i,i}^{\star\star} &= \bar{d}_{i-1,i-1} - d_{i,i-1}, \quad i = 1,2,\ldots,k.
\end{aligned}$$

From this we obtain the following form

$$
M_{k-1}^{\star\star} = \frac{1}{(2k+\alpha-2)(2k+\alpha-1)^2}
\begin{pmatrix}
a_1 & c_k & 0 & 0 & \cdots & 0 \\
b_2 & a_2 & c_{k-1} & 0 & \cdots & 0 \\
0 & b_3 & a_3 & c_{k-2} & \cdots & 0 \\
 & \ddots & \ddots & \ddots & & \\
0 & \cdots & \cdots & b_{k-1} & a_{k-1} & c_2 \\
0 & \cdots & 0 & 0 & b_k & a_k
\end{pmatrix},
$$

where

$$
\begin{aligned}
a_i &= i(i+\gamma)(2k-2i+\beta+1)-(k-i+1)(k-i+\beta+1)(2i+\gamma-1), && i=1,2,\ldots,k, \\
b_i &= (i-1)(i+\gamma-1)(2k-2i+\alpha-\gamma), && i=2,3,\ldots,k, \\
c_i &= -(i-1)(i+\beta-1)(2k-2i+\alpha-\beta), && i=2,3,\ldots,k.
\end{aligned}
$$

The entries in the sub-diagonals of $M_{k-1}^{\star\star}$ do not vanish. Hence, rank $M_{k-1}^{\star\star} \geq k-1$, since $M_{k-1}^{\star}$ is skew symmetric the rank of $M_{k-1}^{\star\star}$ is even. So for k even we obtain rank k, while for k odd we obtain rank $k-1$. This completes the proof of

Theorem 2 The number of nodes N of cubature formulae of degree $2k-1$ for integrals of type $I_{\alpha,\beta,\gamma}$ satisfies the following lower bound

$$
N \geq k(k+1)/2 + [k/2].
$$

The bound will be attained for small k; it is open what happens if k is increasing. The following table - which is taken from the data base on cubature formulae being compiled by R. Cools and P. Rabinowitz – illustrates the gap between the lower bound and the number of nodes attained by known cubature formulae for $I_{2,0,0}$.

degree	lower bound	number of nodes attained	quality	references
3	4	4	NI	[22]
5	7	7	PI	[4, 5, 12, 13, 14, 22]
7	12	12	PI	[2, 8, 9]
9	17	19	PI	[5, 12, 13]
11	24	27	PO	[13]
13	31	37	PI	[3]

References

[1] Appell, P., Kampé de Fériet, J. (1926) Fonctions hypergéometriques et hypersphériques – Polynomes d'Hermite, Gauthiers-Villars et Cie., Paris.

[2] Becker, T. (1987) 'Konstruktion von interpolatorischen Kubaturformeln mit Anwendungen in der Finit-Element-Methode', thesis, Technische Hochschule Darmstadt.

[3] Berntsen, J., T. O. Espelid (1990) 'Degree 13 symmetric quadrature rules for the triangle', Reports in Informatics, Report No. 44, Department of Informatics, Univ. Bergen, Bergen, Norway, March 1990.

[4] Cowper, G. R. (1973) 'Gaussian quadrature formulas for triangles', Internat. J. Numer. Methods Engrg., 7, 405–408.

[5] Dunavant, D.A. (1985) 'High degree efficient symmetrical gaussian quadrature rules for the triangle', Internat. J. Numer. Methods Engrg., 21, 1129–1148.

[6] Engels, H. (1970) 'Über gleichgewichtete Kubaturformeln für ein Dreiecksgebiet', Elek. Daten. 12, 535–539.

[7] Engels, H. (1980) Numerical Quadrature and Cubature, Academic Press, London, New York, Toronto, Sydney, San Francisco.

[8] Gatermann, K. (1986) 'Die Konstruktion von Gaußformeln und symmetrischen Kubaturformeln', Diplomarbeit, Universität Hamburg.

[9] Gatermann, K. (1988) 'The construction of symmetric cubature formulas for the square and the triangle', Computing, 40, 229–240.

[10] Kowalski, M.A. (1982) 'The recursion formulas for orthogonal polynomials in n variables', SIAM J. Math. Anal. 13, 309 – 315.

[11] Kowalski, M.A. (1982) 'Orthogonality and recursion formulas for orthogonal polynomials in n variables', SIAM J. Math. Anal. 13, 316 – 323.

[12] Laursen, M.E. and M. Gellert (1978) 'Some criteria for numerically integrated matrices and quadrature formulas for triangles', Internat. J. Numer. Methods Engrg. 12, 67–76.

[13] Lyness, J.N. and D. Jespersen (1975) 'Moderate degree symmetric quadrature rules for the triangle', J. Inst. Maths Applics. 15, 19–32.

[14] Moan, T. (1974) 'Experiences with orthogonal polynomials and "best" numerical integration formulas on a triangle; with particular reference to finite element approximations', ZAMM, 54, 501–508.

[15] Möller, H.M. (1976) 'Kubaturformeln mit minimaler Knotenzahl', Numerische Mathematik 25, 185 – 200.

[16] Mysovskikh, I.P. (1981) Interpolatory cubature formulae (Russian), Nauka, Moscow.

44

[17] Mysovskikh, I.P. (1976) 'Numerical characteristics of orthogonal polynomials in two variables', Vestnik Leningrad. Univ. Math. 3, 323 – 332.

[18] Radon, J. (1948) 'Zur mechanischen Kubatur', Monatshefte Mathematik 52, 286 – 300.

[19] Rasputin, G.G. (1983) 'On the question of numerical characteristics for orthogonal polynomials of two variables' (Russian), Metody Vychislenij 13, 145 – 154.

[20] Renner, G. (1986) 'Darstellung von strikt quadratpositiven linearen Funktionalen auf endlichdimensionalen Polynomräumen', thesis, Univ. Erlangen-Nürnberg, 93 p.

[21] Schmid, H.J. (1983) 'Interpolatorische Kubaturformeln', Diss. Math. CCXX, 1 – 122.

[22] Stroud, A.H. (1971) Approximate calculation of multiple integrals, Prentice Hall.

[23] Verlinden, P. (1986) 'Expliciete uitdrukkingen en rekursie-betrekkingen voor meerdimensionale invariante orthogonale veeltermen', master's thesis, Katholieke Universiteit Leuven.

ON QUADRATURE FORMULAE NEAR GAUSSIAN QUADRATURE

K.-J. FÖRSTER
Institut für Mathematik
Universität Hildesheim
3200 Hildesheim
Germany

ABSTRACT. In this paper, for product integration on the finite interval $[a, b]$, we consider the class of n–point quadrature formulae Q_n of at least algebraic degree $2n - 3$. We study a new approach for their characterization using the simple fact that such a quadrature formula is uniquely determined by one node y and its associated weight b. For a given node $y \in \mathbb{R}$, we characterize all weights b for which the pair (y, b) generates a positive quadrature formula. Furthermore, for each fixed node $y \in \mathbb{R}$ not a node of the Gaussian formula Q_{n-1}^G and each weight $b < 0$, we show that there exists a quadrature formula Q_n of degree $2n - 3$ using the node y and the associated weight b. With these results, conditions for convergence of such quadrature rules in classes of differentiable functions are obtained. In particular, if all nodes are contained in $[a, b]$, we prove convergence of each such rule for every $f \in C^2[a, b]$, but not for functions in the class $C^1[a, b]$.

1. Introduction and Statement of the Results

For a given (nonnegative) weight function w on $[-1, 1]$, a quadrature formula Q_n of degree $\deg(Q_n) = m \geq 0$ is a linear functional of the type

$$(1.1) \quad Q_n[f] := \sum_{\nu=1}^{n} a_{\nu,n} f(x_{\nu,n}), \qquad a_{\nu,n} \in \mathbb{R}, \quad -\infty < x_{1,n} < x_{2,n} < ... < x_{n,n} < \infty,$$

$$(1.2) \quad \int_{-1}^{1} f(x) w(x) dx = Q_n[f] + R_n[f], \qquad R_n[p_\mu] \begin{cases} = 0 & \text{for } \mu = 0, ..., m, \\ \neq 0 & \text{for } \mu = m + 1 \end{cases}$$

where p_μ denotes the monomial $p_\mu(x) = x^\mu$ and R_n is the so-called error functional of Q_n. A quadrature rule (Q_n) is a sequence of quadrature formulae Q_n $(n = 1, 2, 3, ...)$. We say that Q_n is a positive quadrature formula if and only if all weights $a_{\nu,n}$ are nonnegative. Furthermore, in the following we denote by Q_n^G the Gaussian formula using n nodes, defined uniquely by its maximal degree $\deg(Q_n^G) = 2n - 1$ (cf. here and in the following sections [1], [3] and [8]). Finally, let P_n be the associated orthogonal polynomial,

$$(1.3) \quad P_n(x) = k_n \prod_{\nu=1}^{n} (x - x_{\nu,n}^G), \qquad \int_{-1}^{1} P_n(x) P_m(x) w(x) dx = \delta_{n,m}, \qquad k_n > 0,$$

and let λ_n be the so-called Christoffel function

$$(1.4) \qquad \lambda_n = \frac{k_n}{k_{n-1}} \cdot \frac{1}{P_n' P_{n-1} - P_n P_{n-1}'} = \left\{ \sum_{\nu=0}^{n-1} P_\nu^2 \right\}^{-1} .$$

45

T. O. Espelid and A. Genz (eds.), Numerical Integration, 45–54.
© 1992 Kluwer Academic Publishers.

In this paper, we investigate conditions for convergence of quadrature rules (Q_n) of degree $\deg(Q_n) \geq 2n - 3$. All quadrature formulae Q_n of degree $\deg(Q_n) \geq 2n - 2$ are positive (see, e.g., [6]), while a quadrature formula Q_n of degree $\deg(Q_n) = 2n - 3$ may have (at most) one negative weight (see, e.g., [9]). For a given node y, the following Theorem 1 shows which associated negative weight may occur.

THEOREM 1. *For every $y \in \mathbb{R} \setminus \{x_{1,n-1}^G, x_{2,n-1}^G, ..., x_{n-1,n-1}^G\}$ and every $b < 0$, there exists one and only one quadrature formula Q_n satisfying $\deg(Q_n) \geq 2n - 3$ which has a node in y and the associated weight b. All nodes of Q_n are contained in the interval $[\alpha, \beta]$, where $\alpha = \min\{y, x_{1,n-1}^G\}$ and $\beta = \max\{y, x_{n-1,n-1}^G\}$.*

We see that, for every fixed $n \in \mathbb{N}$, in the class $C[-1, 1]$ of continuous functions the norm $\|Q_n\|_C = \sum_{\nu=1}^n |a_{\nu,n}|$ of a quadrature formula Q_n satisfying $\deg(Q_n) \geq 2n - 3$ may be arbitrarily large, even if we require that all nodes are contained in $[-1, 1]$ and if we prescribe a node $y \in [-1, 1] \setminus \{x_{1,n-1}^G, ..., x_{n-1,n-1}^G\}$. Therefore, by the Theorem of Banach-Steinhaus, we obtain respective results on divergence in the class $C[-1, 1]$. For the class $C^1[-1, 1]$ of differentiable functions having a continuous first derivative, we have the following result.

THEOREM 2. *There exists a quadrature rule (Q_n) satisfying, for every $n \in \mathbb{N}$,*

(i) $\deg(Q_n) \geq 2n - 3$,
(ii) *all nodes $x_{\nu,n}$ are contained in $(-1, 1)$,*

and a function $f \in C^1[-1, 1]$, such that $\liminf\limits_{n \to \infty} |R_n[f]| > 0$.

For the well-known Newton-Cotes quadrature formulae Q_n^{NC}, note that there exist analytic functions f on $[-1, 1]$ for which (Q_n^{NC}) diverges (see, e.g., [1, p.134]). Here, for the class $A_2[-1, 1]$ of functions f having an absolutely continuous first derivative f' on $[-1, 1]$, we have the following result on convergence.

THEOREM 3. *Let (Q_n) be a quadrature rule satisfying, for every $n \in \mathbb{N}$,*

(i) $\deg(Q_n) \geq 2n - 3$,
(ii) *all nodes $x_{\nu,n}$ are contained in $[-1, 1]$.*

Then, for every function $f \in A_2[-1, 1]$, $\lim\limits_{n \to \infty} R_n[f] = 0$.

For different quadrature formulae Q_n, in order to compare the associated error functionals R_n in the class of all functions f having a bounded k-th derivative on $[-1, 1]$, we investigate the so-called error constants of order k,

$$(1.5) \qquad c_k(Q_n) := \sup_{\|f^{(k)}\| \leq 1} |R_n[f]|, \qquad \|g\| := \sup_{|x| \leq 1} |g(x)|,$$

which are the best possible constants c for estimates of the form

$$(1.6) \qquad |R_n[f]| \leq c \cdot \|f^{(k)}\|.$$

The constant $c_k(Q_n)$ is unbounded if and only if $\deg(Q_n) < k - 1$ (see, e.g., [1, p. 245ff.]). We say that a linear functional S on $C^j[-1, 1]$ is positive (negative) definite of order j if $f^{(j)} > 0$ on $[-1, 1]$ implies that $S[f] > 0$ $(S[f] < 0)$. For the error constant of highest order $c_{2n-2}(Q_n)$ of a nonpositive quadrature formula Q_n having $\deg(Q_n) \geq 2n - 3$, we state the following results, which have been proven for the weight function $w \equiv 1$ in [2].

THEOREM 4. *Let* $\deg(Q_n) \geq 2n - 3$ *and let* Q_n *be nonpositive. Then,* R_n *is positive definite of order* $2n - 2$ *and*

$$(1.7) \qquad\qquad c_{2n-2}(Q_n) > c_{2n-2}(Q_{n-1}^G).$$

Furthermore, if additionally $w > 0$ *almost everywhere on* $[-1, 1]$ *and if all nodes of* (Q_n) *are contained in* $[-1, 1]$, *then*

$$(1.8) \qquad\qquad \lim_{n\to\infty} \frac{c_{2n-2}(Q_n)}{c_{2n-2}(Q_{n-1}^G)} = 1.$$

The proof of the theorems above will be given in Section 3, while in Section 2 we present some preliminary lemmas on the structure of quadrature formulae Q_n with $\deg(Q_n) \geq 2n - 3$, which may be of interest in their own.

2. Preliminary Lemmas

In the literature, several investigations on the existence and characterization of (positive) quadrature formulae Q_n of high algebraic degree $\deg(Q_n) \geq 2n - 1 - s$ can be found (see, e.g. , [4], [15], [12], [16], [13], [14]) using the following well-known representation of the node polynomial w_n,

$$(2.1) \qquad w_n(x) := \prod_{\nu=0}^{n} (x - x_{\nu,n}) = \text{const} \sum_{\nu=0}^{s} b_\nu P_{n-\nu}(x), \qquad s \leq n.$$

Here, we are interested in the case $s = 2$. A complete description for those parameters b_0, b_1, b_2, which generate a polynomial having only real zeros and therefore generate quadrature formulae Q_n with $\deg(Q_n) \geq 2n - 3$, has been given by Micchelli and Rivlin [9]. In the following we consider another representation, using the simple fact that a quadrature formula Q_n with $\deg(Q_n) \geq 2n - 3$ is uniquely determined by only one node and its associated weight. Note that one weight of Q_n may have the value zero, i.e., $Q_n = Q_{n-1}^G$.

LEMMA 1. *Let* $\deg(Q_n) \geq 2n-3, \deg(\bar{Q}_n) \geq 2n-3$ *and let* $Q_n[f] = \sum_{\nu=1}^{n} a_{\nu,n} f(x_{\nu,n}), \bar{Q}_n[f] = \sum_{\nu=1}^{n} \bar{a}_{\nu,n} f(\bar{x}_{\nu,n})$. *If one of the following conditions holds, then,* $Q_n = \bar{Q}_n$.

    (i)   *There exist* ν *and* μ *with* $x_{\nu,n} = \bar{x}_{\mu,n}$ *and* $a_{\nu,n} = \bar{a}_{\mu,n}$,
    (ii) *there exist* ν_1, ν_2 ($\nu_1 \neq \nu_2$) *and* μ_1, μ_2 *with* $x_{\nu_1,n} = \bar{x}_{\mu_1,n}$ *and* $x_{\nu_2,n} = \bar{x}_{\mu_2,n}$,
    (iii) $\bar{Q}_n = Q_{n-1}^G$ *and there exist* ν *and* μ *with* $x_{\nu,n} = x_{\mu,n-1}^G$.

For the proof, assume $Q_n \neq \bar{Q}_n$ and consider the functional $H := Q_n - \bar{Q}_n$, which vanishes on the set of all polynomials of degree $2n - 3$. Using one of the above conditions, we have

$$(2.2) \qquad H[f] := \sum_{\nu=1}^{2n-2} d_\nu f(y_\nu), \qquad d_\nu \neq 0,$$

which is impossible (see, e.g., [5]).

Using the same method, we consider, for $Q_n \neq Q_{n-1}^G$, the functional $G = Q_n - Q_{n-1}^G = R_{n-1}^G - R_n$, which vanishes as above on the set of all polynomials of degree $2n - 3$. We have

$$(2.3) \qquad G[f] = Q_n[f] - Q_{n-1}^G[f] := \sum_{\nu=1}^{2n-1} d_\nu f(y_\nu), \quad y_1 < y_2 < ... < y_{2n-1},$$

and therefore G is a constant multiple of the divided difference using the nodes of Q_n and Q_{n-1}^G. From the well-known fact that the signs of the coefficients d_ν of such a divided difference alternate, and with the notations

$$(2.4) \qquad \begin{aligned} &I_0 := (-\infty, x_{1,n-1}^G), \quad I_{n-1} := (x_{n-1,n-1}^G, \infty), \\ &I_\nu := (x_{\nu,n-1}^G, x_{\nu+1,n-1}^G), \quad \nu = 1, 2, ..., n-2, \end{aligned}$$

we directly have the following result.

LEMMA 2. (Micchelli and Rivlin[9]). Let $\deg(Q_n) \geq 2n - 3$ and $Q_n \neq Q_{n-1}^G$. Then

    (i) Q_n has a node in I_ν, for each $\nu = 1, 2, ..., n - 2$,
    (ii) Q_n is positive if and only if $x_{1,n} \in I_0$ and $x_{n,n} \in I_{n-1}$,
    (iii) Q_n is nonpositive if and only if one of the following conditions holds:

        (a) $x_{1,n}, x_{2,n} \in I_0$ and $a_{1,n} < 0$,
        (b) $x_{n-1,n}, x_{n,n} \in I_{n-1}$ and $a_{n,n} < 0$,
        (c) there exists a $\nu \in \{1, 2, ..., n - 2\}$ such that $x_{\nu,n}, x_{\nu+1,n}, x_{\nu+2,n} \in I_\nu$ and $a_{\nu+1,n} < 0$.

Here, to represent the node polynomial w_n, we use the monic polynomial $r_{n,a,y}$ with parameters $y \in \mathbb{R} \setminus \{x_{1,n-1}^G, ..., x_{n-1,n-1}^G\}$ and $a \in \mathbb{R} \setminus \{-k_n/k_{n-1}\}$, such that

$$(2.5) \quad (k_n + ak_{n-1})r_{n,a,y}(x) := P_n(x) - \frac{P_n(y)}{P_{n-1}(y)} P_{n-1}(x) + a(x - y)P_{n-1}(x), \quad P_{n-1}(y) \neq 0,$$

which, by a short computation using the well-known recurrence formula

$$(2.6) \quad P_n(x) = (\alpha_n x + \beta_n)P_{n-1}(x) - \gamma_n P_{n-2}(x), \qquad \alpha_n = \frac{k_n}{k_{n-1}}, \quad \gamma_n = \frac{\alpha_n}{\alpha_{n-1}}$$

(see, e.g., [17, p.42]), for $Q_n \neq Q_{n-1}^G$, can be shown to be equivalent to (2.1) for $s = 2$. Note that $r_{n,a,y}(y) = 0$ and, by the separation of the zeros of P_n and P_{n-1}, that

$$(2.7) \qquad \text{sign}\{r_{n,a,y}(x_{\nu,n-1}^G)\} = (-1)^{n+\nu} \text{sign}(k_n + ak_{n-1}).$$

Therefore, $r_{n,a,y}$ has a real zero in each interval I_ν ($\nu = 1, ..., n - 2$).

LEMMA 3. Let $\deg(Q_n) \geq 2n - 3$ and let $r_{n,a,y}$ be the node polynomial of Q_n. Then, for every node $x_{\nu,n}$ and associated weight $a_{\nu,n}$ of Q_n,

$$(2.8) \qquad a_{\nu,n} = \frac{k_n/k_{n-1}}{P_n'(x_{\nu,n})P_{n-1}(x_{\nu,n}) - P_n(x_{\nu,n})P_{n-1}'(x_{\nu,n}) + aP_{n-1}^2(x_{\nu,n})}.$$

Furthermore, if y and z are two different nodes of Q_n, then, for $Q_n \neq Q_{n-1}^G$,

$$(2.9) \qquad a = \frac{P_n(y)P_{n-1}(z) - P_n(z)P_{n-1}(y)}{(z - y)P_{n-1}(y)P_{n-1}(z)}.$$

PROOF. Consider G in (2.3), a constant multiple of the divided difference with the nodes being the zeros of $v := P_{n-1}r_{n,a,y}$. Using the well known representation of the weights of divided differences, we have

$$(2.10) \qquad d_\nu = \text{const}\{v'(y_\nu)\}^{-1}, \qquad \nu = 1, \ldots, 2n - 1,$$

and comparing this with the representation of the Gaussian weights (see, e.g., [3, p.97]), one gets

$$(2.11) \qquad a_{\nu,n-1}^G = -\frac{k_n/k_{n-1}}{P_n(x_{\nu,n-1})P_{n-1}'(x_{\nu,n-1})} \cdot$$

The assertion (2.8) now follows by a simple calculation. If y and z are two nodes of Q_n, then there exist a and $\bar{a}$ with $w_n = r_{n,a,y} = r_{n,\bar{a},z}$, i.e.,

$$(2.12) \quad \begin{aligned} & (k_n + \bar{a}k_{n-1})\{P_n(x) - \frac{P_n(y)}{P_{n-1}(y)}P_{n-1}(x) + a(x - y)P_{n-1}(x)\} \\ & = (k_n + ak_{n-1})\{P_n(x) - \frac{P_n(z)}{P_{n-1}(z)}P_{n-1}(x) + \bar{a}(x - z)P_{n-1}(x)\} \end{aligned}$$

for every $x \in \mathbb{R}$. The use of $x = x_{\nu,n-1}^G$ in (2.12) yields $a = \bar{a}$, and putting $x = z$ in (2.5) proves (2.9).

LEMMA 4. The polynomial $r_{n,a,y}$ is a node polynomial of a positive quadrature formula Q_n with $\deg(Q_n) \geq 2n - 3$ and $Q_n \neq Q_{n-1}^G$ if and only if $a > -k_n/k_{n-1}$ and $y \in \mathbb{R}\backslash\{x_{1,n-1}^G, \ldots, x_{n-1,n-1}^G\}$.

PROOF. (i) If $a > -k_n/k_{n-1}$, then it follows from (2.7) that $r_{n,a,y}(x_{n-1,n-1}^G) < 0$ and $(-1)^n r_{n,a,y}(x_{1,n-1}^G) < 0$. Since $r_{n,a,y}$ is a monic of degree n, it has a zero in I_0 and a zero in I_{n-1}. The existence and positivity of Q_n now follows from (2.7) and Lemma 2(ii).
(ii) Assume $a \leq -k_n/k_{n-1}$ and let $r_{n,a,y}$ be the node polynomial of a positive quadrature formula Q_n with $\deg(Q_n) \geq 2n - 3$. Let $y := x_{\nu,n}$. By (2.8) and (1.4) it follows that

$$(2.13) \quad \begin{aligned} a_{\nu,n} & \geq \frac{k_n/k_{n-1}}{P_n'(x_{\nu,n})P_{n-1}(x_{\nu,n}) - P_n(x_{\nu,n})P_{n-1}'(x_{\nu,n}) - (k_n/k_{n-1})P_{n-1}^2(x_{\nu,n})} \\ & = \lambda_{n-1}(x_{\nu,n}). \end{aligned}$$

Since Q_n is positive, $\deg(Q_n) \geq 2n - 3$ and $Q_n \neq Q_{n-1}^G$, this is a contradiction to Lemma 5 below.

LEMMA 5. (see, e.g., [7, p. 51]). Let Q_m be positive with $\deg(Q_m) \geq 2n - 4$. Then,

$$(2.14) \qquad a_{\nu,m} \leq \lambda_{n-1}(x_{\nu,m})$$

and the equality sign in (2.13) is attained only if $m = n - 1$.

We immediately have the following results.

COROLLARY 1. *There exists a positive quadrature formula $Q_n \neq Q_{n-1}^G$ with $\deg(Q_n) \geq 2n-3$, which has a node in y and an associated weight b, if and only if the two following conditions hold:*

(i) $y \in \mathbb{R} \setminus \{x_{1,n-1}^G, ..., x_{n-1,n-1}^G\}$,
(ii) $0 < b < \lambda_{n-1}(y)$.

COROLLARY 2. *Let Q_n be a nonpositive quadrature formula with $\deg(Q_n) \geq 2n - 3$. Let $a_{\nu,n}$ be a positive weight of Q_n. Then,*

$$(2.15) \qquad\qquad a_{\nu,n} \geq \lambda_{n-1}(x_{\nu,n}).$$

With the help of Lemma 4, (2.9) and the formula of Christoffel-Darboux, a characterization of positive Q_n with $\deg(Q_n) \geq 2n - 3$ involving two nodes follows.

COROLLARY 3. *Let $z > y$. There exists a positive quadrature formula $Q_n \neq Q_{n-1}^G$ with $\deg(Q_n) \geq 2n - 3$, which has a node in y and a node in z, if and only if*

$$(2.16) \qquad\qquad \frac{P_{n-2}(y)}{P_{n-1}(y)} < \frac{P_{n-2}(z)}{P_{n-1}(z)}.$$

Now, let

$$(2.17) \qquad a^*(y) := -\frac{P_n'(y)P_{n-1}(y) - P_n(y)P_{n-1}'(y)}{P_{n-1}^2(y)}, \qquad P_{n-1}(y) \neq 0.$$

Using (1.4), we have

$$(2.18) \qquad a^*(y) = -\frac{k_n}{k_{n-1}} \frac{1}{P_{n-1}^2(y)} \sum_{\nu=0}^{n-1} P_\nu^2(y) < -\frac{k_n}{k_{n-1}} < 0.$$

Differentiation of $r_{n,a,y}$ yields

$$(2.19) \qquad
\begin{aligned}
(k_n + ak_{n-1})r_{n,a,y}'(y) &= P_n'(y) - \frac{P_n(y)}{P_{n-1}(y)}P_{n-1}'(y) + aP_{n-1}(y) \\
&= (1 - \frac{a}{a^*(y)})\{P_n'(y) - \frac{P_n(y)}{P_{n-1}(y)}P_{n-1}'(y)\}.
\end{aligned}$$

Using again (1.4) we conclude that

$$(2.20) \qquad \operatorname{sign}\{r_{n,a,y}'(y)\} = \operatorname{sign}\{(k_n + ak_{n-1})(1 - \frac{a}{a^*(y)})P_{n-1}(y)\},$$

where

$$(2.21) \qquad \operatorname{sign}\{P_{n-1}(y)\} = (-1)^{n+\mu-1} \text{ for } y \in I_\mu.$$

Now we consider three cases for the values of $a^*(y)$ and y :

(i) Let $a < a^*(y), y \in I_\mu, \mu \in \{1, 2, ..., n - 2\}$. Then (2.20/2.21) and (2.7) imply that $r_{n,a,y}$ has at least three real and separated zeros in I_μ. Therefore, all zeros are real, separated and contained in $(x_{1,n-1}^G, x_{n-1,n-1}^G)$.

(ii) Let $a < a^*(y), y \in I_0$ resp. $y \in I_{n-1}$. By (2.20/2.21) and (2.7) the existence of two real and seperated zeros of $r_{n,a,y}$ in I_0 resp. I_{n-1} follows. Therefore, all zeros are real, separated and contained in $[y, x_{n-1,n-1}^G)$ resp. $(x_{1,n-1}^G, y]$.

(iii) Let $a \in (a^*(y), -k_n/k_{n-1}), y \in I_0$ resp. $y \in I_{n-1}$. Since $(k_n + ak_{n-1}) < 0$ and $r_{n,a,y}$ is monic, there follows, as above, the existence of a second separated real zero $\bar{y}$ in I_0 resp. I_{n-1}. Therefore, all zeros are real, separated and contained in $[\bar{y}, x_{n-1,n-1}^G)$ resp. $(x_{n-1,n-1}^G, \bar{y}]$.

Now, (2.8) yields the following result.

LEMMA 6. *The polynomial $r_{n,a,y}$ is a node polynomial of a quadrature formula Q_n with $\deg(Q_n) \geq 2n - 3$ having a negative weight associated with the node y if and only if*

$$(2.22) \qquad a < -\frac{P_n'(y)P_{n-1}(y) - P_n(y)P_{n-1}'(y)}{\{P_{n-1}(y)\}^2}.$$

Considering additionally the case (iii) above, we have:

LEMMA 7. *Let $y \in \mathbb{R}\backslash[x_{1,n-1}^G, x_{n-1,n-1}^G]$ and $b \in \mathbb{R}$. Then there exists one and only one quadrature formula Q_n with $\deg(Q_n) \geq 2n-3$ which has a node in y and the associated weight b.*

Finally, we consider $G = Q_n - Q_{n-1}^G = R_{n-1}^G - R_n$, using the notations in (2.3). Lemma 2 (iii) yields $d_{2n-1} < 0$. By the positive definiteness of divided differences and the positivity of the weight associated with the greatest node of a divided difference, the negative definiteness of the order $2n - 2$ for G follows. Therefore the following result is valid:

LEMMA 8. *Let $\deg(Q_n) \geq 2n - 3$ and let Q_n be nonpositive. Then, for every $f \in C^{2n-2}[-1,1]$ with $f^{(2n-2)} > 0$ on $[-1,1]$,*

$$(2.23) \qquad R_n[f] > R_{n-1}^G[f].$$

3. Proof of the Theorems

PROOF OF THEOREM 1. This is a direct consequence of Lemma 4 and Lemma 6 noting the cases (i) and (ii) of the proof of Lemma 6.

PROOF OF THEOREM 2. We first require the following lemma.

LEMMA 9. (cf., [8, p. 268]). *Let all nodes of (Q_n) be contained in $[-1,1]$. In order that the quadrature rule (Q_n) converges for any $f \in C^1[-1,1]$, the following conditions are necessary and sufficient:*

1. *(Q_n) converges for each polynomial;*
2. *there exists a number M for which*

$$(3.1) \qquad \sum_{\nu=1}^{n} \Big| \sum_{\mu=1}^{\nu} a_{\mu,n} \Big| (x_{\nu+1,n} - x_{\nu,n}) \leq M \qquad \text{for } n \in \mathbb{N}, \ x_{n+1,n} := 1.$$

Consider $r_{n,a,y}$ defined in (2.5) and $a^*(y)$ defined in (2.16). Then

$$r_{n,a,y}(y) = r'_{n,a,y}(y) = r''_{n,a,y}(y) = 0$$
$$\Leftrightarrow a = a^*(y) \text{ and } h(y) = 0, \text{ where } h = P_{n-1}(P''_n P_{n-1} - P_n P''_{n-1}) - 2P'_{n-1}(P'_n P_{n-1} - P_n P'_{n-1}).$$

If $P_{n-1}(y) = 0$, then $h(y) > 0$ for $P'_{n-1}(y) < 0$ and $h(y) < 0$ for $P'_{n-1}(y) > 0$. Therefore, there exist at least $n - 2$ points $y \in (x^G_{1,n-1}, x^G_{n-1,n-1})$ for which $r_{n,a^*(y),y}$ has a triple zero in y. Let $\bar{y}$ be one of these points and let $a < a^*(\bar{y})$. Lemma 6 and Lemma 2 show that $r_{n,a,\bar{y}}$ is the node polynomial of a nonpositive quadrature formula Q_n having all nodes contained in $(x^G_{1,n-1}, x^G_{n-1,n-1})$. Let $\bar{y} := x_{\nu,n}$. From (2.10) we have, by rough estimation,

$$(3.2) \qquad a_{\nu,n} < -e_{n,a,\bar{y}}\{(x_{\nu+1,n} - x_{\nu,n})(x_{\nu,n} - x_{\nu-1,n})2^{2n-4}\}^{-1}$$

where $e_{n,\bar{a},\bar{y}} > e_{n,\tilde{a},\bar{y}} > 0$ for $\bar{a} < \tilde{a} < a^*(y)$. For $a < a^*(y)$ and $a \to a^*(y)$, we have $x_{\nu+1,n} \to x_{\nu,n}$ and $x_{\nu,n-1} \to x_{\nu,n}$. By Lemma 9 and the fact that $\sum_{\nu=1}^n a_{\nu,n} = \int_{-1}^1 w(x)dx$, the result follows.

PROOF OF THEOREM 3. For every positive subsequence of (Q_n) the result is valid (see, e.g., [8, p.264]). Therefore, let Q_n be nonpositive. From (2.3) we have

$$(3.3) \qquad R_n[f] = R^G_{n-1}[f] - G[f],$$

which yields

$$(3.4) \qquad |R_n[f]| \leq |R^G_{n-1}[f]| + \|f''\| \int_{-1}^1 |K_{2,n}(x)|dx,$$

where $K_{2,n}$ is the Peano kernel of second order of G,

$$(3.5) \qquad K_{2,n}(x) = \sum_{\nu=1}^{2n-1} d_\nu(x - y_\nu)_+ = \sum_{\nu=1}^{2n-1} d_\nu(y_\nu - x)_+.$$

The functional G is definite of order $2n - 2$, therefore $K_{2,n}$ has at least $2n - 4$ changes of sign (see, e.g., [5]), showing that there is exactly one change of sign in every interval $(y_\nu, y_{\nu+1})$ for $\nu = 2, ..., 2n - 3$ and no change of sign in (y_1, y_2) and (y_{2n-2}, y_{2n-1}). Lemma 8 yields $d_\nu < 0$ for $y_\nu = x^G_{\nu,n-1}$. Let $a_{k,n} < 0$ for some k. At first, we consider $x_{k,n} \in (x^G_{1,n-1}, x^G_{n-1,n-1})$. Lemma 2 shows that $y_1 = x^G_{1,n-1}, y_{2n-1} = x^G_{n-1,n-1}$. We have

$$(3.6) \qquad K_{2,n}(x) \geq \min\{-a^G_{\nu,n-1}(x - x^G_{\nu,n-1}), -a^G_{\nu+1,n-1}(x^G_{\nu+1,n-1} - x)\}$$

for $x \in [x^G_{\nu,n-1}, x^G_{\nu+1,n-1}]$ and $\nu \in \{1, ..., n - 2\}$, and therefore

$$2\int_{-1}^1 K_{2,n}(x)dx > -\sum_{\nu=1}^{n-2} a^G_{\nu,n-1}(x^G_{\nu+1,n-1} - x^G_{\nu,n-1})^2 - \sum_{\nu=2}^{n-1} a^G_{\nu,n-1}(x^G_{\nu,n-1} - x^G_{\nu-1,n-1})^2$$

$$(3.7) \qquad > -2 \max_{\mu\in\{1,...,n-1\}} a^G_{\mu,n-1} \sum_{\nu=1}^{n-2}(x^G_{\nu+1,n-1} - x^G_{\nu,n-1})^2.$$

Since $\int_{-1}^{1} K_{2,n}(x)dx = 0$ it follows that

$$(3.8) \qquad \int_{-1}^{1} |K_{2,n}(x)|dx < 2 \max_{\mu \in \{1,...,n-1\}} a_{\mu,n-1}^G \sum_{\nu=1}^{n-2} (x_{\nu+1,n-1}^G - x_{\nu,n-1}^G)^2.$$

Now let $x_{k,n} \notin (x_{1,n-1}^G, x_{n-1,n-1}^G)$. Lemma 2 shows that $k = 1$ or $k = n$. We have

$$(3.9) \qquad K_{2,n}(x) < (x+1)a_{1,n-1}^G \text{ for } x \in [-1, x_{1,n-1}^G],$$

$$(3.10) \qquad K_{2,n}(x) < \max\{a_{\nu,n-1}^G(x_{\nu+1,n-1}^G - x), a_{\nu+1,n-1}^G(x - x_{\nu-1}^G)\}$$
$$\text{for } x \in [x_{\nu,n-1}^G, x_{\nu+1,n-1}^G] \quad \text{and} \quad \nu \in \{1, ..., n-2\},$$

$$(3.11) \qquad K_{2,n}(x) < (1-x)a_{n-1,n-1}^G \text{ for } x \in [x_{n-1,n-1}^G, 1].$$

Since $\int_{-1}^{1} K_{2,n}(x)dx = 0$ it follows, as above, that

$$(3.12) \qquad \int_{-1}^{1} |K_{2,n}(x)|dx < 2 \max_{\mu \in \{1,...,n-1\}} a_{\mu,n-1}^G \sum_{\nu=0}^{n-1} (x_{\nu+1,n-1}^G - x_{\nu,n-1}^G)^2,$$

using the notation $x_{0,n-1}^G := -1$, $x_{n,n-1}^G := 1$. By (3.8) and (3.12) for nonpositive Q_n having all nodes contained in $[-1, 1]$, we have

$$(3.13) \qquad \int_{-1}^{1} |K_{2,n}(x)|dx < 2 \max_{\mu \in \{1,...,n-1\}} a_{\mu,n-1}^G \sum_{\nu=0}^{n-1} (x_{\nu+1,n-1}^G - x_{\nu,n-1}^G)^2,$$

$$(3.14) \qquad \qquad\qquad < 4 \max_{\mu \in \{1,...,n-1\}} a_{\mu,n-1}^G.$$

Since $\lim_{n \to \infty} \max_{1 \leq \nu \leq n-1} \{a_{\nu,n-1}^G\} = 0$, by (3.4) the result follows.

REMARK. With the methods above, estimating $|R_{n-1}^G[f]|$ and using for the Peano kernels of G the identity $K_{s-1,n} = K_{s,n}'$ as well as the fact that $K_{s,n}$ has $2n - 2 - s$ changes of sign in (y_1, y_{2n-1}), for special weight functions w sharper results can be obtained, e.g., for $w \equiv 1$ and (Q_n) fulfilling the conditions (i) and (ii) of Theorem 3,

$$(3.15) \qquad \limsup_{n \to \infty} n^s c_s(Q_n) < \infty \qquad \text{for fixed } s > 1.$$

PROOF OF THEOREM 4. First note that $R_{n-1}^G[f] = f^{(2n-2)}(\xi)/\{(2n - 2)!k_{n-1}^2\}, |\xi| \leq 1$, for $f \in C^{2n-2}[-1, 1]$ (see, e.g., [3, p.98]). Therefore, the first part of Theorem 4 follows from Lemma 8. Let $r_{n,a,y}$ be the node polynomial of Q_n and let the negative weight be associated with y. Since R_n is positive definite and $\deg(Q_n) \geq 2n-3$ we have $c_{2n-2}(Q_n) = R_n[p_{2n-2}]/(2n-2)!$ (see, e.g., [1, p.57]). Therefore, using $a < -\{P_n'(y)P_{n-1}(y) - P_n(y)P_{n-1}'(y)\}/\{P_{n-1}(y)\}^2$ (see Lemma 6),

$$\frac{c_{2n-2}(Q_n)}{c_{2n-2}(Q_{n-1}^G)} = k_{n-1}^2 R_n[p_{2n-2}] = \frac{k_{n-1}^2}{k_{n-2}} \int_{-1}^{1} P_{n-2}(x) r_{n,a,y}(x) w(x) dx$$

$$= \frac{ak_{n-1}^2}{k_{n-2}(k_n + ak_{n-1})} \int_{-1}^{1} x P_{n-2}(x) P_{n-1}(x) w(x) dx = \left\{1 + \frac{k_n}{k_{n-1}} \frac{1}{a}\right\}^{-1} < \{1 - P_{n-1}^2(y)\lambda_n(y)\}^{-1}$$

Now, since $\lambda_n \leq \lambda_{n-1}$, the proof is completed by using Lemma 10 below.

54

LEMMA 10. *Let $w > 0$ a.e. on $[-1,1]$. Then, for $x \in [-1,1]$,*

$$(3.16) \qquad \lim_{n \to \infty} \lambda_n(x) P_n^2(x) = 0,$$

and the convergence is uniform for $x \in [-1,1]$.

PROOF. We use a method described in the monograph of Nevai[10]. Let $\epsilon = \frac{2}{m}, m \in \mathbb{N} \backslash \{1\}$ be fixed, let $U_\nu := (-1 + (\nu - 1)\epsilon, -1 + \nu\epsilon]$ and let $f_{\epsilon,\nu} := 1$ for $x \in U_\nu, f_{\epsilon,\nu}(x) := 0$ for $x \notin U_\nu$. Then, for every $\nu \in \{1, ..., m\}$,

$$(3.17) \qquad \lim_{n \to \infty} Q_n^G[f_{\epsilon,\nu} P_{n-1}^2] = \frac{2}{\pi} \int_{U_\nu} \sqrt{1 - t^2} dt > \frac{2}{\pi} \int_{1-\epsilon}^1 \sqrt{1-t} = \frac{4}{3\pi} \epsilon^{3/2}$$

(see [10, p.17]). Therefore, there exist $n_\epsilon \in \mathbb{N}$ with $Q_n^G[f_{\epsilon,\nu} P_{n-1}^2] > \frac{1}{3} \epsilon^{3/2}$ for every $n > n_\epsilon$. Let $x \in [0,1], x \in U_\nu$. If $x \notin [0,1]$ consider $\bar{w}(x) := w(-x)$. Then,

$$(3.18) \qquad Q_n^G[\frac{1}{(\cdot - x)^2} P_{n-1}^2] \geq \frac{1}{\epsilon^2} \sum_{x_{k,n}^G \in U_\nu} a_{\nu,n}^G P_{n-1}^2(x_{\nu,n}^G) = \epsilon^{-2} Q_n^G[f_{\epsilon,\nu} P_{n-1}^2] > \frac{1}{3\sqrt{\epsilon}}$$

for $n > n_\epsilon$ and every $\nu \in \{1, ..., m\}$. We have (see [10, p.26])

$$(3.19) \qquad \lambda_n(x) P_n^2(x) = \frac{\gamma_n^2}{\gamma_{n-1}^2} Q_n^G[\frac{1}{(\cdot - x)^2} P_{n-1}^2]^{-1} < \frac{\gamma_n^2}{\gamma_{n-1}^2} 3\sqrt{\epsilon} \qquad \text{for } n > n_\epsilon.$$

With $\lim_{n \to \infty} \gamma_n / \gamma_{n-1} = \frac{1}{2}$ (see, e.g., [11, p.20]) the result follows.

References

1. BRASS, H. : Quadraturverfahren. Vandenhoeck & Ruprecht, Göttingen, (1977).
2. BRASS, H.; FÖRSTER, K.-J. : Error bounds for quadrature formulas near Gaussian quadrature. J. Comp. Appl. Math. 28, 145–154, (1989).
3. DAVIS, P.J.; RABINOWITZ, P. : Methods of Numerical Integration, 2nd ed. Academic Press, London, (1984).
4. FEJÉR, L.: Mechanische Quadraturen mit positiven Cotesschen Zahlen. Math. Z., 37, 287–309, (1933).
5. FÖRSTER, K.-J. : A comparison theorem for linear functionals and its application in quadrature. In G. Hämmerlin (Ed.): Numerical Integration, ISNM 57, Birkhäuser Verlag, 66–76, (1982).
6. FREUD, G. : Orthogonale Polynome. Birkhäuser Verlag, Basel, (1969).
7. KARLIN, S.J.; STUDDEN, W.J. : Tchebycheff systems. J. Wiley & Sons, New York, (1966).
8. KRYLOV, V.I. : Approximate Calculation of Integrals. (Translated from the Russian by A.H. Stroud) Macmillan Company, New York, (1962).
9. MICCHELLI, C.A.; RIVLIN, T.J. : Numerical integration rules near Gaussian quadrature. Isreal J. Math. 16, 287–299, (1973).
10. NEVAI, P.G. : Orthogonal Polynomials. Memoirs Amer. Math. Soc. Vol. 18, Rhode Island, (1979).
11. NEVAI, P.G. : Géza Freud, orthogonal polynomials and Christoffel functions. A case study. J. Approx. Theory 48, 3–167, (1986).
12. PEHERSTORFER, F. : Characterization of positive quadrature formulas. SIAM J. Math. Anal. 12, 935–942, (1981).
13. PEHERSTORFER, F. : Characterization of quadrature formula II. SIAM J. Math. Anal. 15, 1021–1030 (1984).
14. SCHMID, H.-J. : A note on positive quadrature rules. Rocky Mt. J. Math. 19, 395–404, (1989).
15. SHOHAT, J. : On mechanical quadratures, in particular with positive coefficients. Trans. Amer. Math. Soc. 42, 461–496, (1937).
16. SOTTAS, G.; WANNER, G. : The number of positive weights of a quadrature formula. BIT 22, 339–352, (1982).
17. SZEGÖ, G.: Orthogonal Polynomials. AMS Providence, Rhode Island, 4-th edition 1975, (1939).

NUMERICAL INTEGRATION IN HIGH DIMENSIONS
– THE LATTICE RULE APPROACH

Ian H. SLOAN
School of Mathematics
University of New South Wales
Kensington, NSW 2033 Australia
Email: sloan@hydra.maths.unsw.oz.au

ABSTRACT. Of the few methods available for numerical integration in high dimensions, one of the most interesting is the number–theoretic method of good lattice points originated by Korobov and Hlawka. This paper introduces the subject of lattice rules, which may be thought of as a generalisation of the method of good lattice points, and reviews recent developments. After introducing the concept of the rank of a lattice rule (the method of good lattice points has rank 1, the product–rectangle rule in s dimensions has rank s, and rules with every rank between 1 and s also exist) the paper will discuss some theoretical and practical grounds for believing that certain rules with rank greater than 1 might be of practical value.

1 Introduction

Lattice rules for numerical integration in many dimensions have recently attracted considerable interest. In this paper we review recent developments, after a sketch of their origins, definition, and general properties. We describe a way in which they may be classified, introduce the concept of rank, and show at the same time how lattice rules may be presented in a form suitable for computer study. And then we examine, using mainly theoretical tools, whether some of the lattice rules which were previously unknown might hold some promise for practical computations.

An important special case, developed thirty years ago, is the method of good lattice points, reviewed by Niederreiter (1978, 1991). For a review of the early development of lattice rules in the more general sense see Sloan (1984). For contemporary reviews from different perspectives see Lyness (1989) and Niederreiter (1991).

2 Lattice Rules – Special Cases

It is useful to begin with the 1–dimensional case. Consider the problem of approximating

$$If = \int_0^1 f(x)dx, \tag{1}$$

where f is a continuous function with a smooth 1–periodic extension to the real line. For a given number N of quadrature points on $[0,1)$ it is well known that the N–point rectangle

T. O. Espelid and A. Genz (eds.), Numerical Integration, 55–69.
© 1992 *Kluwer Academic Publishers.*

rule

$$Q_N f = \frac{1}{N} \sum_{j=0}^{N-1} f\left(\frac{j}{N}\right) \tag{2}$$

cannot in general be bettered. Certainly a composite Newton–Cotes rule such as Simpson's rule cannot generally offer any advantage over the rectangle rule, since if one takes the Simpson's rule result with a repeat length of $2h$ and averages it with the result obtained from the h–shifted Simpson's rule (which, given the periodicity assumption, surely has an equal claim), then one obtains just the rectangle rule with spacing h; a similar argument applies to any composite Newton–Cotes rule.

The rectangle–rule error in this periodic setting is usually studied by means of the Euler–Maclaurin expansion (see, for example, Davis and Rabinowitz (1975)). For our present purposes, however, a Fourier analysis turns out to be more instructive. Suppose f has an absolutely convergent Fourier series

$$f(x) = \sum_{h=-\infty}^{\infty} \hat{f}(h) e^{2\pi i h x}, \tag{3}$$

where

$$\hat{f}(h) = \int_0^1 e^{-2\pi i h x} f(x) dx, \quad h \in . \tag{4}$$

Then it is easily seen that the error in the rectangle rule (2) applied to f is

$$Q_N f - I f = \sum_{k=-\infty}^{\infty} \hat{f}(kN) - \hat{f}(0) = \sum_{k=-\infty}^{\infty}{}' \hat{f}(kN), \tag{5}$$

where the prime indicates that the term with $k = 0$ is to be omitted from the sum. The result (5) expresses the fact, easily established by summing a geometric series, that the only Fourier components $\hat{f}(h) \exp(2\pi i h x)$ for which the quadrature sum is not zero are those for which h is a multiple of N.

The s–dimensional generalization of the problem (1) is

$$I f = \int_0^1 \cdots \int_0^1 f(x_1, \ldots, x_s) dx_1 \ldots dx_s = \int f(\boldsymbol{x}) d\boldsymbol{x}, \tag{6}$$

where the integration region is understood to be the unit cube in s dimensions, and f is assumed to have a (more or less) smooth 1–periodic extension with respect to each of $x_1, \ldots, x_s$. (We should concede that periodicity of this kind does not often occur naturally with respect to all variables, even if it occurs naturally for some of them. In practice it will usually be necessary to force some degree of smooth periodisation by an initial non–linear change of variables, of the form

$$\int_0^1 g(x) dx = \int_0^1 g(\phi(t)) \phi'(t) dt, \tag{7}$$

for each variable for which a smooth periodic extension does not occur naturally. Here ϕ is a monotone increasing function which maps $[0, 1]$ onto $[0, 1]$, and which has one or more derivatives vanishing at both 0 and 1. For example. if g in (7) is continuous on $[0,1]$, the choice

$$\varphi(t) = 3t^2 - 2t^3, \quad t \in [0,1], \tag{8}$$

which gives $\phi'(t) = 6t(1 - t)$, ensures that the integrand on the right of (7) has a 1-periodic extension that is at least continuous. Greater smoothing can of course be achieved, assuming that g is appropriately smooth on $(0, 1)$, by using transformations ϕ for which ϕ' has higher–order zeros at the end–points. From now on we shall simply assume that all necessary transformations of this kind have already been carried out.)

A lattice rule for the integral (6) is an equal–weight generalization of the 1–dimensional rectangle rule (2). (We shall come to the formal definition in the next section.) The most obvious such generalization is a tensor product of 1–dimensional rectangle rules (hereafter called the 'product–rectangle rule'),

$$Qf = \frac{1}{n^s} \sum_{k_1=0}^{n-1} \cdots \sum_{k_s=0}^{n-1} f\left(\frac{k_1}{n}, \ldots, \frac{k_s}{n}\right). \tag{9}$$

This is a valid lattice rule. but a very costly one when the dimensionality is high, since the total number of quadrature points is $N = n^s$.

A more interesting generalization of the 1–dimensional rectangle rule is the 'method of good lattice points', developed around 1960 by Korobov (1960) and Hlawka (1962). Other notable contributors have been Zaremba (1966), Niederreiter (1978) and Hua and Wang (1981), and the method was rediscovered by Conroy (1967), a physical chemist.

The method of good lattice points employs a quadrature rule of the form

$$Q(z)f = \frac{1}{N} \sum_{j=0}^{N-1} f\left(\left\{j\frac{z}{N}\right\}\right), \tag{10}$$

where z is a (well chosen) integer vector, and the braces indicate that each component of the vector is to be replaced by its fractional part: that is

$$\{(x_1, \ldots, x_s)\} = (\{x_1\}, \ldots, \{x_s\}), \tag{11}$$

with

$$\{x\} = x - \lfloor x \rfloor, \tag{12}$$

$\lfloor x \rfloor$ denoting the largest integer which does not exceed x. Since f is 1–periodic with respect to each component, the braces can be omitted without loss. Thus in future we shall write, for simplicity,

$$Q(z)f = \frac{1}{N} \sum_{j=0}^{N-1} f\left(j\frac{z}{N}\right). \tag{13}$$

A simple 2 dimensional example is the 5 point rule

$$\frac{1}{5} \sum_{j=0}^{4} f\left(j\frac{(1,2)}{5}\right) = \frac{1}{5}\left[f(0,0) + f\left(\frac{1}{5}, \frac{2}{5}\right) + f\left(\frac{2}{5}, \frac{4}{5}\right) + f\left(\frac{3}{5}, \frac{1}{5}\right) + f\left(\frac{4}{5}, \frac{3}{5}\right)\right].$$

A typical early theorem is the following. In this theorem E_α, for $\alpha > 1$, denotes the class of 1–periodic functions f whose Fourier coefficients are bounded by

$$|\hat{f}(h_1, \ldots, h_s)| \leq \frac{1}{(\overline{h_1} \ldots \overline{h_s})^\alpha}, \tag{14}$$

where

$$\overline{h} = \begin{cases} 1 & \text{if } h = 0, \\ |h| & \text{if } h \neq 0. \end{cases} \tag{15}$$

Theorem 2.1 *Given $\alpha > 1$, there exist positive numbers C and β, depending only on s and α, such that for each prime number N there is a $z \in {}^s$ for which, for all $f \in E_\alpha$,*

$$|Q(z)f - If| \leq C \frac{(\log N)^\beta}{N^\alpha}.$$

The order of convergence in this theorem is wonderfully good, if one is prepared to ignore the powers of log N, in that the product–rectangle rule (9) can achieve an error of order only $O(n^{-\alpha}) = O(N^{-\alpha/s})$ for general $f \in E_\alpha$.

Much more precise results for the method of good lattice points are now available, particularly with respect to the constants C and β in the theorem, and the restriction that N be prime has been removed. For the best results and an up–to–date survey, see Niederreiter (1992b).

Only in the two–dimensional case is an explicit construction of a good z for use in (13) known (see Zaremba (1966)). In all other dimensions the existence proofs in the literature are non–constructive, and good choices of z must in practice be found by computer search.

3 Lattice Rules

The more general notion of a lattice rule, anticipated by Frolov (1977) and rediscovered and developed by Sloan and Kachoyan (1987), includes the product–rectangle rule and the method of good lattice points as special cases.

Definition. A *lattice rule* for approximate integration over the unit s–dimensional cube is an equal–weight rule of the form

$$Qf = \frac{1}{N} \sum_{j=0}^{N-1} f(x_j), \tag{16}$$

where $x_0, \ldots, x_{N-1}$ are all the points of an 'integration lattice' that lie in the half–open unit cube $[0, 1)^s$.

Definition. An *integration lattice* in ^{s} is a subset of ^{s} which is discrete and closed under addition and subtraction (and hence is a *lattice*), and which contains ^{s} as a subset.

In the two examples of lattice rules we have met already, the corresponding integration lattices are

$$\left\{ \left(\frac{k_1}{n}, \ldots, \frac{k_s}{n} \right) \ : \ k_i \in , \quad 1 \le i \le s \right\}$$

for the product–rectangle rule, and

$$\left\{ j\frac{z}{N} + u \ : \ j \in , \ u \in {}^s \right\}$$

for the method of good lattice points, with z in the latter case a prescribed integer vector.

Let us denote the lattice corresponding to the lattice rule Q by $L = L(Q)$. Because the lattice is invariant under translation by any point of the lattice, it turns out that the error expression (5) for the rectangle rule has a natural generalization. Let the Fourier coefficients of f be denoted by

$$\hat{f}(h) = \int e^{-2\pi i h \cdot x} f(x) dx, \quad h \in {}^s,$$

where again the integration is over the unit cube. The following theorem, from Sloan and Kachoyan (1987), shows that the only Fourier components of f not integrated exactly by the lattice rule L are those for which h is a point in the 'reciprocal lattice' $L^{\perp}$, defined by

$$\begin{aligned} L^{\perp} &= \{ h \in {}^s \ : \ h \cdot x \in \quad \forall \, x \in L \} \\ &= \{ h \in {}^s \ : \ h \cdot x_j \in \text{ for } j = 0, \ldots, N-1 \}. \end{aligned} \tag{17}$$

For the the method of good lattice points the reciprocal lattice is

$$L^{\perp} = \{ h \in {}^s \ : \ h \cdot z \equiv 0 \pmod{N} \}.$$

For this particular case the error expression in the following theorem is well known.

Theorem 3.1 *If f has the absolutely convergent Fourier series*

$$f(x) = \sum_{h} \hat{f}(h) e^{2\pi i h \cdot x}. \tag{18}$$

then the error in the lattice rule (16) applied to f is

$$Qf - If = {\sum_{h \in L^{\perp}}}' \hat{f}(h). \tag{19}$$

Again the prime indicates that the term $h = 0$ is to be omitted from the sum. It is easy to understand the appearance of the reciprocal lattice in the error expression, if we note that for $h \in L^{\perp}$ the Fourier component $\hat{f}(h) \exp(2\pi i h \cdot x)$ has the same value $\hat{f}(h)$ at each quadrature point x_j, because $h \cdot x_j$ is an integer. ¿From this it follows that the quadrature rule applied to a component with $h \in L^{\perp}$ and $h \neq 0$ yields $\hat{f}(h)$, instead of the correct value 0. The non–trivial part of the proof is the demonstration that for all values of h *not* in $L^{\perp}$ the quadrature sum applied to the Fourier component does yield the correct value 0. This and other general properties of lattice rules are discussed in Sloan and Kachoyan (1987).

We note in passing that the error expression (19) is known to hold in a certain sense even for functions which, though well behaved on $[0,1]^s$, do not have a continuous periodic extension, and for which therefore the hypothesis of absolute convergence of the Fourier series cannot hold. In this case the right side of (19) must be interpreted as the limit of the square partial sums, i.e. as the limit as $\ell \to \infty$ of the sums over $h \in L^{\perp}$, $h \neq 0$, with $|h_1|, \ldots, |h_s| \leq \ell$. This result is established in Price and Sloan (1991). The key feature is that the lattice rule is applied not to f itself but to a 1-periodic function $\overline{f}$ which coincides with f in $(0,1)^s$, and whose values at points of discontinuity are defined in a special way: for example, in the 1-dimensional case as the mean of the right-hand and left-hand limits. We do not consider this extension any further in this paper.

4   What Lattice Rules Are There?

The considerations in the preceding section do not tell us whether lattice rules exist, beyond those we have already met in section 2. Nor, since the lattice–rule definition is geometrical in nature, do they tell us how to present a general lattice rule in a form suitable for computer study.

In this section we introduce briefly a classification scheme which answers these questions. The following theorem of Sloan and Lyness (1989) shows that every lattice rule can be written in a similar way to that seen already for the method of good lattice points, but with, in general, more than one sum. The 'rank' of the lattice rule is the minimum possible number of such sums.

Theorem 4.1 *Every lattice rule may be written as a non–repetitive expression of the form*

$$Qf = \frac{1}{n_1 \ldots n_r} \sum_{j_1=0}^{n_1-1} \cdots \sum_{j_r=0}^{n_r-1} f\left(j_1 \frac{z_1}{n_1} + \ldots + j_r \frac{z_r}{n_r}\right), \tag{20}$$

where r (the 'rank') and $n_1, \ldots, n_r$ (the 'invariants') are uniquely determined natural numbers satisfying

$$n_{i+1} | n_i, \qquad i = 1, \ldots, r-1, \tag{21}$$

and

$$n_r > 1, \tag{22}$$

and where $z_1, \ldots, z_r$ are linearly independent vectors in ^{s}.

The rule (20) is non–repetitive if it has $n_1 \ldots n_r$ distinct quadrature points after reduction to $[0,1)^s$ by the subtraction of appropriate integer vectors.

The theorem is proved by recognising that the quadrature points in the lattice rule (16) form a finite abelian group, and then appealing to a standard theorem on the classification of abelian groups.

We have already met lattice rules of ranks 1 and s: the method of good lattice points is manifestly of rank 1 (there is only one sum), and the product–rectangle rule (9) with $n > 1$ has rank s and invariants $n, n, \ldots, n$. The following theorem, again from Sloan and Lyness

(1989), shows that these values of the rank are at the extremes of the range of possible values. The theorem follows immediately from the linear independence of $z_1, \ldots, z_r$.

Theorem 4.2 *The rank r of an s-dimensional lattice rule satisfies $1 \leq r \leq s$.*

Lattice rules exist for every rank in this range, and for every possible set of invariants, i.e. for every set of positive integers $n_1, \ldots, n_r$ satisfying (21) and (22): for if we choose z_i in (20) to be the vector with a 1 in the ith position and 0's everywhere else then we obtain a rule which is not repetitive, and which therefore, by the uniqueness part of the theorem, really is a rule with the designated rank and invariants.

An interesting question, at present unanswered, is the total number of distinct lattice rules with a given rank r and given invariants $n_1, \ldots, n_r$. On the other hand the total number of lattice rules with a given number of quadrature points N is known, explicit formulas being stated by Lyness and Sørevik (1989). It is fair to say that the numbers, even for moderate values of s and N, are truly astronomical. For example, if $N = p^t$ where p is a prime then the quoted number of rules is

$$\prod_{i=1}^{s-1} \frac{p^{i+t} - 1}{p^i - 1}.$$

Thus if $s = 7$ and $N = 3^3 = 27$ (a modest number of quadrature points indeed) then the number of lattice rules already exceeds 6×10^8. On the other side, however, it should be said that many rules differ only in uninteresting ways, for example by a permutation of the coordinates, or by reflections of the form $x_i \to 1 - x_i$. Little or nothing is known about the number of rules that are distinct when one take this 'geometrical equivalence' of rules into account.

Rules which have the maximal rank s will play an important role in the next section when we turn to questions of merit. It turns out (see Sloan and Lyness (1989)) that rules of rank s have a very simple characterisation:

Theorem 4.3 *An s-dimensional lattice rule is of rank s if and only if it is the n^s copy, with $n > 1$, of a lattice rule of lower rank.*

Definition. The n^s *copy* of a quadrature rule on $[0, 1]^s$ is the rule obtained by shrinking the given rule to a cube of side $1/n$, and then applying the rule to each of the n^s cubes of side $1/n$ into which the cube is naturally divided.

Note that the product–rectangle rule (9) is the n^s copy of the 1–point rule $Tf = f(\mathbf{0})$.

For a more detailed version of theorem 4.3 in which the invariant structure of the copied rule is specified, and for theorems on questions such as the invariant structure of rules obtained by omitting some of the components of a lattice rule, we refer to the original paper of Sloan and Lyness (1989).

An unattractive feature of the lattice–rule representation given by theorem 4.1 is that the vectors $z_1, \ldots, z_r$ are not uniquely determined. The problem of obtaining a unique representation of lattice rules has been solved for an important sub–class of lattice rules (the so–called 'projection regular' rules) by Sloan and Lyness (1990), but has not yet been solved for all lattice rules.

A different approach to the specification of lattice rules which does lead to a unique representation has been emphasized by Lyness and colleagues (see Lyness (1989)). In this approach the lattice L is specified indirectly, through a 'generator matrix' of the reciprocal lattice $L^\perp$. (A generator matrix of $L^\perp$ is simply an $s \times s$ matrix with the property that $L^\perp$ is the set of all integer linear combinations of the rows.) Since $L^\perp$ is an integer lattice in the present context, its generator matrices have integer entries, and a canonical form can be obtained by appeal to the theory of integer matrices. It is this approach which leads to the explicit formulas of Lyness and Sørevik (1989), mentioned earlier, for the total number of all lattice rules.

5   Are Rules With Rank > 1 Of Value?

In this section we ask whether well chosen lattice rules with rank > 1 might be competitive with the method of good lattice points (which of course uses a rule with rank $= 1$). Following the number theorists, we use as a measure of numerical accuracy the quantity

$$P_\alpha = \sum_{h \in L^\perp}' \frac{1}{(\overline{h_1} \ldots \overline{h_s})^\alpha}, \tag{23}$$

where $\alpha > 1$ is a fixed real number (in practice often 2 or 4). The significance of this measure is easily stated: if $f \in E_\alpha$, where E_α is (as in theorem 2.1) the set of all 1–periodic functions f whose Fourier coefficients satisfy (14), then it follows from theorem 3.1 that

$$|Qf - If| = \left| \sum_{h \in L^\perp}' \hat{f}(h) \right| \leq \sum_{h \in L^\perp}' \frac{1}{(\overline{h_1} \ldots \overline{h_s})^\alpha} = P_\alpha.$$

Thus P_α is a bound on the error for $f \in E_\alpha$. Moreover, it is an achievable bound, since, again by theorem 3.1,

$$P_\alpha = Qf_\alpha - If_\alpha = Qf_\alpha - 1, \tag{24}$$

where $f_\alpha \in E_\alpha$ is defined by

$$f_\alpha(\boldsymbol{x}) = \sum_{h \in \mathscr{s}} \frac{1}{(\overline{h_1} \ldots \overline{h_s})^\alpha} e^{2\pi i h \cdot \boldsymbol{x}}. \tag{25}$$

Equation (24) provides a practical way of computing P_α if α is an even integer, since

$$f_\alpha(\boldsymbol{x}) = \prod_{i=1}^{s} f_\alpha(x_i), \tag{26}$$

where

$$f_\alpha(x) = \sum_{h=-\infty}^{\infty} \frac{1}{\overline{h}^\alpha} e^{2\pi i h x} = 1 + 2 \sum_{h=1}^{\infty} \frac{1}{h^\alpha} \cos 2\pi h x. \tag{27}$$

For α even, the latter sum is expressible on $[0, 1]$ in terms of Bernouilli polynomials: for example

$$f_2(x) = 1 + 2\pi^2(x^2 - x + 1/6), \quad 0 \le x \le 1. \tag{28}$$

For simplicity we shall often fix $\alpha = 2$, but a different choice would make little difference to our considerations.

How can we decide which of two classes of quadrature rules gives the best value of P_α, when the number of rules in each class is enormous? Until very recently no theoretical results were available. In the absence of such results Sloan and Walsh (1990) carried out an extensive computer search of certain classes of rules of rank 2 and small second invariant n_2 (e.g. $n_2 = 2$ or 3), seeking always rules that gave relatively small values of P_2 (or of P_6, or both). Various devices were used to limit the searches to manageable proportions. The results, when compared to similar searches of rank 1 rules, were encouraging, in that for $n_2 = 2$ the best of the rank 2 rules almost always gave a smaller value of P_2 or P_6 than the best of the rank 1 rules. While the reductions (being typically perhaps 15% for P_2) were not dramatic, the pattern seemed clear. On the other hand larger values of n_2 were almost invariably less satisfactory than $n_2 = 2$.

A simple but useful theoretical observation in that paper is that a rank 2 rule with fixed smaller invariant n_2 can always be found that is as good as the best rank 1 rule, if we are concerned only about order of convergence and not with the multiplying constant that stands in front. The result, stated formally in Sloan and Walsh (1990), is based on the rather obvious fact that a rank 2 rule with invariants mn_2 and n_2 contains as a subset the points of an m–point rank 1 rule. The value of P_α can only be larger for this m–point rule than for the mn_2^2–point rule, since $L_1 \subset L_2$ implies $L_1^\perp \supset L_2^\perp$. A much more precise result has recently been obtained by Niederreiter (1992a), which shows that P_α for the best rank 2 rule is of the same order as for the best known rank 1 rule even when n_2 is allowed some slow growth.

While the theoretical results indicated in the preceding paragraph are consistent with the empirical observations of Sloan and Walsh (1990), they cannot provide a basis for a quantitative comparison between rank 1 rules and rules of higher rank, because they are concerned only with orders of convergence, and make no attempt to compare the constant factors in front. (In fact the constant factor may appear very bad if, as above, the argument is merely that a rule of higher rank gives as good a P_α value as a rank 1 rule with many fewer points.)

The first successful attempt at a quantitative comparison between rank 1 rules and rules of higher rank concerns not rules of rank 2, but rules of the maximal rank s. For most of the remainder of this section we describe results of Disney and Sloan (1991a) and (1991b), always foregoing generality in favour of simplicity of statement. The first of these papers is concerned with rules of the rank 1 form (13), while the rules considered by Disney and Sloan (1991b) are of the form

$$Q^{(n)}(\boldsymbol{w})f = \frac{1}{n^s a} \sum_{j=0}^{a-1} \sum_{k_1=0}^{n-1} \cdots \sum_{k_s=0}^{n-1} f\left(j\frac{\boldsymbol{w}}{na} + \frac{(k_1, \ldots, k_s)}{n}\right), \tag{29}$$

with $w \in {}^s$. It is clear that $Q^{(n)}(w)f$ is the n^s copy of the a- point rule

$$Q(w)f = Q^{(1)}(w)f = \frac{1}{a} \sum_{j=0}^{a-1} f\left(j\frac{w}{a}\right).$$ (30)

The considerations above would lead us to expect that n should be small. In fact we shall see that in a certain sense the best value is 2.

For fixed α and fixed a and n, the ideal theorem would compare $P_\alpha(Q^{(n)}(w))$ for the best possible choice of the vector w (best, that is, in the sense of minimising P_α) with $P_\alpha(Q(z))$ for the best choice of z in the rank 1 rule (13) for a similar number of points. No such theorem is known. What *is* known is a theorem which compares certain *means* of rules of each type. For simplicity we again fix $\alpha = 2$. The mean of $P_2(Q^{(n)}(w))$ for fixed n and a, and with a assumed to be prime, we denote by $m(n,a)$. It is the mean of $P_2(Q^{(n)}(w))$ over the $(a-1)^s$ values of w obtained by allowing the components of w to take values in the range

$$-\frac{a}{2} < w_i \le \frac{a}{2}, \quad w_i \ne 0, \quad 1 \le i \le s.$$

The mean of $P_2(Q(z))$ for the rank 1 rules (13) when N is prime we denote by $M(N)$. It is the mean of $P_2(Q(z))$ over the $(N-1)^s$ values of z obtained by allowing the components of z to take values in the range

$$-\frac{N}{2} < z_i \le \frac{N}{2}, \quad z_i \ne 0, \quad 1 \le i \le s.$$

Explicit expressions for both $m(n,a)$ and $M(N)$ are known. An explicit expression is also known, see Disney (1990), for an appropriate mean of $P_\alpha(Q(z))$ even when N is not prime. However, we prefer to use always the expression for N prime, because it is more favourable to the rank 1 rules. To do otherwise would introduce an unfair bias against the rank 1 rules, given that one could always choose, when using rank 1 rules, to use only prime values of N. In the following theorem M occurs with the non–prime argument $2^s a$, because we want to compare rules with the same number of points. One may think of $M(2^s a)$ as the mean of $P_2(Q(z))$ for a prime number of points close to $2^s a$. The following theorem is from Disney and Sloan (1991b).

Theorem 5.1 *If a is prime and $s \ge 3$ than*

$$\frac{m(2,a)}{M(2^s a)} \le \left(\frac{2 + \pi^2/6}{1 + \pi^2/3}\right)^s \approx 0.85^s.$$ (31)

This favourable result for the means of P_2 for the two kinds of rule suggests that a similar result might hold for the *best* rules of each class.

Some remarks about the proof of (31) may give useful insight. For the denominator we have, for N prime,

$$M(N) = \operatorname*{mean}_z P_2(Q(z)) = \operatorname*{mean}_z (Q(z)f_2 - 1)$$

$$= \operatorname*{mean}_{\boldsymbol{z}} \left(\frac{1}{N} f_2(\boldsymbol{0}) + \frac{1}{N} \sum_{j=1}^{N-1} f_2 \left(j \frac{\boldsymbol{z}}{N} \right) \right) - 1$$

$$= \frac{1}{N} \left(1 + \frac{\pi^2}{3} \right)^s + \operatorname*{mean}_{\boldsymbol{z}} \left(\frac{1}{N} \sum_{j=1}^{N-1} f_2 \left(j \frac{\boldsymbol{z}}{N} \right) \right) - 1, \tag{32}$$

in which we have sensibly separated the two terms which do not depend on $\boldsymbol{z}$, and used, from (26) and (27),

$$f_2(\boldsymbol{0}) = (1 + 2\zeta(2))^s = (1 + \pi^2/3)^s.$$

The middle term of (32) is evaluated by Disney and Sloan (1991a), to give finally

$$M(N) = \frac{1}{N} \left(1 + \frac{\pi^2}{3} \right)^s + \frac{N-1}{N} \left(1 - \frac{\pi^2}{3N} \right)^s - 1. \tag{33}$$

It turns out that it is the first term, which is simply $f_2(\boldsymbol{0})/N$, that dominates the results: The other terms when combined are negative, and relatively small when s is large.

The numerator $m(2, a)$, or more generally $m(n, a)$, is evaluated in Disney and Sloan (1991b) by the use of a trick that reduces the computation practically to that carried out already for the rank 1 case. Noting that the lattice corresponding to $Q^{(n)}(\boldsymbol{w})$ is just a scaled version of that corresponding to $Q(\boldsymbol{w}) = Q^{(1)}(\boldsymbol{w})$, we may write

$$L(Q^{(n)}(\boldsymbol{w})) = \frac{1}{n} L(Q(\boldsymbol{w})),$$

from which follows

$$L(Q^{(n)}(\boldsymbol{w}))^{\perp} = n L(Q(\boldsymbol{w}))^{\perp},$$

so that

$$\begin{aligned} P_\alpha(Q^{(n)}(\boldsymbol{w})) &= \sideset{}{'}\sum_{\boldsymbol{h} \in L(Q^{(n)})^{\perp}} \frac{1}{(\overline{h_1} \ldots \overline{h_s})^\alpha} \\ &= \sideset{}{'}\sum_{\boldsymbol{h} \in L(Q(\boldsymbol{w}))^{\perp}} \frac{1}{(\overline{nh_1} \ldots \overline{nh_s})^\alpha} \\ &= Q(\boldsymbol{w}) f_\alpha^{(n)} - 1, \end{aligned} \tag{34}$$

where

$$\begin{aligned} f_\alpha^{(n)}(\boldsymbol{x}) &= \sum_{\boldsymbol{h} \in^s} \frac{1}{(\overline{nh_1} \ldots \overline{nh_s})^\alpha} e^{2\pi i \boldsymbol{h} \cdot \boldsymbol{x}} \\ &= \prod_{i=1}^{s} f_\alpha^{(n)}(x_i), \end{aligned} \tag{35}$$

with

$$f_\alpha^{(n)}(x) = \sum_{h=-\infty}^{\infty} \frac{1}{(\overline{nh})^\alpha} e^{2\pi i h x}$$

$$= 1 + \frac{2}{n^\alpha} \sum_{h=1}^{\infty} \frac{1}{h^\alpha} \cos 2\pi h x$$

$$= 1 + \frac{1}{n^\alpha}(f_\alpha(x) - 1). \tag{36}$$

Notice that (34) expresses $P_\alpha(Q^{(n)}(\boldsymbol{w}))$ as the error in an a-point rank 1 rule applied to the function $f_\alpha^{(n)}$, which by (35) and (36) is related in a simple way to a function which we know already.

The arguments outlined above for the rank 1 case now go through with little change, to give for our means of $P_2(Q^{(n)}(\boldsymbol{w}))$

$$m(n,a) = \frac{1}{a} f_2^{(n)}(\boldsymbol{0}) + \operatorname*{mean}_{\boldsymbol{w}} \left(\frac{1}{a} \sum_{j=1}^{a-1} f_2^{(n)} \left(j \frac{\boldsymbol{w}}{a} \right) \right) - 1$$

$$= \frac{1}{n^s a} \left(n + \frac{\pi^2}{3n} \right)^s + \frac{a-1}{a} \left(1 - \frac{\pi^2}{3an^2} \right)^s - 1, \tag{37}$$

in which again it is the first term that is dominant, and which leads to the result in the theorem. Disney and Sloan (1991b) also point out that the best value of n, in the sense of minimising the dominant term of $m(n,a)$ for fixed $N = n^s a$, is $n = 2$. In effect $n = 2$ gives the best compromise between the rapidly growing first term of $n + \pi^2/(3n)$ and the initially large but declining second term.

We conclude this section by remarking that theorem 5.1 has been extended from rules of rank s to certain rules of general rank r and smallest invariant 2 by Joe and Disney (1991). A similar bound to that in theorem 5.1 is obtained, but with the exponent s on the right replaced by r. Interestingly, the result for the means of selected rank 2 rules compared to the means of rank 1 rules is reasonably consistent with the empirical results of Sloan and Walsh (1990) for the best rules of each class. For further details we refer to the paper of Joe and Disney.

6   Implementation

The arguments in the preceding section, and in particular theorem 5.1, give good reasons for believing that lattice rules which are 2^s copies of well chosen rank 1 rules may be competitive in accuracy with rank 1 rules with a similar number of points.

There are other reasons, apart from accuracy, why one might find 2^s copy rules attractive. One is that it is relatively easy to find 'good' rules of this class, since the computation of $P_\alpha(Q^{(n)})$ can be achieved, via (34)–(36), by a quadrature sum which uses just $N/2^s$ points, compared to N points for an equivalent rank 1 rule. The effect can be dramatic. Thus Joe and Sloan (1992) quote a 20–dimensional rule with approximately 4×10^9 points, obtained by a search using a quadrature rule with just a few thousand points.

Another attraction of the 2^s copy rules is that it is possible, following Joe and Sloan (1992), to obtain at no extra cost not just one estimate, but a whole family. This raises the possibility of obtaining an error estimate that is essentially cost–free. Joe and Sloan calculate the finite sequence $Q_0 f, \ldots, Q_s f$ given by

$$Q_r f = \frac{1}{2^r a} \sum_{j=0}^{a-1} \sum_{k_1=0}^{1} \cdots \sum_{k_r=0}^{1} f\left(j\frac{w}{a} + \frac{(k_1, \ldots, k_r, 0, \ldots, 0)}{2}\right), \quad 0 \le r \le s, \qquad (38)$$

where a is odd, and present an algorithm by which all members of the sequence may be calculated simultaneously. Obviously the sequence is 'imbedded' (each point of Q_{r-1} is also a point of Q_r), with the number of points doubling from one member to the next. The rule $Q_r f$ is shown in the cited paper to be of rank r for $r = 1, \ldots, s$, and the last rule $Q_s f$ is shown to be the 2^s copy of the rank 1 rule

$$Q_0 f = \frac{1}{a} \sum_{j=0}^{a-1} f\left(j\frac{w}{a}\right). \qquad (39)$$

It is reasonable to choose the vector w to minimise

$$P_\alpha(Q_s) = Q_0 f_\alpha^{(2)} - 1. \qquad (40)$$

where we have used (34). That still leaves the order of the components of w undetermined, since $P_\alpha(Q_s)$ is invariant under a permutation of the components of w. Joe and Sloan choose the components $w_s, w_{s-1}, \ldots, w_2$, in that order, from the $s, s-1, \ldots, 2$ possible choices at each stage, so as to minimise successively $P_\alpha(Q_{s-1}), P_\alpha(Q_{s-2}), \ldots, P_\alpha(Q_1)$.

As an example, in table 1 we show a 6–dimensional rule with $5,003 \times 2^6 = 320,192$ points obtained by Joe and Sloan in this way, with the search restricted to vectors w of the form $w = (1, b, b^2, \ldots, b^{s-1})$ (mod a), and $\alpha = 2$. We conclude with a numerical example of a

Table 1: A set of parameters for the lattice rule sequence (38), from Joe and Sloan (1992).

s	=	6
a	=	5,003
w	=	(4718, 4538, 1, 1229, 162, 3981)

lattice rule applied to a 6–dimensional integral. The given integral is

$$c \int_0^1 \cdots \int_0^1 \exp[-5(|x_1 - 0.6| + \ldots + |x_6 - 0.6|)] dx_1 \ldots dx_6, \qquad (41)$$

with c chosen so that the exact value of the integral is 1. As it stands the integrand does not have a continuous periodic extension, thus as a first step we make the transformation $x_i = \phi(t_i)$ for each variable, where ϕ is given by (8). (Higher order periodisation would in this case be pointless, given the discontinuities in the first derivatives that already occur in the interior.) After this transformation the family of rules in table 1 was applied, giving the results shown in table 2. Recalling that the rules are optimised to make $Q_6 f$ the best estimate, we might conclude, given the information in the table above, that

$$If \approx 1.000011.$$

As a rough estimate of the error we might use

$$|Q_6 f - Q_5 f| = 0.000952,$$

which at least in this case turns out to be comfortingly conservative. Further investigation is needed to see if an error estimate of this kind is useful in general.

Table 2: The family of rules in table 1 applied to the integral (41), after making the transformation (8) in each variable. The last entry is a 2^6 copy rule.

Rule	No. of points	$Q_r f$	Error ($\times 10^6$)
Q_0	5,003	1.016987	16987
Q_1	10,006	0.998598	-1402
Q_2	20,012	0.997869	-2131
Q_3	40,024	0.996252	-3748
Q_4	80,048	0.997973	-2027
Q_5	160,096	0.999059	-941
Q_6	320,192	1.000011	11

Acknowledgement

The writer acknowledges the support of the Australian Research Council and the Technical University of Vienna.

References

Conroy, H. (1967) 'Molecular Schrödinger equation VIII: A new method for the evaluation of multidimensional integrals', *J. Chem. Phys.* **47**, 5307–5318.

Davis, P. and Rabinowitz, P. (1975) 'Methods of Numerical Integration', Academic Press, New York.

Disney, S.A.R. (1990), 'Error bounds for the rank 1 lattice rules modulo composites', *Monatsh. Math.*, **110**, 89–100.

Disney, S.A.R. and Sloan, I.H. (1991a) 'Error bounds for the method of good lattice points', *Math. Comp.* **56**, 257–266.

Disney, S.A.R. and Sloan, I.H. (1991b) 'Lattice integration rules of maximal rank', *SIAM J. Numer. Anal.*, to appear.

Frolov, K.K. (1977) 'On the connection between quadrature formulas and sublattices of integral vectors', (Russian), *Dokl. Akad. Nauk SSSR*, **232**, 40–43.

Hlawka, E. (1962) 'Zur angenäherten Berechnung mehrfacher Integrale', *Monatsh. Math.*, **66**, 140–151.

Hua Loo Keng and Wang Yuan (1981) 'Applications of Number Theory to Numerical Analysis', Springer–Verlag, Berlin, Science Press, Beijing.

Joe, S. and Disney, S.A.R. (1991) 'Intermediate rank lattice rules for multiple integration', submitted.

Joe, S. and Sloan, I.H. (1992) 'Imbedded lattice rules for multidimensional integration', *SIAM J. Numer. Anal.*, to appear.

Korobov, N.M. (1960) 'Properties and calculation of optimal coefficients', *Dokl. Akad. Nauk SSSR*, **132**, 1009–1012 (Russian), *Soviet Math. Dokl.*, **1**, 696–700 (English transl.).

Lyness, J.N. (1989) 'An introduction to lattice rules and their generator matrices', *IMA J. Numer. Anal.*, **9**, 405–419.

Lyness, J.N. and Sørevik, T. (1989) 'The number of lattice rules', *BIT*, **29**, 527–534.

Niederreiter, H. (1978) 'Quasi–Monte Carlo methods and pseudo–random numbers', *Bull. Amer. Math. Soc.*, **84**, 957–1041.

Niederreiter, H. (1991) 'Random Number Generation and Quasi–Monte Carlo Methods', *SIAM*, Philadelphia, to appear.

Niederreiter, H. (1992a) 'The existence of efficient lattice rules for multidimensional numerical integration', *Math. Comp.*, to appear.

Niederreiter, H. (1992b) 'Improved error bounds for lattice rules', submitted.

Price, J. and Sloan, I.H. (1991) 'Pointwise convergence of multiple Fourier series and an application to numerical integration', *J. Math. Anal. and Applics.*, to appear.

Sloan, I.H. (1984) 'Lattice methods for multiple integration', *J. Comput. Appl. Math.* **12 & 13**, 131–143.

Sloan, I.H. and Kachoyan, P.J. (1987) 'Lattice methods for multiple integration: theory, error analysis and examples', *SIAM J. Numer. Anal.*, **24**, 116–128.

Sloan, I.H. and Lyness, J.N. (1989) 'The representation of lattice quadrature rules as multiple sums', *Math. Comp.*, **52**, 81–94.

Sloan, I.H. and Lyness, J.N. (1990) 'Lattice rules: projection regularity and unique representations', *Math. Comp.*, **54**, 649–660.

EXISTENCE THEOREMS FOR EFFICIENT LATTICE RULES

HARALD NIEDERREITER
Institute for Information Processing
Austrian Academy of Sciences
Sonnenfelsgasse 19
A-1010 Vienna
Austria

ABSTRACT. Lattice rules form a special class of methods for the numerical integration of multivariate functions over multidimensional unit cubes. The success of these methods depends on the proper choice of an integration lattice. In this paper we present results that guarantee the existence of integration lattices for which the corresponding lattice rules yield small error bounds. These results improve or generalize earlier existence theorems in this area. The description of the results uses the Sloan-Lyness classification of lattice rules in terms of ranks and invariants.

1. Introduction

For $s \geq 2$ an s-dimensional *lattice* is a set consisting of all integer linear combinations of s linearly independent vectors in $\mathbb{R}^s$. An s-dimensional *integration lattice* is an s-dimensional lattice containing the integer lattice $\mathbb{Z}^s$. If L is an integration lattice, then $L \cap [0,1)^s$ is a finite set consisting, say, of the distinct points $\mathbf{x}_0, \ldots, \mathbf{x}_{N-1}$. The s-dimensional *lattice rule* corresponding to L (or simply the lattice rule L) approximates the integral $I(f)$ of a function f over $[0,1]^s$ by

$$Q(L; f) := \frac{1}{N} \sum_{k=0}^{N-1} f(\mathbf{x}_k).$$

If we want to emphasize that the number of nodes in a lattice rule is N, then we speak of an *N-point lattice rule*. To rule out a trivial case, we always assume $N \geq 2$.

Lattice rules are particularly well suited for periodic integrands with period interval $[0,1]^s$, and for such integrands they can be considered as multidimensional analogs of the trapezoidal rule. However, lattice rules can also be applied to nonperiodic functions, but the error analysis proceeds along different lines (see Niederreiter and Sloan [14]). The first steps in the direction of lattice rules were taken by Frolov [4] and a systematic theory was developed by Sloan and Kachoyan [16]. Special classes of lattice rules, namely lattice rules of rank 1, were introduced much earlier by Korobov [6] and Hlawka [5]. Lattice rules have been intensively studied over the last few years, and a summary of these developments can be found in a survey article by Lyness [8] and in the forthcoming book [12].

71

T. O. Espelid and A. Genz (eds.), Numerical Integration, 71–80.
© *1992 Kluwer Academic Publishers.*

The error analysis for lattice rules with periodic integrands uses regularity classes defined in terms of the rate of decrease of Fourier coefficients. For $\mathbf{h} = (h_1, \ldots, h_s) \in \mathbf{Z}^s$ we put

$$r(\mathbf{h}) = \prod_{i=1}^{s} \max(1, |h_i|).$$

Definition 1. Let $\alpha > 1$ and $C > 0$ be real. Then $f \in \mathcal{E}_\alpha^s(C)$ if f is continuous and periodic on $\mathbf{R}^s$ with period interval $[0, 1]^s$ and if its Fourier coefficients satisfy

$$|\hat{f}(\mathbf{h})| \leq Cr(\mathbf{h})^{-\alpha} \qquad \text{for all nonzero } \mathbf{h} \in \mathbf{Z}^s.$$

We write $f \in \mathcal{E}_\alpha^s$ if $f \in \mathcal{E}_\alpha^s(C)$ for some $C > 0$.

Sufficient conditions for $f \in \mathcal{E}_\alpha^s$ in terms of smoothness properties are known (see [9, pp. 984–985]). For an integration lattice L we define its *dual lattice*

$$L^\perp = \{\, \mathbf{h} \in \mathbf{Z}^s : \mathbf{h} \cdot \mathbf{x} \in \mathbf{Z} \text{ for all } \mathbf{x} \in L \,\},$$

where $\mathbf{h} \cdot \mathbf{x}$ denotes the standard inner product of $\mathbf{h}$ and $\mathbf{x}$. In the following we use an asterisk to indicate deletion of the origin from a range of summation.

Definition 2. For $\alpha > 1$ and any integration lattice L let

$$P_\alpha(L) = \sum_{\mathbf{h} \in L^\perp}^{*} r(\mathbf{h})^{-\alpha}.$$

Sloan and Kachoyan [16] have established a best possible error bound for lattice rules with integrands $f \in \mathcal{E}_\alpha^s(C)$, namely that for any $\alpha > 1$ and $C > 0$ and any lattice rule L we have

$$\max_{f \in \mathcal{E}_\alpha^s(C)} |Q(L; f) - I(f)| = C P_\alpha(L).$$

Thus, the quantity $P_\alpha(L)$ plays a crucial role in the assessment of the suitability of lattice rules.

It is convenient to use the following classification of lattice rules due to Sloan and Lyness [17]. If L is an s-dimensional integration lattice, then $L/\mathbf{Z}^s$ is a subgroup of $\mathbf{R}^s/\mathbf{Z}^s$ of order N, where N is the number of nodes. Like any finite abelian group, $L/\mathbf{Z}^s$ is isomorphic to a direct sum of cyclic groups of orders $n_1, \ldots, n_r \geq 2$, with the additional property that n_{i+1} divides n_i for $1 \leq i < r$. The uniquely determined integer r satisfies $1 \leq r \leq s$ and is called the *rank* of L, and the uniquely determined integers $n_1, \ldots, n_r$ are called the *invariants* of L. Moreover, there exist $\mathbf{g}_1, \ldots, \mathbf{g}_r \in \mathbf{Z}^s$ (not uniquely determined) such that the nodes of the lattice rule L are exactly the fractional parts

$$\left\{ \sum_{i=1}^{r} \frac{k_i}{n_i} \mathbf{g}_i \right\} \qquad \text{with } 0 \leq k_i < n_i \text{ for } 1 \leq i \leq r,$$

where the fractional part $\{\mathbf{x}\}$ of $\mathbf{x} \in \mathbf{R}^s$ is obtained by reducing all coordinates of $\mathbf{x}$ modulo 1. In particular, we have $N = n_1 \cdots n_r$.

Section 2 is devoted to lattice rules of rank 1. For this case we describe two different methods which lead to improvements on earlier results. The results for rank 1 form the basis for an extension to higher ranks, as will be explained in Section 3. In this section we also present another approach to existence theorems for efficient lattice rules which so far has been carried out in detail for lattice rules of rank 2 only.

2. Results for Rank 1

Before stating our new results, we review previous theorems guaranteeing small values of $P_\alpha(L)$ for lattice rules of rank 1. First we recall the definition of the *figure of merit*

$$\rho(L) = \min_{\substack{h \in L^\perp \\ h \neq 0}} r(h)$$

for any lattice rule L. We have the general bound

$$P_\alpha(L) = O\left(\rho(L)^{-\alpha}\left(1 + \log\rho(L)\right)^{s-1}\right) \tag{1}$$

due to Sloan and Kachoyan [16]. Here and in the sequel it is understood that, unless stated otherwise, the implied constant in a Landau symbol depends only on s and α.

From the Sloan-Lyness classification of lattice rules described in Section 1 it is clear that a lattice rule L of rank 1 is completely specified by prescribing the number N of nodes and a point $g \in \mathbf{Z}^s$. The node set of L consists then exactly of all fractional parts $\{(k/N)g\}$, $k = 0, 1, \ldots, N - 1$, and we have $L^\perp = \{h \in \mathbf{Z}^s : h \cdot g \equiv 0 \bmod N\}$. Instead of $P_\alpha(L)$ and $\rho(L)$ we also write $P_\alpha(g, N)$ and $\rho(g, N)$, respectively. To get N distinct nodes, we have to assume that $g = (g_1, \ldots, g_s)$ satisfies $\gcd(g_1, \ldots, g_s, N) = 1$. Let $C_s(N)$ be the set of all g with $-N/2 < g_i \leq N/2$ for $1 \leq i \leq s$, let $G_s(N)$ be the set of all $g \in C_s(N)$ with $\gcd(g_i, N) = 1$ for $1 \leq i \leq s$, and let $E_s(N)$ be the set of all $g \in C_s(N)$ with $g_1 = 1$.

Zaremba [20] proved that for every $s \geq 2$ and $N \geq 2$ there exists a $g \in E_s(N)$ with

$$\rho(g, N) > \frac{c_s N}{(\log N)^{s-1}},$$

where $c_s > 0$ depends only on s. Together with (1) this shows that with a suitable $g \in E_s(N)$ we can achieve

$$P_\alpha(g, N) = O\left(N^{-\alpha}(\log N)^{(\alpha+1)(s-1)}\right) \qquad \text{for all } \alpha > 1. \tag{2}$$

Another method of obtaining existence theorems for small values of $P_\alpha(g, N)$ is based on the following argument. For any $g \in \mathbf{Z}^s$ and $N \geq 2$ define

$$R_\alpha(g, N) = \sum_{\substack{h \in C_s(N) \\ h \cdot g \equiv 0 \bmod N}}^{*} r(h)^{-\alpha} \qquad \text{for } \alpha \geq 1. \tag{3}$$

Then it was shown by Niederreiter [10] that for every $s \geq 2$ and $N \geq 2$ there exists a $\mathbf{g} \in G_s(N)$ with

$$R_1(\mathbf{g}, N) < \frac{1}{N}\left(2\log N + \frac{7}{5}\right)^s.$$

Furthermore, $P_\alpha(\mathbf{g}, N)$ can be bounded in terms of $R_1(\mathbf{g}, N)$ either by [9, eq. (4.6)] or by the more refined inequality

$$P_\alpha(\mathbf{g}, N) < R_1(\mathbf{g}, N)^\alpha + O(N^{-\alpha})$$

established in [12, Theorem 5.5] which is valid for any $\mathbf{g} \in G_s(N)$ and $\alpha > 1$. This yields the existence of a $\mathbf{g} \in G_s(N)$ with

$$P_\alpha(\mathbf{g}, N) = O\left(N^{-\alpha}(\log N)^{\alpha s}\right) \qquad \text{for all } \alpha > 1. \tag{4}$$

Here the implied constant grows exponentially with s. More recently, Disney [2] and Disney and Sloan [3] have developed a different approach which yields the bound in (4) for a $\mathbf{g}$ depending also on α, but with the coefficient of the leading term being $(2e/s)^{\alpha s}$, which decreases superexponentially as s increases and α is fixed.

For the special case where N is prime, Bakhvalov [1] has shown that for every $s \geq 2$ and $\alpha > 1$ there exists a $\mathbf{g} \in \mathbf{Z}^s$ with

$$P_\alpha(\mathbf{g}, N) = O\left(N^{-\alpha}(\log N)^{\alpha(s-1)}\right), \tag{5}$$

where the implied constant is not specified. Although (5) yields the best order of magnitude among the bounds above, it is important for applications to higher-rank lattice rules to have good results for composite N (compare with Section 3). As to lower bounds, Sharygin [15] proved that $P_\alpha(\mathbf{g}, N)$ is always at least of the order of magnitude $N^{-\alpha}(\log N)^{s-1}$.

We now have the following new result. We use ϕ to denote Euler's totient function, i.e., $\phi(N)$ is the number of integers n with $1 \leq n \leq N$ and $\gcd(n, N) = 1$.

Theorem 1. *For every $s \geq 2$, $N \geq 2$, and $\alpha > 1$ there exists a $\mathbf{g} \in G_s(N)$ with*

$$P_\alpha(\mathbf{g}, N) \leq c(s, \alpha)\frac{(\log N)^{\alpha(s-1)+1}}{N^\alpha}\left(\frac{N}{\phi(N)}\right)^{(\alpha-1)(s-1)}\left(1 + O\left(\frac{(\log\log(N+1))^{b(s)}}{\log N}\right)\right),$$

where

$$c(s, \alpha) = 2^{\alpha(s-1)+1}\alpha\left(\frac{\alpha}{(s-1)!\,(\alpha-1)}\right)^{\alpha-1},$$

$b(s) = 3$ *for* $s = 2$, *and* $b(s) = s - 1$ *for* $s \geq 3$.

Since $N/\phi(N) = O(\log\log(N+1))$ for all $N \geq 2$ with an absolute implied constant (see [7, Theorem 6.26]), Theorem 1 improves on (4) for all s and on (2) for $s \geq 3$. The case $s = 2$ is special since an explicit construction by Bakhvalov [1] (see also Zaremba [19]) shows that there are infinitely many N for which $P_\alpha(\mathbf{g}, N) = O(N^{-\alpha}\log N)$ can be achieved for all $\alpha > 1$.

We give a sketch of the proof of Theorem 1; the details can be found in [13]. For fixed s, N, and α we consider

$$M := \sum_{\mathbf{g} \in G_s(N)} \rho(\mathbf{g}, N)^{\alpha-1} R_\alpha(\mathbf{g}, N)$$

with the notation in (3). Inserting the definition of $R_\alpha(\mathbf{g}, N)$ and interchanging the order of summation, we get

$$M = \sum_{\mathbf{h} \in C_s(N)}^{*} r(\mathbf{h})^{-\alpha} \sum_{\substack{\mathbf{g} \in G_s(N) \\ \mathbf{h} \cdot \mathbf{g} \equiv 0 \bmod N}} \rho(\mathbf{g}, N)^{\alpha-1}.$$

If for a nonzero $\mathbf{h} \in C_s(N)$ we have $\mathbf{h} \cdot \mathbf{g} \equiv 0 \bmod N$, then $\rho(\mathbf{g}, N) \leq r(\mathbf{h})$ by the definition of the figure of merit. Thus, for every $\mathbf{g}$ in the inner sum we have $\rho(\mathbf{g}, N)^{\alpha-1} \leq r(\mathbf{h})^{\alpha-1}$, and so

$$M \leq \sum_{\mathbf{h} \in C_s(N)}^{*} r(\mathbf{h})^{-1} \sum_{\substack{\mathbf{g} \in G_s(N) \\ \mathbf{h} \cdot \mathbf{g} \equiv 0 \bmod N}} 1 = \sum_{\mathbf{g} \in G_s(N)} R_1(\mathbf{g}, N).$$

By the inequality (24) in Niederreiter [10] we therefore get

$$M < \frac{\phi(N)^s}{N} \left(2 \log N + \frac{7}{5} \right)^s. \tag{6}$$

From the definition of M it follows that for any $t > 0$ we have

$$M \geq t^{\alpha-1} \sum_{\substack{\mathbf{g} \in G_s(N) \\ \rho(\mathbf{g}, N) \geq t}} R_\alpha(\mathbf{g}, N).$$

Together with (6) this yields

$$\sum_{\substack{\mathbf{g} \in G_s(N) \\ \rho(\mathbf{g}, N) \geq t}} R_\alpha(\mathbf{g}, N) < \frac{\phi(N)^s}{N t^{\alpha-1}} \left(2 \log N + \frac{7}{5} \right)^s. \tag{7}$$

From the work of Zaremba [20] we obtain the existence of at least $c_1 \phi(N)^s$ points $\mathbf{g} \in G_s(N)$ with

$$\rho(\mathbf{g}, N) \geq t_0 = c_2 \left(\frac{\phi(N)}{N} \right)^{s-1} \frac{N}{(\log N)^{s-1}},$$

where $c_1, c_2 > 0$ are constants depending only on s and α. Thus, applying (7) with $t = t_0$, we deduce the existence of a $\mathbf{g} \in G_s(N)$ with

$$R_\alpha(\mathbf{g}, N) = O\left(\frac{(\log N)^s}{N t_0^{\alpha-1}} \right) = O\left(N^{-\alpha} (\log N)^{\alpha(s-1)+1} \left(\frac{N}{\phi(N)} \right)^{(\alpha-1)(s-1)} \right).$$

By [12, eq. (5.8)] we have

$$P_\alpha(\mathbf{g}, N) \leq R_\alpha(\mathbf{g}, N) + O(N^{-\alpha})$$

for any $\mathbf{g} \in G_s(N)$ and $\alpha > 1$, and so we arrive at Theorem 1 without a specified constant and error term.

Another new existence theorem for small values of $P_\alpha(\mathbf{g}, N)$ is stated below. We denote by $\tau(N)$ the number of positive divisors of N.

Theorem 2. *For every $s \geq 3$, $N \geq 2$, and $\alpha > 1$ there exists a $\mathbf{g} \in E_s(N)$ with*

$$P_\alpha(\mathbf{g}, N) \leq \frac{(\log N)^{\alpha(s-1)}}{N^\alpha} \left(c_1(s, \alpha) + \frac{c_2(s, \alpha)\tau(N)}{(\log N)^{s-1}} \right) \left(1 + O\left(\frac{(\log \log(N+1))^{s-1}}{\log N} \right) \right),$$

where

$$c_1(s, \alpha) = 2\zeta(s-1)\alpha \left(\frac{2^{s-1}\alpha}{(s-1)!(\alpha-1)} \right)^\alpha,$$

$$c_2(s, \alpha) = 2s\alpha \left(\frac{\alpha}{\alpha-1} + \zeta(\alpha) \right) \left(\frac{2^{s-1}\alpha}{(s-1)!(\alpha-1)} \right)^{\alpha-1},$$

and ζ is the Riemann zeta-function.

There is also a similar type of result for $s = 2$ (see [13]). If N is prime, then $\tau(N) = 2$ and Theorem 2 again yields Bakhvalov's result (5), but with a specified constant. If $s \geq 3$ and $\tau(N) \leq c_3(s, \alpha)(\log N)^s$ with a suitable constant $c_3(s, \alpha)$, then Theorem 2 is better than Theorem 1; this will happen for most values of N since the average order of magnitude of $\tau(N)$ is $\log N$ by [7, Theorem 6.30]. Note, however, that by [7, p. 164] there exist sequences of values of N through which $\tau(N)$ grows faster than any given power of $\log N$; for such values of N Theorem 1 yields the better result. The coefficients of the leading terms of the bounds in Theorems 1 and 2 decrease at basically the same superexponential rate as the coefficient $(2e/s)^{\alpha s}$ of Disney [2] and Disney and Sloan [3] as s increases and α is fixed.

The proof of Theorem 2 is rather technical and so we have to refer the reader to [13]. The main idea is to consider the average of $P_\alpha(\mathbf{g}, N)$ over those $\mathbf{g} \in E_s(N)$ for which $\rho(\mathbf{g}, N)$ is sufficiently large.

3. Results for Higher Ranks

We first establish a general principle of extending a lattice rule of rank 1 to a lattice rule of higher rank.

Lemma 1. *For $s \geq 2$ let a rank r with $1 \leq r \leq s$ and invariants $n_1, \ldots, n_r \geq 2$ with n_{i+1} dividing n_i for $1 \leq i < r$ be given, and let L_1 be an s-dimensional lattice rule of rank 1 with n_1 nodes which is generated by a point $\mathbf{g}_1 = (g_1^{(1)}, \ldots, g_s^{(1)}) \in \mathbf{Z}^s$ with $\gcd(g_1^{(1)}, n_1) = 1$. Then there exists an s-dimensional lattice rule L with rank r and invariants $n_1, \ldots, n_r$ such that the node set of L contains the node set of L_1.*

Proof. We may assume $r \geq 2$. Construct L by putting

$$\mathbf{g}_i = (g_1^{(i)}, \ldots, g_s^{(i)}) \in \mathbf{Z}^s \quad \text{for } 2 \leq i \leq r$$

with $g_j^{(i)} = 0$ for $1 \leq j \leq i-1$ and $\gcd(g_i^{(i)}, n_i) = 1$ and letting the nodes of L be all fractional parts

$$\left\{ \sum_{i=1}^r \frac{k_i}{n_i} \mathbf{g}_i \right\} \quad \text{with } 0 \leq k_i < n_i \text{ for } 1 \leq i \leq r.$$

Consider the mapping ψ on the finite abelian group $A = (\mathbf{Z}/n_1\mathbf{Z}) \oplus \cdots \oplus (\mathbf{Z}/n_r\mathbf{Z})$ defined by

$$\psi : (k_1 + n_1\mathbf{Z}, \ldots, k_r + n_r\mathbf{Z}) \longmapsto \sum_{i=1}^{r} \frac{k_i}{n_i}\mathbf{g}_i + \mathbf{Z}^s \in L/\mathbf{Z}^s.$$

Then ψ is a surjective group homomorphism. We claim that the kernel of ψ is trivial, i.e., that

$$\sum_{i=1}^{r} \frac{k_i}{n_i}\mathbf{g}_i \equiv \mathbf{0} \bmod \mathbf{Z}^s \tag{8}$$

implies $k_i \equiv 0 \bmod n_i$ for $1 \leq i \leq r$. By comparing the first coordinates in (8) we get $(k_1/n_1)g_1^{(1)} \equiv 0 \bmod 1$, hence $k_1 g_1^{(1)} \equiv 0 \bmod n_1$, and thus $k_1 \equiv 0 \bmod n_1$ since $\gcd(g_1^{(1)}, n_1) = 1$. Thus (8) reduces to $\sum_{i=2}^{r}(k_i/n_i)\mathbf{g}_i \equiv \mathbf{0} \bmod \mathbf{Z}^s$, and comparing second coordinates we get $k_2 \equiv 0 \bmod n_2$. By continuing in this manner we establish the claim. Consequently, $L/\mathbf{Z}^s$ is isomorphic to A, and so L has rank r and invariants $n_1, \ldots, n_r$. Clearly, the node set of L contains the node set of L_1. $\square$

Theorem 3. *For $s \geq 2$ let a rank r with $1 \leq r \leq s$ and invariants $n_1, \ldots, n_r \geq 2$ with n_{i+1} dividing n_i for $1 \leq i < r$ be given. Then for every $\alpha > 1$ there exists an s-dimensional lattice rule L with this rank and these invariants such that*

$$P_\alpha(L) \leq c(s, \alpha)\frac{(\log n_1)^{\alpha(s-1)+1}}{n_1^\alpha} \left(\frac{n_1}{\phi(n_1)}\right)^{(\alpha-1)(s-1)} \left(1 + O\left(\frac{(\log\log(n_1+1))^{b(s)}}{\log n_1}\right)\right),$$

where $c(s, \alpha)$ and $b(s)$ are as in Theorem 1.

Proof. By Theorem 1 there exists a $\mathbf{g}_1 = (g_1^{(1)}, \ldots, g_s^{(1)}) \in G_s(n_1)$ such that $P_\alpha(\mathbf{g}_1, n_1)$ satisfies the bound in Theorem 1 with $N = n_1$. Let L_1 be the corresponding lattice rule of rank 1 with n_1 nodes and let L be as in Lemma 1. Then $L \supseteq L_1$, hence $L^\perp \subseteq L_1^\perp$, and so $P_\alpha(L) \leq P_\alpha(L_1) = P_\alpha(\mathbf{g}_1, n_1)$ follows immediately from Definition 2. By the construction of $\mathbf{g}_1$ we get the desired result. $\square$

Theorem 4. *For $s \geq 3$ let a rank r and invariants $n_1, \ldots, n_r$ be given as in Theorem 3. Then for every $\alpha > 1$ there exists an s-dimensional lattice rule L with this rank and these invariants such that*

$$P_\alpha(L) \leq \frac{(\log n_1)^{\alpha(s-1)}}{n_1^\alpha}\left(c_1(s, \alpha) + \frac{c_2(s, \alpha)\tau(n_1)}{(\log n_1)^{s-1}}\right)\left(1 + O\left(\frac{(\log\log(n_1+1))^{s-1}}{\log n_1}\right)\right),$$

where $c_1(s, \alpha)$ and $c_2(s, \alpha)$ are as in Theorem 2.

Proof. Proceed as in the proof of Theorem 3, but use Theorem 2 instead of Theorem 1. $\square$

For $s = 2$ we get a result analogous to Theorem 4 by using the version of Theorem 2 for $s = 2$ (see [13]). It should be noted that according to a general lower bound in [11, Proposition 2], $P_\alpha(L)$ is at least of the order of magnitude $n_1^{-\alpha}$ for any lattice rule L with first invariant n_1.

When $r \geq 2$, Theorems 3 and 4 are only of interest if n_1 is large. Since the number N of nodes is $N = n_1 \cdots n_r$, this means equivalently that $n_2, \ldots, n_r$ should be small compared to N (and compared even to n_1). The fact that efficient lattice rules of rank $r \geq 2$ can only be obtained for small invariants $n_2, \ldots, n_r$ was already observed in computer searches (see Sloan and Walsh [18]) and also in theoretical work (see Niederreiter [11]). In this connection it also becomes clear why Theorems 1 and 2 are needed for composite moduli N in the application to higher-rank lattice rules. Note that in the proofs of Theorems 3 and 4 we apply Theorems 1 and 2, respectively, with the modulus n_1. If n_1 is a prime, then by the divisibility property of invariants the invariants $n_2, \ldots, n_r$ must be equal to n_1. Thus $n_1 = N^{1/r}$ in terms of the number N of nodes. It follows then from the lower bound for $P_\alpha(L)$ mentioned above that for any lattice rule L with these identical prime invariants, $P_\alpha(L)$ is at least of the order of magnitude $N^{-\alpha/r}$. On the other hand, if $n_2, \ldots, n_r$ are chosen as small invariants and n_1 is large (and necessarily composite), so that n_1 may be considered to be of the order of magnitude N, then Theorem 3 guarantees the existence of a lattice rule L with these prescribed invariants such that $P_\alpha(L)$ is of the order of magnitude $N^{-\alpha}$, up to logarithmic factors.

Theorems 3 and 4 have been obtained by a method which is in a sense indirect, since in the proof we first consider efficient lattice rules of rank 1. There is also a direct method of establishing existence theorems for efficient lattice rules in which we immediately look at lattice rules of the desired rank. For a dimension $s \geq 2$ let a rank r with $1 \leq r \leq s$ and invariants $n_1, \ldots, n_r \geq 2$ with the usual divisibility property be given. We put

$$Z_i = \{ g \in \mathbf{Z} : 0 \leq g < n_i \text{ and } \gcd(g, n_i) = 1 \} \quad \text{for } 1 \leq i \leq r.$$

Let $\mathcal{L} = \mathcal{L}(s; n_1, \ldots, n_r)$ be the family of all s-dimensional lattice rules L for which the node set consists exactly of all fractional parts

$$\left\{ \sum_{i=1}^{r} \frac{k_i}{n_i} \mathbf{g}_i \right\} \quad \text{with } 0 \leq k_i < n_i \text{ for } 1 \leq i \leq r,$$

where the $\mathbf{g}_i$ have the special form $\mathbf{g}_i = (g_1^{(i)}, \ldots, g_s^{(i)})$ with $g_j^{(i)} = 0$ for $1 \leq j \leq i - 1$ and $g_j^{(i)} \in Z_i$ for $i \leq j \leq s$. It follows from the proof of Lemma 1 that each $L \in \mathcal{L}$ has rank r and invariants $n_1, \ldots, n_r$. In analogy with (3) we define

$$R_1(L) = \sum_{\mathbf{h} \in E(L)}^{*} r(\mathbf{h})^{-1} \quad \text{with } E(L) = C_s(N) \cap L^{\perp},$$

where $N = n_1 \cdots n_r$ is the number of nodes of L. Let

$$M(\mathcal{L}) = \frac{1}{\operatorname{card}(\mathcal{L})} \sum_{L \in \mathcal{L}} R_1(L)$$

be the average value of $R_1(L)$ as L runs through $\mathcal{L}$. Note that $\operatorname{card}(\mathcal{L}) = \prod_{i=1}^{r} \phi(n_i)^{s-i+1}$. Inserting the definition of $R_1(L)$ and interchanging the order of summation, we get

$$M(\mathcal{L}) = \frac{1}{\operatorname{card}(\mathcal{L})} \sum_{\mathbf{h} \in C_s(N)} A(\mathbf{h}) r(\mathbf{h})^{-1} - 1,$$

where $A(\mathbf{h})$ is the number of $L \in \mathcal{L}$ with $\mathbf{h} \in L^{\perp}$. Writing $e(t) = e^{2\pi\sqrt{-1}t}$ for real t and using [16, Theorem 1], we obtain

$$A(\mathbf{h}) = \frac{1}{N} \sum_{k_1=0}^{n_1-1} \cdots \sum_{k_r=0}^{n_r-1} \prod_{j=1}^{s} \prod_{i=1}^{\min(j,r)} \left(\sum_{g \in Z_i} e\left(\frac{k_i}{n_i} h_j g\right) \right)$$

for $\mathbf{h} = (h_1, \ldots, h_s) \in C_s(N)$. Therefore,

$$\sum_{\mathbf{h} \in C_s(N)} \frac{A(\mathbf{h})}{r(\mathbf{h})} = \frac{1}{N} \sum_{k_1=0}^{n_1-1} \cdots \sum_{k_r=0}^{n_r-1} \prod_{j=1}^{s} \left(\sum_{h \in C_1(N)} \frac{1}{\max(1,|h|)} \prod_{i=1}^{\min(j,r)} \left(\sum_{g \in Z_i} e\left(\frac{k_i}{n_i} hg\right) \right) \right).$$

For the innermost sum we have the formula

$$\sum_{g \in Z_i} e\left(\frac{k_i}{n_i} hg\right) = \sum_{\substack{d|n_i \\ d|k_i h}} \mu\left(\frac{n_i}{d}\right) d,$$

where μ is the Möbius function. The resulting expression for $M(\mathcal{L})$ is rather complicated. Even in the simplest higher-rank case $r = 2$, very elaborate number-theoretic arguments are needed to deal with this expression. The final result that is obtained in [11] for this case is

$$M(\mathcal{L}) < c_s \left(\frac{(\log N)^s}{N} + \frac{\log N}{n_1} \right)$$

with a constant c_s depending only on s. Since $P_\alpha(L) = O\left(R_1(L)^\alpha\right)$, we then arrive at the following theorem.

Theorem 5. *For every $s \geq 2$ and any prescribed invariants $n_1, n_2 \geq 2$ with n_2 dividing n_1 there exists an s-dimensional lattice rule L of rank 2 with these invariants such that*

$$P_\alpha(L) = O\left(\left(\frac{(\log N)^s}{N} + \frac{\log N}{n_1} \right)^\alpha \right) \qquad \text{for all } \alpha > 1.$$

References

[1]   Bakhvalov, N.S.: "Approximate computation of multiple integrals" (Russian), *Vestnik Moskov. Univ. Ser. Mat. Mekh. Astr. Fiz. Khim.* **1959**, no. 4, 3–18.

[2]   Disney, S.: "Error bounds for rank 1 lattice quadrature rules modulo composites", *Monatsh. Math.* **110**, 89–100 (1990).

[3]   Disney, S. and Sloan, I.H.: "Error bounds for the method of good lattice points", *Math. Comp.* **56**, 257–266 (1991).

[4]   Frolov, K.K.: "On the connection between quadrature formulas and sublattices of the lattice of integral vectors" (Russian), *Dokl. Akad. Nauk SSSR* **232**, 40–43 (1977).

[5]   Hlawka, E.: "Zur angenäherten Berechnung mehrfacher Integrale", *Monatsh. Math.* **66**, 140–151 (1962).

[6]   Korobov, N.M.: "The approximate computation of multiple integrals" (Russian), *Dokl. Akad. Nauk SSSR* **124**, 1207–1210 (1959).

[7]   LeVeque, W.J.: *Fundamentals of Number Theory*, Addison-Wesley, Reading, Mass., 1977.

[8]   Lyness, J.N.: "An introduction to lattice rules and their generator matrices", *IMA J. Numer. Analysis* **9**, 405–419 (1989).

[9]   Niederreiter, H.: "Quasi-Monte Carlo methods and pseudo-random numbers", *Bull. Amer. Math. Soc.* **84**, 957–1041 (1978).

[10] Niederreiter, H.: "Existence of good lattice points in the sense of Hlawka", *Monatsh. Math.* **86**, 203–219 (1978).

[11] Niederreiter, H.: "The existence of efficient lattice rules for multidimensional numerical integration", *Math. Comp.*, to appear.

[12] Niederreiter, H.: *Random Number Generation and Quasi-Monte Carlo Methods*, SIAM, Philadelphia, to appear.

[13] Niederreiter, H.: "Improved error bounds for lattice rules", preprint, 1991.

[14] Niederreiter, H. and Sloan, I.H.: "Lattice rules for multiple integration and discrepancy", *Math. Comp.* **54**, 303–312 (1990).

[15] Sharygin, I.F.: "A lower estimate for the error of quadrature formulas for certain classes of functions" (Russian), *Zh. Vychisl. Mat. i Mat. Fiz.* **3**, 370–376 (1963).

[16] Sloan, I.H. and Kachoyan, P.J.: "Lattice methods for multiple integration: theory, error analysis and examples", *SIAM J. Numer. Analysis* **24**, 116–128 (1987).

[17] Sloan, I.H. and Lyness, J.N.: "The representation of lattice quadrature rules as multiple sums", *Math. Comp.* **52**, 81–94 (1989).

[18] Sloan, I.H. and Walsh, L.: "A computer search of rank-2 lattice rules for multidimensional quadrature", *Math. Comp.* **54**, 281–302 (1990).

[19] Zaremba, S.K.: "Good lattice points, discrepancy, and numerical integration", *Ann. Mat. Pura Appl.* **73**, 293–317 (1966).

[20] Zaremba, S.K.: "Good lattice points modulo composite numbers", *Monatsh. Math.* **78**, 446-460 (1974).

SINC QUADRATURES FOR
CAUCHY PRINCIPAL VALUE INTEGRALS

BERNARD BIALECKI
Department of Mathematics
University of Kentucky
Lexington, KY 40506-0027 USA
Email: bialecki@ms.uky.edu

ABSTRACT. Three types of SINC quadratures are surveyed for the evaluation of Cauchy principal value integrals $\int_\Gamma F(t)dt/(t-x)$, $x \in \Gamma$, where Γ is an arc in the complex plane. Under suitable assumptions on F, the quadrature errors are of order $O\left(e^{-c\sqrt{N}}\right)$, where N is the number of quadrature nodes and c is a positive constant independent of N. Special consideration is given to SINC quadratures for Cauchy principal value integrals over $(-\infty, \infty)$, $(0, \infty)$, and $(-1, 1)$.

1 Introduction

Let Γ be an arc in the complex plane and let F be a function defined on Γ. For $x \in \Gamma$, the Cauchy principal value (CPV) integral is defined by [7]

$$\int_\Gamma \frac{F(t)}{t-x}\,dt = \lim_{\varepsilon \to 0^+} \int_{\Gamma \setminus \Gamma_\varepsilon} \frac{F(t)}{t-x}\,dt,$$

where Γ_ε is the set of all points of Γ which are within a distance ε from x. Assuming that F is an analytic function in a region containing Γ, the residue theorem and Plemelj formula are used to derive systematically three types of SINC quadratures for the evaluation of CPV integrals. Weights of all SINC quadratures depend on x. In the first quadrature, all nodes depend on x, whereas all nodes of the second quadrature are independent of x. In the third quadrature, x is one of the nodes and all remaining nodes are independent of x. For certain points x, the second and third quadratures are identical. Following terminology used in [4] for Gauss type quadratures, we call the second and third quadrature of this paper strict and modified SINC quadrature, respectively. Under suitable additional assumptions on F, bounds on SINC quadrature errors converge to zero like $e^{-c\sqrt{N}}$, as $N \to \infty$, where N is the number of quadrature nodes and c is a positive constant independent of N. The constant c is the same for the first and modified SINC quadratures and for the strict SINC quadrature it is $1/\sqrt{2}$ times the constant c for the modified SINC quadrature. Special treatment is given to SINC quadratures for CPV integrals over $(-\infty, \infty)$, $(0, \infty)$, and $(-1, 1)$.

This survey paper gives a unified account of the SINC quadratures for CPV integrals presented in [2, 5, 8, 9, 10].

T. O. Espelid and A. Genz (eds.), Numerical Integration, 81–92.
© *1992 Kluwer Academic Publishers.*

2 Preliminaries

Throughout the paper, $\mathcal{Z} = \{k : k = 0, \pm1, \pm2, \ldots\}$, $\mathcal{R} = (-\infty, \infty)$, and $\mathcal{C} = \{x + iy : x \in \mathcal{R}, y \in \mathcal{R}\}$. For $x \in \mathcal{R}$ and $z \in \mathcal{C}$, $\operatorname{sgn} x$ and $\operatorname{Im} z$ are the sign of x and the imaginary part of z, respectively. For $x \in \mathcal{R}$, $\lceil x \rceil$ denotes the smallest integer $\geq x$. A generic positive constant independent of x and N is denoted by c. A generic positive constant independent of N but depending on x is denoted by $c(x)$ ($c(x)$ may become unbounded as x approaches either of the endpoints of Γ).

For $h > 0$ and $k \in \mathcal{Z}$, the SINC function $S(k, h)$ is defined by

$$S(k, h)(z) = \frac{\sin[\pi(z - kh)/h]}{\pi(z - kh)/h}, \quad z \in \mathcal{C}.$$

First, a few important relations are derived. The Plemelj formula [7] and the identity

$$\int_{\mathcal{R}} \frac{\sin(\alpha t)}{t - z} \, dt = \pi e^{i\alpha z \operatorname{sgn} \operatorname{Im} z}, \quad \alpha > 0, \quad z \in \mathcal{C} \setminus \mathcal{R}, \tag{2.1}$$

give

$$\int_{\mathcal{R}} \frac{\sin(\alpha t)}{t - x} \, dt = \pi \cos(\alpha x), \quad \alpha > 0, \quad x \in \mathcal{R},$$

from which it follows that

$$\int_{\mathcal{R}} S(k, h)(t) \, dt = h. \tag{2.2}$$

Writing $1/[(t - kh)(t - z)]$ as $[1/(t - kh) - 1/(t - z)]/(kh - z)$, (2.1) and (2.2) give

$$\int_{\mathcal{R}} \frac{S(k, h)(t)}{t - z} \, dt = \frac{h}{kh - z} \left[1 - (-1)^k e^{i(\pi/h)z \operatorname{sgn} \operatorname{Im} z} \right], \quad z \in \mathcal{C} \setminus \mathcal{R}. \tag{2.3}$$

Assume $d > 0$, and let D_d be the domain defined by

$$D_d = \{z \in \mathcal{C} : |\operatorname{Im} z| < d\}. \tag{2.4}$$

Let D be a simply connected open domain in the complex plane, and let a and b ($a \neq b$) be points on the boundary ∂D of D. Let ϕ be a conformal map of D onto D_d such that $\phi(a) = -\infty$ and $\phi(b) = \infty$. Let ψ denote the inverse map of ϕ, and assume that

$$\Gamma = \{\psi(x) : x \in \mathcal{R}\},$$

where the direction along Γ is from a to b. For $h > 0$, let z_k be the points on Γ given by

$$z_k = \psi(kh), \quad k \in \mathcal{Z}.$$

Let $B(D)$ denote the family of all functions F that are continuous in $D \cup \partial D \setminus \{a, b\}$, analytic in D, such that

$$N(F, D) \equiv \int_{\partial D} |F(z) \, dz| < \infty,$$

and such that

$$\int_{\psi(L_x)} |F(z)\,dz| \to 0 \text{ as } x \to \pm\infty,$$

where $L_x = \{x + iy : -d < y < d\}$.

In the remainder of this section, the SINC quadrature of [8] for the evaluation of $\int_\Gamma F(x)\,dx$ is reviewed. This quadrature plays a key role in obtaining and analyzing SINC quadratures for CPV integrals. Assume $F \in B(D)$. Then

$$\int_\Gamma F(x)\,dx = h \sum_{k \in \mathcal{Z}} \frac{F(z_k)}{\phi'(z_k)} + E_h(F), \tag{2.5}$$

where

$$E_h(F) = \frac{i}{2} \int_{\partial D} \frac{e^{i[\pi\phi(z)/h]\,\mathrm{sgn}\,\mathrm{Im}\,\phi(z)}}{\sin[\pi\phi(z)/h]} F(z)\,dz. \tag{2.6}$$

Since

$$\sinh(\pi d/h) \le |\sin[\pi\phi(z)/h]|, \quad \left| e^{i[\pi\phi(z)/h]\,\mathrm{sgn}\,\mathrm{Im}\,\phi(z)} \right| = e^{-\pi d/h}, \quad z \in \partial D, \tag{2.7}$$

it follows from (2.6) that

$$|E_h(F)| \le N(F, D) \frac{e^{-2\pi d/h}}{1 - e^{-2\pi d/h}}. \tag{2.8}$$

(The SINC quadrature (2.5) for $\Gamma = (-1, 1)$ with $\phi = \log[(1 + z)/(1 - z)]$ is also known as the tanh rule [3]). In order to replace the infinite sum in (2.5) by the finite sum

$$Q_{N_1, N_2}(F) = h \sum_{-N_1 \le k \le N_2} \frac{F(z_k)}{\phi'(z_k)} \tag{2.9}$$

for positive integers N_1, N_2, assume that there are positive constants $c_j, \alpha_j, j = 1, 2$, such that

$$|F(z)| \le c_j |\phi'(z)| e^{-\alpha_j |\phi(z)|}, \quad z \in \Gamma_j, \quad j = 1, 2, \tag{2.10}$$

where

$$\Gamma_1 = \{\psi(x) : x \in (-\infty, 0)\}, \quad \Gamma_2 = \{\psi(x) : x \in (0, \infty)\}.$$

Condition (2.10) and the inequalities $e^{\alpha_j h} - 1 \ge \alpha_j h, j = 1, 2$, yield

$$h \left(\sum_{k < -N_1} + \sum_{k > N_2} \right) \left| \frac{F(z_k)}{\phi'(z_k)} \right| \le \frac{c_1}{\alpha_1} e^{-\alpha_1 N_1 h} + \frac{c_2}{\alpha_2} e^{-\alpha_2 N_2 h}. \tag{2.11}$$

It follows from (2.5), (2.9), (2.8), (2.11), and the triangle inequality that

$$\left| \int_\Gamma F(x)\,dx - Q_{N_1, N_2}(F) \right| \le N(F, D) \frac{e^{-2\pi d/h}}{1 - e^{-2\pi d/h}} + \frac{c_1}{\alpha_1} e^{-\alpha_1 N_1 h} + \frac{c_2}{\alpha_2} e^{-\alpha_2 N_2 h}. \tag{2.12}$$

For fixed h, the integers N_1 and N_2 are selected so that the three terms on the right-hand side in the inequality (2.12) are approximately of the same order of magnitude. Thus, if

$$N_1 = \left\lceil 2\pi d/(\alpha_1 h^2) \right\rceil, \quad N_2 = \left\lceil 2\pi d/(\alpha_2 h^2) \right\rceil,$$

which are derived from $e^{-2\pi d/h} = e^{-\alpha_1 N_1 h} = e^{-\alpha_2 N_2 h}$, then

$$\left| \int_\Gamma F(x)\, dx - Q_{N_1,N_2}(F) \right| \le ce^{-2\pi d/h}.$$

In particular, if $\alpha_1 = \alpha_2 = \alpha$, and $h = [2\pi d/(\alpha N)]^{1/2}$, where N is a positive integer, then

$$\left| \int_\Gamma F(x)\, dx - Q_{N,N}(F) \right| \le ce^{-(2\pi d\alpha N)^{1/2}}.$$

3 First SINC quadrature

Assume $F \in B(D)$, $x \in \Gamma$, and set $w = \phi(x)$. Then by the change of variable $t = \psi(w + u)$,

$$\int_\Gamma \frac{F(t)}{t - x}\, dt = \int_{\mathcal{R}} \frac{F[\psi(w + u)]\psi'(w + u)}{\psi(w + u) - \psi(w)}\, du$$

$$= \frac{1}{2} \int_{\mathcal{R}} \left[\frac{F[\psi(w + u)]\psi'(w + u)}{\psi(w + u) - \psi(w)} + \frac{F[\psi(w - u)]\psi'(w - u)}{\psi(w - u) - \psi(w)} \right] du, \quad (3.1)$$

since $\int_{\mathcal{R}} g(t)\, dt = \int_{\mathcal{R}} g_e(t)\, dt$, where g_e is the even component of g (cf. [6]). Again, by the change of variable $u = \phi(t) + h/2$ applied to the last integral in (3.1),

$$\int_\Gamma \frac{F(t)}{t - x}\, dt = \int_\Gamma [G^+(t) + G^-(t)]\, dt,$$

where

$$G^\pm(z) = \frac{1}{2} \frac{F(\psi[\phi(x) \pm \phi(z) \pm h/2])\phi'(z)}{\phi'(\psi[\phi(x) \pm \phi(z) \pm h/2])(\psi[\phi(x) \pm \phi(z) \pm h/2] - x)}.$$

Since $G^+ + G^- \in B(D)$, (2.5) and (2.6) give

$$\int_\Gamma \frac{F(t)}{t - x}\, dt = h \sum_{k \in \mathcal{Z}} \frac{F[\hat{z}_k(x)]}{\phi'[\hat{z}_k(x)][\hat{z}_k(x) - x]} + E_h^{(1)}(F, x), \quad (3.2)$$

where

$$\hat{z}_k(x) = \psi[\phi(x) + kh + h/2],$$

and

$$E_h^{(1)}(F, x) = \frac{i}{2} \int_{\partial D} \frac{e^{i[\pi\phi(z)/h]\,\mathrm{sgn}\,\mathrm{Im}\,\phi(z)}}{\sin[\pi\phi(z)/h]} [G^+(z) + G^-(z)]\, dz. \quad (3.3)$$

By the change of variable $\zeta = \psi[\phi(x) \pm \phi(z) \pm h/2]$,

$$\int_{\partial D} |G^\pm(z)\, dz| = \frac{1}{2} \int_{\partial D} \left| \frac{F(\zeta)}{\zeta - x}\, d\zeta \right|. \quad (3.4)$$

It follows from (3.3), (2.7), and (3.4) that

$$\left| E_h^{(1)}(F, x) \right| \le N(F, D, x) \frac{e^{-2\pi d/h}}{1 - e^{-2\pi d/h}}, \quad (3.5)$$

where

$$N(F, D, x) = \int_{\partial D} \left| \frac{F(z)}{z - x} \, dz \right|.$$

To replace the infinite sum in (3.2) by the finite sum

$$Q^{(1)}_{N_1, N_2}(F, x) = h \sum_{-N_1 \leq k \leq N_2} \frac{F[\hat{z}_k(x)]}{\phi'[\hat{z}_k(x)][\hat{z}_k(x) - x]}, \tag{3.6}$$

where $N_1, N_2 \in \mathcal{Z}$, assume $\Gamma \subset \mathcal{R}$, $a < b$, and that (2.10) holds. Let N_1, N_2 be such that

$$N_1 \geq \max(\phi(x)/h - 1/2, -1/2), \quad N_2 \geq \max(-\phi(x)/h - 3/2, -3/2). \tag{3.7}$$

Since $N_1 \geq -1/2$, $N_2 \geq -3/2$, $x = \psi[\phi(x)]$, and since ψ is increasing on $\mathcal{R}$,

$$|\hat{z}_k(x) - x| \geq x - \hat{z}_{-N_1-1}(x) > 0, \quad k \leq -N_1 - 1, \quad |\hat{z}_k(x) - x| \geq \hat{z}_{N_2+1}(x) - x > 0, \quad k \geq N_2 + 1. \tag{3.8}$$

Also, since $N_1 \geq \phi(x)/h - 1/2$ and $N_2 \geq -\phi(x)/h - 3/2$, (2.10) and (3.8) imply

$$h \left(\sum_{k < -N_1} + \sum_{k > N_2} \right) \left| \frac{F[\hat{z}_k(x)]}{\phi'[\hat{z}_k(x)][\hat{z}_k(x) - x]} \right|$$

$$\leq \frac{c_1}{\alpha_1} \frac{e^{-\alpha_1[(N_1 - 1/2)h - \phi(x)]}}{x - \hat{z}_{-N_1-1}(x)} + \frac{c_2}{\alpha_2} \frac{e^{-\alpha_2[(N_2+1/2)h + \phi(x)]}}{\hat{z}_{N_2+1}(x) - x}. \tag{3.9}$$

It follows from (3.2), (3.6), (3.5), and (3.9) that

$$\left| \int_{\Gamma} \frac{F(t)}{t - x} \, dt - Q^{(1)}_{N_1, N_2}(F, x) \right| \leq N(F, D, x) \frac{e^{-2\pi d/h}}{1 - e^{-2\pi d/h}}$$

$$+ \frac{c_1}{\alpha_1} \frac{e^{-\alpha_1[(N_1 - 1/2)h - \phi(x)]}}{x - \hat{z}_{-N_1-1}(x)} + \frac{c_2}{\alpha_2} \frac{e^{-\alpha_2[(N_2+1/2)h + \phi(x)]}}{\hat{z}_{N_2+1}(x) - x}.$$

Thus, if

$$N_1 = \left\lceil \max \left(\frac{2\pi d}{\alpha_1 h^2} + \frac{\phi(x)}{h} + \frac{1}{2}, -\frac{1}{2} \right) \right\rceil, \quad N_2 = \left\lceil \max \left(\frac{2\pi d}{\alpha_2 h^2} - \frac{\phi(x)}{h} - \frac{1}{2}, -\frac{3}{2} \right) \right\rceil,$$

which are derived from $e^{-2\pi d/h} = e^{-\alpha_1[(N_1-1/2)h - \phi(x)]} = e^{-\alpha_2[(N_2+1/2)h + \phi(x)]}$ and (3.7), then

$$\left| \int_{\Gamma} \frac{F(t)}{t - x} \, dt - Q^{(1)}_{N_1, N_2}(F, x) \right| \leq c(x) e^{-2\pi d/h}.$$

Also, if $\alpha_1 = \alpha_2 = \alpha$, and $h = [2\pi d/(\alpha N)]^{1/2}$, where N is a sufficiently large positive integer, then

$$\left| \int_{\Gamma} \frac{F(t)}{t - x} \, dt - Q^{(1)}_{N,N}(F, x) \right| \leq c(x) e^{-(2\pi d\alpha N)^{1/2}}.$$

The SINC quadrature of this section and its error bound were presented in [8] for arbitrary arc Γ, and in [5] for $\Gamma = \mathcal{R}$.

4 Strict SINC quadrature

Assume $F \in B(D), \zeta \in D \setminus \Gamma, t \in \Gamma$ but $t \neq z_k$ for all $k \in \mathcal{Z}$. Then by the residue theorem,

$$\frac{F(t)[\phi(t) - \phi(\zeta)]}{\phi'(t)(t - \zeta)} = \sum_{k \in \mathcal{Z}} \frac{F(z_k)}{\phi'(z_k)} \frac{kh - \phi(\zeta)}{z_k - \zeta} S(k,h)[\phi(t)]$$

$$+ \frac{\sin[\pi\phi(t)/h]}{2\pi i} \int_{\partial D} \frac{F(z)[\phi(z) - \phi(\zeta)] \, dz}{[\phi(z) - \phi(t)](z - \zeta) \sin[\pi\phi(z)/h]}. \tag{4.1}$$

Multiply (4.1) through by $\phi'(t)/[\phi(t) - \phi(\zeta)]$, integrate both sides over Γ with respect to t and use (2.1), (2.3) to get

$$\int_\Gamma \frac{F(t)}{t - \zeta} \, dt = h \sum_{k \in \mathcal{Z}} \frac{1 - (-1)^k e^{i[\pi\phi(\zeta)/h] \operatorname{sgn} \operatorname{Im} \phi(\zeta)}}{\phi'(z_k)(z_k - \zeta)} F(z_k)$$

$$+ \frac{1}{2i} \int_{\partial D} \frac{e^{i[\pi\phi(\zeta)/h] \operatorname{sgn} \operatorname{Im} \phi(\zeta)} - e^{i[\pi\phi(z)/h] \operatorname{sgn} \operatorname{Im} \phi(z)}}{(z - \zeta) \sin[\pi\phi(z)/h]} F(z) \, dz. \tag{4.2}$$

Assume $x \in \Gamma$ and $x \neq z_k$ for all $k \in \mathcal{Z}$. Taking the arithmetic average of the two equations which are obtained from (4.2) on letting ζ approach x from the left- and right-hand sides of Γ, respectively, Plemelj's formula gives

$$\int_\Gamma \frac{F(t)}{t - x} \, dt = h \sum_{k \in \mathcal{Z}} \frac{1 - \cos[\pi(kh - \phi(x))/h]}{\phi'(z_k)(z_k - x)} F(z_k) + E_h^{(2)}(F, x), \tag{4.3}$$

where

$$E_h^{(2)}(F, x) = \frac{1}{2i} \int_{\partial D} \frac{\cos[\pi\phi(x)/h] - e^{i[\pi\phi(z)/h] \operatorname{sgn} \operatorname{Im} \phi(z)}}{(z - x) \sin[\pi\phi(z)/h]} F(z) \, dz. \tag{4.4}$$

By a continuity argument, (4.3) holds also for $x = z_l$, $l \in \mathcal{Z}$. It follows from (4.4), (2.7), and $|\cos[\pi\phi(x)/h]| \leq 1$ that

$$\left| E_h^{(2)}(F, x) \right| \leq N(F, D, x) e^{-\pi d/h} \frac{1 + e^{-\pi d/h}}{1 - e^{-2\pi d/h}}. \tag{4.5}$$

To replace the infinite sum in (4.3) by the finite sum

$$Q_{N_1, N_2}^{(2)}(F, x) = h \sum_{-N_1 \leq k \leq N_2} \frac{1 - \cos[\pi(kh - \phi(x))/h]}{\phi'(z_k)(z_k - x)} F(z_k), \tag{4.6}$$

where $N_1, N_2 \in \mathcal{Z}$, assume $\Gamma \subset \mathcal{R}$, $a < b$, and that (2.10) holds. Let N_1, N_2 be such that

$$N_1 \geq \max(-\phi(x)/h, -1), \qquad N_2 \geq \max(\phi(x)/h, -1). \tag{4.7}$$

Since $N_1 \geq -\phi(x)/h$, $N_2 \geq \phi(x)/h$, $x = \psi[\phi(x)]$, and since ψ is increasing on $\mathcal{R}$,

$$|z_k - x| \geq x - z_{-N_1 - 1} > 0, \quad k \leq -N_1 - 1, \qquad |z_k - x| \geq z_{N_2 + 1} - x > 0, \quad k \geq N_2 + 1. \tag{4.8}$$

Also, since $N_1 \geq -1$, $N_2 \geq -1$, and $|1 - \cos[\pi(kh - \phi(x))/h]| \leq 2$, (2.10) and (4.8) imply

$$h\left(\sum_{k<-N_1} + \sum_{k>N_2}\right)\left|\frac{1 - \cos[\pi(kh - \phi(x))/h]}{\phi'(z_k)(z_k - x)}F(z_k)\right| \leq \frac{2c_1}{\alpha_1}\frac{e^{-\alpha_1 N_1 h}}{x - z_{-N_1-1}} + \frac{2c_2}{\alpha_2}\frac{e^{-\alpha_2 N_2 h}}{z_{N_2+1} - x}.$$

$$(4.9)$$

It follows from (4.3), (4.6), (4.5), and (4.9) that

$$\left|\int_\Gamma \frac{F(t)}{t-x}\,dt - Q^{(2)}_{N_1,N_2}(F,x)\right| \leq N(F,D,x)e^{-\pi d/h}\frac{1 + e^{-\pi d/h}}{1 - e^{-2\pi d/h}}$$

$$+ \frac{2c_1}{\alpha_1}\frac{e^{-\alpha_1 N_1 h}}{x - z_{-N_1-1}} + \frac{2c_2}{\alpha_2}\frac{e^{-\alpha_2 N_2 h}}{z_{N_2+1} - x}.$$

Thus, if

$$N_1 = \left\lceil \max(\pi d/(\alpha_1 h^2), -\phi(x)/h) \right\rceil, \qquad N_2 = \left\lceil \max(\pi d/(\alpha_2 h^2), \phi(x)/h) \right\rceil,$$

which are derived from $e^{-\pi d/h} \doteq e^{-\alpha_1 N_1 h} = e^{-\alpha_2 N_2 h}$ and (4.7), then

$$\left|\int_\Gamma \frac{F(t)}{t-x}\,dt - Q^{(2)}_{N_1,N_2}(F,x)\right| \leq c(x)e^{-\pi d/h}.$$

Also, if $\alpha_1 = \alpha_2 = \alpha$, and $h = [\pi d/(\alpha N)]^{1/2}$, where N is a sufficiently large positive integer, then

$$\left|\int_\Gamma \frac{F(t)}{t-x}\,dt - Q^{(2)}_{N,N}(F,x)\right| \leq c(x)e^{-(\pi d\alpha N)^{1/2}}.$$

The strict SINC quadrature and its error bound were presented in [9] and [10].

5   Modified SINC quadrature

Assume $F \in B(D)$, $\zeta \in D \setminus \Gamma$, $t \in \Gamma$ but $t \neq z_k$ for all $k \in \mathcal{Z}$. Then by the residue theorem

$$\frac{F(t)}{\phi'(t)(t-\zeta)} = \sum_{k\in\mathcal{Z}} \frac{F(z_k)}{\phi'(z_k)(z_k - \zeta)}S(k,h)[\phi(t)] + \frac{F(\zeta)}{\sin[\pi\phi(\zeta)/h]}\frac{\sin[\pi\phi(t)/h]}{\phi(t) - \phi(\zeta)}$$

$$+ \frac{\sin[\pi\phi(t)/h]}{2\pi i}\int_{\partial D} \frac{F(z)\,dz}{[\phi(z) - \phi(t)](z - \zeta)\sin[\pi\phi(z)/h]}. \qquad (5.1)$$

Multiply (5.1) through by $\phi'(t)$, integrate both sides over Γ with respect to t and use (2.1), (2.2) to get

$$\int_\Gamma \frac{F(t)}{t-\zeta}\,dt = h\sum_{k\in\mathcal{Z}} \frac{F(z_k)}{\phi'(z_k)(z_k - \zeta)} + \pi\frac{e^{i[\pi\phi(\zeta)/h]\,\mathrm{sgn}\,\mathrm{Im}\,\phi(\zeta)}}{\sin[\pi\phi(\zeta)/h]}F(\zeta)$$

$$- \frac{1}{2i}\int_{\partial D} \frac{e^{i[\pi\phi(z)/h]\,\mathrm{sgn}\,\mathrm{Im}\,\phi(z)}}{(z - \zeta)\sin[\pi\phi(z)/h]}F(z)\,dz. \qquad (5.2)$$

Assume $x \in \Gamma$ and $x \neq z_k$ for all $k \in \mathcal{Z}$. Taking the arithmetic average of the two equations which are obtained from (5.2) on letting ζ approach x from the left- and right-hand sides of Γ, respectively, Plemelj's formula gives

$$
\int_\Gamma \frac{F(t)}{t - x}\,dt =
\begin{cases}
h \displaystyle\sum_{k \in \mathcal{Z}} \frac{F(z_k)}{\phi'(z_k)(z_k - x)} + \pi \cot\left[\frac{\pi\phi(x)}{h}\right] F(x) + E_h^{(3)}(F, x) & \text{if } x \neq z_l, \\[2ex]
h \displaystyle\sum_{\substack{k \in \mathcal{Z} \\ k \neq l}} \frac{F(z_k)}{\phi'(z_k)(z_k - x)} - \frac{h\phi''(x)}{2[\phi'(x)]^2} F(x) + \frac{h}{\phi'(x)} F'(x) + E_h^{(3)}(F, x) & \text{if } x = z_l,
\end{cases}
$$

(5.3)

where

$$
E_h^{(3)}(F, x) = -\frac{1}{2i} \int_{\partial D} \frac{e^{i[\pi\phi(z)/h]\,\mathrm{sgn}\,\mathrm{Im}\,\phi(z)}}{(z - x)\sin[\pi\phi(z)/h]} F(z)\,dz.
$$

(5.4)

It follows from (5.4) and (2.7) that

$$
\left| E_h^{(3)}(F, x) \right| \leq N(F, D, x) \frac{e^{-2\pi d/h}}{1 - e^{-2\pi d/h}}.
$$

(5.5)

To replace the infinite sum in (5.3) by the finite sum

$$
Q_{N_1,N_2}^{(3)}(F, x) =
\begin{cases}
h \displaystyle\sum_{-N_1 \leq k \leq N_2} \frac{F(z_k)}{\phi'(z_k)(z_k - x)} + \pi \cot\left[\frac{\pi\phi(x)}{h}\right] F(x) & \text{if } x \neq z_l, \\[2ex]
h \displaystyle\sum_{\substack{-N_1 \leq k \leq N_2 \\ k \neq l}} \frac{F(z_k)}{\phi'(z_k)(z_k - x)} - \frac{h\phi''(x)}{2[\phi'(x)]^2} F(x) + \frac{h}{\phi'(x)} F'(x) & \text{if } x = z_l,
\end{cases}
$$

(5.6)

where $N_1, N_2 \in \mathcal{Z}$, assume $\Gamma \subset \mathcal{R}$, $a < b$, and that (2.10) holds. Let N_1, N_2 be as in (4.7). Then the inequalities in (4.8) are satisfied, and hence by (2.10),

$$
h\left(\sum_{k < -N_1} + \sum_{k > N_2}\right) \left|\frac{F(z_k)}{\phi'(z_k)(z_k - x)}\right| \leq \frac{c_1}{\alpha_1} \frac{e^{-\alpha_1 N_1 h}}{x - z_{-N_1-1}} + \frac{c_2}{\alpha_2} \frac{e^{-\alpha_2 N_2 h}}{z_{N_2+1} - x}.
$$

(5.7)

It follows from (5.3), (5.6), (5.5), and (5.7) that

$$
\left| \int_\Gamma \frac{F(t)}{t - x}\,dt - Q_{N_1,N_2}^{(3)}(F, x) \right| \leq N(F, D, x) \frac{e^{-2\pi d/h}}{1 - e^{-2\pi d/h}} + \frac{c_1}{\alpha_1} \frac{e^{-\alpha_1 N_1 h}}{x - z_{-N_1-1}} + \frac{c_2}{\alpha_2} \frac{e^{-\alpha_2 N_2 h}}{z_{N_2+1} - x}.
$$

Thus, if

$$
N_1 = \left\lceil \max(2\pi d/(\alpha_1 h^2), -\phi(x)/h) \right\rceil, \qquad N_2 = \left\lceil \max(2\pi d/(\alpha_2 h^2), \phi(x)/h) \right\rceil,
$$

which are derived from $e^{-2\pi d/h} = e^{-\alpha_1 N_1 h} = e^{-\alpha_2 N_2 h}$ and (4.7), then

$$
\left| \int_\Gamma \frac{F(t)}{t - x}\,dt - Q_{N_1,N_2}^{(3)}(F, x) \right| \leq c(x) e^{-2\pi d/h}.
$$

Also, if $\alpha_1 = \alpha_2 = \alpha$, and $h = [2\pi d/(\alpha N)]^{1/2}$, where N is a sufficiently large positive integer, then

$$\left| \int_\Gamma \frac{F(t)}{t-x} \, dt - Q^{(3)}_{N_1,N_2}(F,x) \right| \le c(x) e^{-(2\pi d\alpha N)^{1/2}}.$$

The modified SINC quadrature and its error bound were presented in [2].

It follows easily from (4.6) and (5.6) that the strict and modified SINC quadratures coincide for $x = \psi[(l+1/2)h]$, $l \in \mathcal{Z}$. It should also be noted that in floating-point arithmetic severe cancellation is likely to occur in $Q^{(3)}_{N_1,N_2}(F,x)$ for x near a quadrature node because of the subtraction of large numbers. However, this does not happen in the application of the modified SINC quadrature to the approximate solution of Cauchy singular integral equations [2].

6 Special cases of arc Γ

In the following, the parameter d is in the range $0 < d < \pi$.

CASE 1: $\Gamma = \mathcal{R}$; *exponential decay at $\pm\infty$*. Let $D = D_d$, where D_d is defined in (2.4). Then $\phi(z) = z$, and the condition (2.10) is equivalent to

$$|F(x)| \le c \begin{cases} e^{-\alpha_1|x|} & \text{if } x < 0, \\ e^{-\alpha_2 x} & \text{if } x > 0. \end{cases}$$

Quadratures (3.6), (4.6), and (5.6) take on the following forms:

$$Q^{(1)}_{N_1,N_2}(F,x) = \sum_{-N_1 \le k \le N_2} \frac{F(x+kh+h/2)}{k+1/2},$$

$$Q^{(2)}_{N_1,N_2}(F,x) = h \sum_{-N_1 \le k \le N_2} \frac{1 - \cos[\pi(kh-x)/h]}{kh-x} F(kh),$$

$$Q^{(3)}_{N_1,N_2}(F,x) = \begin{cases} h \displaystyle\sum_{-N_1 \le k \le N_2} \frac{F(kh)}{kh-x} + \pi \cot\left(\frac{\pi x}{h}\right) F(x) & \text{if } x \ne lh, \\ h \displaystyle\sum_{\substack{-N_1 \le k \le N_2 \\ k \ne l}} \frac{F(kh)}{kh-x} + hF'(x) & \text{if } x = lh. \end{cases}$$

CASE 2: $\Gamma = \mathcal{R}$; *algebraic decay at $\pm\infty$*. Let $D = \{x+iy \in \mathcal{C} : y^2/\sin^2 d - x^2/\cos^2 d \le 1\}$ (see Fig. 2 in [1]). Then $\phi(z) = \log\left(z + \sqrt{1+z^2}\right)$, $z_k = \sinh(kh)$, and the condition (2.10) is equivalent to

$$|F(x)| \le c \begin{cases} |x|^{-\alpha_1-1} & \text{if } x < -1, \\ x^{-\alpha_2-1} & \text{if } x > 1. \end{cases}$$

Quadratures (3.6), (4.6), and (5.6) take on the following forms:

$$Q^{(1)}_{N_1,N_2}(F,x) = h \sum_{-N_1 \le k \le N_2} \frac{\sqrt{1+\hat{z}_k^2(x)}}{\hat{z}_k(x)-x} F\left[\hat{z}_k(x)\right],$$

where $\hat{z}_k(x) = xe^{(k+1/2)h} + \sinh[(k+1/2)h]/\left(x + \sqrt{1+x^2}\right)$,

$$Q^{(2)}_{N_1,N_2}(F,x) = h \sum_{-N_1 \leq k \leq N_2} \sqrt{1+z_k^2}\frac{1 - \cos[\pi(kh - \phi(x))/h]}{z_k - x}F(z_k),$$

$$Q^{(3)}_{N_1,N_2}(F,x) = \begin{cases} h \displaystyle\sum_{-N_1 \leq k \leq N_2} \frac{\sqrt{1+z_k^2}}{z_k - x}F(z_k) + \pi \cot\left[\frac{\pi\phi(x)}{h}\right]F(x) & \text{if } x \neq z_l, \\[4mm] h \displaystyle\sum_{\substack{-N_1 \leq k \leq N_2 \\ k \neq l}} \frac{\sqrt{1+z_k^2}}{z_k - x}F(z_k) + \frac{h}{2}\frac{x}{\sqrt{1+x^2}}F(x) + h\sqrt{1+x^2}F'(x) & \text{if } x = z_l. \end{cases}$$

CASE 3: $\Gamma = (0,\infty)$; *exponential decay at* ∞. Let $D = \{z \in \mathcal{C} : |\arg\sinh z| < d\}$ (see Fig. 3 in [1]). Then $\phi(z) = \log\sinh z$, $z_k = \log\left(e^{kh} + \sqrt{1+e^{2kh}}\right)$, and the condition (2.10) is equivalent to

$$|F(x)| \leq c \begin{cases} x^{\alpha_1 - 1} & \text{if } 0 < x < 1, \\ e^{-\alpha_2 x} & \text{if } 1 < x < \infty. \end{cases}$$

Quadratures (3.6), (4.6), and (5.6) take on the following forms:

$$Q^{(1)}_{N_1,N_2}(F,x) = h \sum_{-N_1 \leq k \leq N_2} \frac{\tanh\hat{z}_k(x)}{\hat{z}_k(x) - x}F[\hat{z}_k(x)],$$

where $\hat{z}_k(x) = \log\left[e^{(k+1/2)h}\sinh x + \sqrt{1 + e^{(2k+1)h}\sinh^2 x}\right]$,

$$Q^{(2)}_{N_1,N_2}(F,x) = h \sum_{-N_1 \leq k \leq N_2} \tanh z_k \frac{1 - \cos[\pi(kh - \phi(x))/h]}{z_k - x}F(z_k),$$

$$Q^{(3)}_{N_1,N_2}(F,x) = \begin{cases} h \displaystyle\sum_{-N_1 \leq k \leq N_2} \frac{\tanh z_k}{z_k - x}F(z_k) + \pi \cot\left[\frac{\pi\phi(x)}{h}\right]F(x) & \text{if } x \neq z_l, \\[4mm] h \displaystyle\sum_{\substack{-N_1 \leq k \leq N_2 \\ k \neq l}} \frac{\tanh z_k}{z_k - x}F(z_k) + \frac{h}{2}\operatorname{sech}^2(x)F(x) + h\tanh(x)F'(x) & \text{if } x = z_l. \end{cases}$$

CASE 4: $\Gamma = (0,\infty)$; *algebraic decay at* ∞. Let $D = \{z \in \mathcal{C} : |\arg(z)| < d\}$. Then $\phi(z) = \log z$, $z_k = e^{kh}$, and the condition (2.10) is equivalent to

$$|F(x)| \leq c \begin{cases} x^{\alpha_1 - 1} & \text{if } 0 < x < 1, \\ x^{-\alpha_2 - 1} & \text{if } 1 < x < \infty. \end{cases}$$

Quadratures (3.6), (4.6), and (5.6) take on the following forms:

$$Q^{(1)}_{N_1,N_2}(F,x) = h \sum_{-N_1 \leq k \leq N_2} \frac{z_{k+1/2}}{z_{k+1/2} - 1}F\left(xz_{k+1/2}\right),$$

where $z_{k+1/2} = e^{(k+1/2)h}$,

$$Q^{(2)}_{N_1,N_2}(F,x) = h \sum_{-N_1 \leq k \leq N_2} z_k \frac{1 - \cos[\pi(kh - \phi(x))/h]}{z_k - x} F(z_k),$$

$$Q^{(3)}_{N_1,N_2}(F,x) = \begin{cases} h \displaystyle\sum_{-N_1 \leq k \leq N_2} \frac{z_k}{z_k - x} F(z_k) + \pi \cot\left[\frac{\pi\phi(x)}{h}\right] F(x) & \text{if } x \neq z_l, \\ h \displaystyle\sum_{\substack{-N_1 \leq k \leq N_2 \\ k \neq l}} \frac{z_k}{z_k - x} F(z_k) + \frac{h}{2} F(x) + hx F'(x) & \text{if } x = z_l. \end{cases}$$

CASE 5: $\Gamma = (-1,1)$. Let $D = \{z \in C : |\arg[(1+z)/(1-z)]| < d\}$ (see Fig. 5 in [1]). Then $\phi(z) = \log[(1+z)/(1-z)]$, $z_k = \left(e^{kh} - 1\right) / \left(e^{kh} + 1\right)$, and the condition (2.10) is equivalent to

$$|F(x)| \leq c \begin{cases} (1+x)^{\alpha_1 - 1} & \text{if } -1 < x < 0, \\ (1-x)^{\alpha_2 - 1} & \text{if } 0 < x < 1. \end{cases}$$

Quadratures (3.6), (4.6), and (5.6) take on the following forms:

$$Q^{(1)}_{N_1,N_2}(F,x) = \frac{h}{2} \sum_{-N_1 \leq k \leq N_2} \frac{1 - z_{k+1/2}^2}{z_{k+1/2}(1 + xz_{k+1/2})} F\left(\frac{x + z_{k+1/2}}{1 + xz_{k+1/2}}\right),$$

where $z_{k+1/2} = \left[e^{(k+1/2)h} - 1\right] / \left[e^{(k+1/2)h} + 1\right]$,

$$Q^{(2)}_{N_1,N_2}(F,x) = \frac{h}{2} \sum_{-N_1 \leq k \leq N_2} (1 - z_k^2) \frac{1 - \cos[\pi(kh - \phi(x))/h]}{z_k - x} F(z_k),$$

$$Q^{(3)}_{N_1,N_2}(F,x) = \begin{cases} \dfrac{h}{2} \displaystyle\sum_{-N_1 \leq k \leq N_2} \frac{1 - z_k^2}{z_k - x} F(z_k) + \pi \cot\left[\frac{\pi\phi(x)}{h}\right] F(x) & \text{if } x \neq z_l, \\ \dfrac{h}{2} \displaystyle\sum_{\substack{-N_1 \leq k \leq N_2 \\ k \neq l}} \frac{1 - z_k^2}{z_k - x} F(z_k) - \frac{h}{2} x F(x) + \frac{h}{2}(1 - x^2) F'(x) & \text{if } x = z_l. \end{cases}$$

References

[1] B. Bialecki, *A modified* SINC *quadrature for functions with poles near the arc of integration*, BIT, 29 (1989), 464–476.

[2] B. Bialecki, *A Sinc-Hunter quadrature rule for Cauchy principal value integrals*, Math. Comp., 55 (1990), 665-681.

[3] P.J. Davis and P. Rabinowitz, *Methods of Numerical Integration*, Academic Press, INC., Orland, Florida, 1984.

[4] W. Gautschi, *A survey of Gauss-Christoffel quadrature formulae*, in: *E.B. Christoffel: The Influence of his Work in Mathematics and the Physical Sciences* (P.L. Butzer and F. Fehér, eds.), Birkhäuser, Basel, 1981, 72–147.

[5] R. Kress and E. Martensen, *Anwendung der Rechteckregel auf die reelle Hilberttranformation mit unendlichem Intervall*, ZAMM 50 (1970), T61–64.

[6] I.M. Longman, *On the numerical evaluation of Cauchy principal values of integrals*, Math. Comp., 12 (1958), 205–207.

[7] N.I. Muskhelishvili, *Singular Integral Equations*, P. Noordhoff N.V., Groningen-Holland, 1953.

[8] F. Stenger, *Approximations via Whittaker's cardinal function*, J. Approx. Theory, 17 (1976), 222-240.

[9] F. Stenger, *Numerical methods based on Whittaker cardinal, or* SINC functions, SIAM Review, 23 (1981), 165–224.

[10] F. Stenger and D. Elliott, *Sinc method of solution of singular integral equations*, IMACS Symposium on Numerical Solution of Singular Integral Equations, IMACS, 1984, 27–35.

Interpolatory Product Integration in the Presence of Singularities: L_p Theory

Philip Rabinowitz

Department of Applied Mathematics and Computer Science
Weizmann Institute of Science
Rehovot 76100, Israel

William E. Smith

Department of Applied Mathematics
University of New South Wales
Kensington, NSW 2033, Australia

Abstract

We study interpolatory product integration rules based on Gauss, Radau and Lobatto points with respect to a generalized Jacobi weight function as applied to the integration of functions with singularities. We give sufficient conditions for convergence of such rules for both endpoint and interior singularities.

1. Introduction

In the study of interpolatory product integration rules (PIR's) for evaluating the integral

$$I(kf) := \int_{-1}^{1} k(x)f(x)dx, \tag{1}$$

there are two theories, an L_2 theory and an L_p theory. In both cases, the sets of integration points $\{\bar{x}_{in}\}$ consist of Gauss (G), Radau (R) or Lobatto (L) points with respect to a nonnegative integrable weight function w on $J := [-1, 1]$ such that $I(w) > 0$. However, in the L_2 case, there are no other restrictions on w

T. O. Espelid and A. Genz (eds.), Numerical Integration, 93–109.
© 1992 Kluwer Academic Publishers.

94

whereas, to apply the L_p theory, we must restrict ourselves to generalized smooth Jacobi (GSJ) weight functions of the form

$$w(x) := \psi(x) \prod_{k=0}^{\rho+1} |x - t_k|^{\gamma_k}, \qquad \gamma_k > -1, \quad k = 0, 1, \ldots, \rho+1 \qquad (2)$$

where $\rho \geq 0$, $-1 = t_0 < t_1 < \cdots < t_{\rho+1} = 1$, $\psi \in C(J)$, $\psi > 0$ and

$$\int_0^2 \omega(\psi; t) t^{-1} dt < \infty.$$

Here, $\omega(\psi; t)$ denotes the modulus of continuity of ψ.

The L_2 theory of product integration is based on generalizations of the Erdös-Turan theorem on mean square convergence of Lagrange interpolation and on the Cauchy-Schwartz inequality and restricts both f and $K := k/w$ to belong to $L_{2,w}$. This case was studied in detail for G points in [4] and extended to R and L points by the authors [12]. The L_p theory is based on the work of Nevai [5] on weighted mean convergence of Lagrange interpolation and uses the Hölder inequality. Nevai's work was applied by the authors [7,11] to prove convergence of PIR's applied to (bounded) Riemann-integrable functions f when certain conditions hold between w and $k \in L_1(J)$. In this work, we shall extend these results to functions f which are unbounded at a finite set of points $Y := \{y_j : j = 1, \ldots, \ell\} \subset J$ where we shall set $f(y_j) = 0$, $j = 1, \ldots, \ell$. We shall consider both endpoint and interior singularities. It will turn out that in the former case, we can ignore the singularities. In the latter case, we shall generally have to avoid the singularities except in the case when w is a Jacobi weight function

$$w_{\alpha\beta}(x) := (1 - x)^\alpha (1 + x)^\beta, \quad \alpha, \beta > -1$$

in which case we shall be able to ignore the singularities in certain situations similar to those considered in [8] and [15].

In Section 2, we provide the necessary background information. In Sections 3 and 4, we deal with endpoint and interior singularities respectively. In Section 5, we make some concluding remarks.

2. Background Material

We shall be concerned with product integration rules of the form

$$I_n(f; k) := \sum_{i=1-r}^{n+s} w_{in}(k) f(\overline{x}_{in}) \qquad (3)$$

where $r, s \in \{0, 1\}$,

$$-1 = \overline{x}_{0n} < \overline{x}_{1n} < \cdots < \overline{x}_{nn} < \overline{x}_{n+1,n} = 1, \tag{4}$$

and the weights $w_{in}(k)$ are chosen such that

$$I_n(f; k) = I(kf)$$

when $f \in P_{n+r+s-1}$, where P_m denotes the set of all polynomials of degree $\leq m$. The integration points $\{\overline{x}_{in}\}$ are G, R or L points with respect to some $w \in GSJ$, i.e. they are the zeros of

$$q_{n+r+s}(x; w) := w_{sr}(x)p_n(x; \overline{w}) \tag{5}$$

where

$$\overline{w}(x) := w_{sr}(x)w(x) \tag{6}$$

and, for any $v \in GSJ$,

$$\{p_n(x; v) := k_n(v)x^n + \ldots; \quad k_n(v) > 0; \quad n = 0, 1, 2, \ldots\}$$

is the sequence of orthonormal polynomials with respect to the inner product $(f, g) := I(vfg)$. We shall denote the zeros of $p_n(x; v)$ by $x_{in}(v)$, $i = 1, \ldots, n$ and the corresponding Gauss weights by $\mu_{in}(v)$ so that

$$I(vg) = \sum_{i=1}^{n} \mu_{in}(v)g(x_{in}(v)) + E_n g \tag{7}$$

with $E_n g = 0$ if $g \in P_{2n-1}$. If we write the G, R and L rules with respect to w in a uniform fashion, then we have that

$$I(wg) = \sum_{i=1-r}^{n+s} \overline{\mu}_{in} g(\overline{x}_{in}) + \overline{E}_n g \tag{8}$$

where

$$\overline{\mu}_{in} = \mu_{in}(\overline{w})/w_{sr}(\overline{x}_{in}) > 0, \qquad i = 1, \ldots, n \tag{9}$$

and $\overline{E}_n g = 0$ if $g \in P_{2n+r+s-1}$. If $g \in C^{2n+r+s}(J)$, then

$$\overline{E}_n g = (-1)^s \frac{g^{(2n+r+s)}(\xi)}{(2n + r + s)!} k_n(\overline{w})^{-2}, \qquad -1 < \xi < 1. \tag{10}$$

Of course, in the G case, $r = s = 0$, $\overline{w} = w$, $\overline{\mu}_{in} = \mu_{in}(w)$ and $\overline{x}_{in} = x_{in}(w)$. The R cases correspond to $r + s = 1$ and the L case, to $r + s = 2$. The weights $\overline{\mu}_{0n}$, when $r = 1$, and $\overline{\mu}_{n+1,n}$, when $s = 1$, are positive and $o(1)$ as $n \to \infty$. For the points $\overline{x}_{in}$, we have that if $\overline{x}_{in} := \cos \theta_{in}$, $0 \le \theta_{in} \le \pi$, then

$$\theta_{in} - \theta_{i+1,n} \sim n^{-1} \tag{11}$$

uniformly for $0 \le i \le n$, $n = 1, 2, \ldots$. As for the weights $\overline{\mu}_{in}$, we have that in any closed subinterval of $J - T$,

$$\overline{\mu}_{in} \sim n^{-1} \tag{12}$$

where $T := \{t_j : j = 0, \ldots, \rho + 1\}$ [5].

We now introduce the Lagrange interpolating polynomial f at the zeros of $p_n(x; w)$,

$$L_n(x; w; f) := \sum_{i=1}^{n} \ell_{in}(x; w) f(x_{in}(w)) \tag{13}$$

where

$$\ell_{in}(x; w) := p_n(x; w) / \left((x - x_{in}(w)) p_n'(x_{in}(w); w) \right).$$

Another expression for $\ell_{in}(x; w)$ is given in [5 (3)] by

$$\ell_{in}(x; w) = \frac{k_{n-1}(w)}{k_n(w)} \mu_{in}(w) p_{n-1}(x_{in}(w); w) \frac{p_n(x; w)}{x - x_{in}(w)}. \tag{14}$$

The Lagrange polynomials interpolating f at the zeros of $q_{n+r+s}(x; w)$ are given by

$$\overline{L}_n(x; \overline{w}; f) := \sum_{i=1-r}^{n+s} \overline{\ell}_{in}(x; \overline{w}) f(\overline{x}_{in}) \tag{15}$$

for certain polynomials $\overline{\ell}_{in}(w) \in P_{n+r+s-1}$ satisfying $\overline{\ell}_{in}(\overline{x}_{jn}; \overline{w}) = \delta_{ij}$. For $i = 1, \ldots, n$

$$\overline{\ell}_{in}(x; \overline{w}) = \frac{w_{sr}(x)}{w_{sr}(\overline{x}_{in})} \ell_{in}(x; \overline{w}).$$

By a well-known result in interpolatory integration [2, p. 74],

$$I_n(f; k) = I\left(k \overline{L}_n(\overline{w}; f) \right), \tag{16}$$

so that

$$w_{in}(k) = I\left(k \overline{\ell}_{in}(\overline{w}) \right). \tag{17}$$

We now derive another expression for the weights, $w_{in}(k)$, $i = 1, \ldots, n$, which generalizes that given in [3] for the G case. If we expand $\bar{\ell}_{in}(x; w)$ in an orthogonal series, we have that

$$\bar{\ell}_{in}(x; \overline{w}) = \sum_{j=0}^{n+r+s-1} (\bar{\ell}_{in}(\overline{w}), p_j(w)) p_j(x; w). \tag{18}$$

We now introduce the discrete inner product

$$[f, g] := \sum_{i=1-r}^{n+s} \overline{\mu}_{in} f(\overline{x}_{in}) g(\overline{x}_{in}) \tag{19}$$

and observe that

$$(\bar{\ell}_{in}(\overline{w}), p_j(w)) = [\bar{\ell}_{in}(\overline{w}), p_j(w)] = \overline{\mu}_{in} p_j(\overline{x}_{in}; w) \tag{20}$$

whenever $n + r + s - 1 + j \leq 2n + r + s - 1$. It follows that

$$(\bar{\ell}_{in}(\overline{w}), p_j(w)) \neq [\bar{\ell}_{in}(\overline{w}), p_j(w)]$$

only when $r + s = 2$ and $j = n + 1$. Hence

$$\bar{\ell}_{in}(x; \overline{w}) = \sum_{j=0}^{n+r+s-1} \overline{\mu}_{in}(\overline{x}_{in}; w) p_j(x; w) + \epsilon_{in}(x) \tag{21}$$

where

$$\epsilon_{in}(x) := \delta_{r+s,2} \overline{E}_n(\bar{\ell}_{in}(\overline{w}) p_{n+1}(w)) p_{n+1}(x; w). \tag{22}$$

Inserting (21) into (17), we find that

$$w_{in}(k) = \overline{\mu}_{in} S^K_{n+r+s-1}(w; \overline{x}_{in}) + \eta_{in} \tag{23}$$

where, as before, $K := k/w$, $\eta_{in} := I(k\epsilon_{in})$, and, for any weight function v and any function h,

$$S^h_m(v; x) := \sum_{j=0}^{m} b^h_j p_j(x; v) \tag{24}$$

with $b^h_j := (h, p_j(v)) = I(vhp_j(v))$. For $h = K$ and $v = w$, we have that

$$b^K_j := I(wKp_j(w)) = I(kp_j(w))$$

is the modified moment of k with respect to $p_j(w)$. The additional term η_{in} in (23) makes things more complicated in the L case. However, there is a certain benefit from treating the L case as we shall see (cf. [12]).

The weights $w_{0n}(k)$, when $r = 1$, and $w_{n+1,n}(k)$, when $s = 1$, are $o(1)$ as $n \to \infty$ for the cases we shall be considering, namely, the cases when

$$I_n(f; k) \to I(kf) \quad \text{as} \quad n \to \infty \tag{25}$$

for all $f \in R(J)$, the set of all Riemann-integrable functions on J. The conditions on k and w to insure (25), given in [11], are that for some $p > 1$

$$\begin{cases} k \in L_p(J) \\ w_{-s+1/4,-r+1/4}\overline{w}^{1/2} \in L_1(J) \\ kw_{s-1/4,r-1/4}\overline{w}^{-1/2} \in L_p(J). \end{cases} \tag{26}$$

We shall need the following result which is a special case of [5, (24)] with $v = 1/w_{rs}$ and $w = \overline{w}$. If C is a fixed positive number, then for every $r \in P_{\lfloor cn \rfloor}$ and $p \in (0, \infty)$

$$\sum_{i=1}^n \overline{\mu}_{in} |r(\overline{x}_{in})|^p \leq C \int_{-1}^1 w(x)|r(x)|^p \, dx. \tag{27}$$

We now define the concept of monotone integrability with respect to a weight function v and a set of points Y as in [12].

Definition 1. We say that g is monotone integrable (MI) on J with respect to v and Y if

(a) there exists a nonnegative integer ℓ and points

$$-1 = y_0 < y_1 < \cdots < y_\ell < y_{\ell+1} = 1$$

such that $v > 0$ in deleted neighborhoods of y_i, $i = 0, \ldots, \ell+1$ and g is Riemann-integrable in each compact subinterval of (y_i, y_{i+1}), $i = 0, 1, \ldots, \ell$ with

$$\lim_{\substack{A \to y_i^+ \\ B \to y_{i+1}^-}} \int_A^B v(x)|g(x)|dx < \infty, \qquad i = 0, 1, \ldots, \ell.$$

Whenever $\overline{\lim}_{x \to y}|g(x)| = \infty$, we set $g(y) = 0$.

(b) there exist functions $G_i(x)$, $i = 0, 1, \ldots, \ell+1$ such that $I(vG_i) < \infty$, G_i is infinitely differentiable in $J - \{y_i\}$, $G_i^{(j)} \geq 0$ in $[-1, y_i)$, $(-1)^j \cdot G_i^{(j)} \geq 0$ in $(y_i, 1]$, $j = 0, 1, 2, \ldots$, and

$$\overline{\lim}_{\substack{x \to y_i \\ x \in (-1,1)}} |g(x)/G_i(x)| < \infty, \qquad i = 0, 1, \ldots, \ell+1.$$

If g has only endpoint singularities, then $Y = \phi$.

If we now assume that g is MI with respect to w and Y and define $\tau(n, g)$ to be the subset of $\{1, 2, \ldots, n\}$ such that $j \in \tau(n, g)$ if either $Y \cap (\overline{x}_{j-1,n}, \overline{x}_{j+1,n}) = \phi$ or if for any i, $1 \leq i \leq \ell$ such that $y_i \in (\overline{x}_{j-1,n}, \overline{x}_{j+1,n})$, $|\int_{\overline{x}_{jn}}^{y_i} w(x) dx| \geq \overline{\mu}_{jn}$, then

$$\sum_{j \in \tau(n,g)} \overline{\mu}_{jn} g(\overline{x}_{jn}) \to I(wg). \tag{28}$$

Note that if $Y = \phi$, the case of endpoint singularities, $\tau(n, g) = \{1, 2, \ldots, n\}$, i.e. the endpoints are not used.

3. Endpoint Singularities

Our first result is an immediate consequence of the special case, $p = 1$, of Theorem 5 in [5].

Theorem 1. Let $f \in C(J)$, $w \in GSJ$ and $g = w_{\gamma\delta}f$ where γ and δ are arbitrary and need not satisfy the conditions $\gamma, \delta > -1$. Let $v^* := w_{\lfloor\gamma\rfloor,\lfloor\delta\rfloor}$. Then

$$\int_{-1}^{1} k(x)[g(x) - L_n(x; w; g)] dx \to 0 \quad \text{as} \quad n \to \infty \tag{29}$$

if

(a) $k \in L_1(J)$

(b) $kv^* \in L \log^+ L(J)$

(c) $kw^{-1/2}(1 - x^2)^{-1/4} \in L_1(J)$

(d) $v^* w^{1/2}(1 - x^2)^{1/4} \in L_1(J)$

where we say that $h \in L \log^+ L(J)$ if $I(|h(\log |h|)|) < \infty$.

Now, this theorem is restricted to G rules and to functions g continuous in $\overset{\circ}{J}$. On the other hand, it accommodates strong singularities in g provided that k and w satisfy suitable conditions. The strongest results occur when γ and δ are negative integers. We illustrate this theorem with several examples.

Example 1. $w(x) := (1 - x^2)^{-1/2}$ and $f \equiv 1$. The points $\{x_{in}(w)\}$ are the zeros of T_n, the Chebyshev polynomial of the first kind. If $g(x) := (1 - x^2)^{-m}$ so that $v^* = g$, then condition (b) requires that $k(x)(1 - x^2)^{-m} \in L \log^+ L(J)$, i.e. that k behave like $(1 - x^2)^{m-1+\epsilon}$. Then (a) and (c) will hold. However (d) can only hold if $m = 0$. Hence, Theorem 1 cannot deal with the case of Chebyshev points as integration points when there are endpoint singularities.

Example 2. $w(x) := 1$ and $f \equiv 1$. The points $\{x_{in}(w)\}$ are the Gauss-Legendre points. If $g(x) := v^*(x) = (1 - x^2)^{-1}$, then condition (d) is satisfied. If $k(x)(1 - x^2)^{-1} \in L \log^+ L(J)$, then all other conditions of Theorem 1 are satisfied so that essentially, we require that k behave like $(1 - x^2)^\epsilon$ near the endpoints, which we need in any case since we require that $kg \in L_1(J)$.

Example 3. $w(x) := (1 - x^2)^{1/2}$ and $f \equiv 1$. The points $\{x_{in}(w)\}$ are the zeros of U_n, the Chebyshev polynomial of the second kind. If $g(x) := v^*(x) = (1 - x^2)^{-m}$, then condition (d) again limits m to be ≤ 1. If we set $m = 1$, we are back to the situation of Example 2.

Example 4. $w(x) := (1 - x^2)^{3/2}$ and $f \equiv 1$. The points $\{x_{in}(w)\}$ are the zeros of the Jacobi polynomial $P_n^{(3/2,3/2)}$. If $g(x) := v^*(x) = (1 - x^2)^{-m}$, then condition (d) requires now that $m \leq 2$. For $m = 2$, condition (b) requires that $k(x)(1 - x^2)^{-2} \in L \log^+ L(J)$, essentially that k behave like $(1 - x^2)^{1+\epsilon}$ near the endpoints. Conditions (a) and (b) hold for this choice of k so that we have convergence of the PIR for the strongly singular function $(1 - x^2)^{-2}$.

Note that if g has a weaker singularity than in the example, the same requirements hold for k. For example, when $g(x) := (1 - x^2)^{-1/2}$ in Example 2, we still require that $k(x)(1-x^2)^{-1} \in L \log^+ L$ even though we only require that k behave like $(1 - x^2)^{-1/2+\epsilon}$ to insure that $kg \in L_1(J)$. Our next theorem overcomes this limitation. In addition it allows the use of R and L points as well as G points and it does not require continuity of g. Instead it requires that g be MI. Before we state the theorem, we quote a lemma from Badkov [1] about mean convergence of orthogonal series.

Lemma 1. Let $v \in GSJ$ and define

$$\widehat{v}(x) := \prod_{k=0}^{\rho+1} |x - t_k|^{\Gamma_k}. \tag{30}$$

Assume that $\widehat{v}h \in L_q(J)$ for some q, $1 < q < \infty$. Then

$$\|\widehat{v}S_m^h(v)\|_q \leq C \|\widehat{v}h\|_q \quad \text{for all} \quad m \tag{31}$$

if

$$\left| \Gamma_k + q^{-1} - \frac{\gamma_k + 1}{2} \right| < \min\left(\mu_k, \frac{\gamma_k + 1}{2} \right), \qquad k = 0, \ldots, \rho + 1 \tag{32}$$

where

$$\mu_0 = \mu_{\rho+1} = \frac{1}{4}, \quad \mu_k = \frac{1}{2}, \qquad 1 \leq k \leq \rho.$$

Theorem 2. Let $w \in GSJ$ and let $K := k/w \in L_{q,w}(J)$ where q is such that

$$|(\gamma_0 + 1)(1/q - 1/2)| < \min(1/4, (\gamma_0 + 1)/2)$$
$$|(\gamma_{\rho+1} + 1)(1/q - 1/2)| < \min(1/4, (\gamma_{\rho+1} + 1)/2) \qquad (33)$$
$$|(\gamma_k + 1)(1/q - 1/2)| < \min(1/2, (\gamma_k + 1)/2), \qquad k = 1, \ldots, \rho.$$

If $|f|^p$ is MI with respect to w and ϕ, where $p^{-1} + q^{-1} = 1$ and if we set $f(\pm 1) = 0$, then (25) holds.

Proof: We first note that the hypotheses of the theorem imply that $kf \in L_i(J)$, since by the Hölder inequality.

$$|I(kf)| = |I(wKf)| \leq \|f\|_{p,w}\|K\|_{q,w} < \infty \text{ since } |f|^p \text{ is } MI.$$

Next, we remark that to establish (25) we need only show that for any g such that $|g|^p$ is MI for some $p > 1$ and such that $g(\pm 1)$ is finite

$$\left|I(k\overline{L}_n(\overline{w}; g))\right| \leq C\|g\|_{p,w}\|K\|_{q,w} \qquad (34)$$

since, if (34) holds, we have that for any polynomial $r \in P_m$ and all $n \geq m$

$$\left|I\big(k(f - \overline{L}_n(\overline{w}; f))\big)\right| \leq \left|I\big(k(f - r)\big)\right| + \left|I\big(k(\overline{L}_n(\overline{w}; r - f))\big)\right|$$
$$\leq \|f - r\|_{p,w}\|K\|_{q,w} + C\|f - r\|_{p,w}\|K\|_{q,w}.$$

This is true since if $|f|^p$ is MI, so is $|f - r|^p$ and $f - r$ is finite at ± 1.

Since for any $\epsilon > 0$, we can find a polynomial $\widehat{r}$ such that $\|f - \widehat{r}\|_{p,w} < \epsilon/(C + 1)\|K\|_{q,w}$, our theorem will follow.

To prove (34), we assume first that $r + s \leq 1$ so that $\eta_{in} = 0$ in (23). Since $|g(\pm 1)| \leq A$ and $w_{0n}(k)$, $w_{n+1,n}(k) = o(1)$ as $n \to \infty$, we have that

$$\left|I\big(k\overline{L}_n(\overline{w}; g)\big)\right| = \left|\sum_{i=1}^n w_{in}(k)g(\overline{x}_{in})\right| + o(1).$$

By (23), $\left|\sum_{i=1}^n w_{in}(k)g(\overline{x}_{in})\right| = \left|\sum_{i=1}^n \overline{\mu}_{in}S_{n+r+s-1}^K(w; \overline{x}_{in})g(\overline{x}_{in})\right|$
$$\leq \left(\sum_{i=1}^n \overline{\mu}_{in}|g(x_{in})|^p\right)^{1/p}\left(\sum_{i=1}^n \overline{\mu}_{in}|S_{n+r+s-1}^K(w; \overline{x}_{in})|^q\right)^{1/q}$$
$$\leq C_1\|g\|_{p,w}I(w|S_{n+r+s-1}^K|^q)^{1/q} \text{ by (27) and (28)}$$
$$= C_1\|g\|_{p,w}I(|w^{1/q}S_{n+r+s-1}^K|^q)^{1/q}$$
$$\leq C_2\|g\|_{p,w}\|K\|_{q,w} \text{ by the hypotheses on } K \text{ and Lemma 1 with } \overline{v} = w^{1/q}.$$

This proves (25) for $r + s \leq 1$. To deal with the case $r + s = 2$, we have to introduce the error term η_{in} from (23) and consider the expression

$$\left| \sum_{i=1}^{n} (\overline{\mu}_{in} S_{n+1}^K(w; \overline{x}_{in}) + \eta_{in}) g(\overline{x}_{in}) \right|$$

$$\leq \left| \sum_{i=1}^{n} \overline{\mu}_{in} S_{n+1}^K(w; x_{in}) g(\overline{x}_{in}) \right| + \left| \sum_{i=1}^{n} \eta_{in} g(\overline{x}_{in}) \right|$$

$$\leq C_1 \|g\|_{p,w} \|K\|_{q,w} + C_2 A_n \|g\|_{p,w}$$

where

$$A_n := \left(\sum_{i=1}^{n} \overline{\mu}_{in} |\eta_{in}/\overline{\mu}_{in}|^q \right)^{1/q}.$$

We see from this that to prove Theorem 2 for $r + s = 2$, we must show that A_n is uniformly bounded in n.

Recalling that $\eta_{in} = I(k\epsilon_{in})$ where ϵ_{in} is given by (22), we have that

$$\eta_{in} = (K, p_{n+1}(w)) \cdot \overline{E}_n\left(\overline{\ell}_{in}(\overline{w}) p_{n+1}(w)\right). \tag{35}$$

Since

$$\overline{\ell}_{in}(x; \overline{w}) = (1 - x^2) \ell_{in}(x; \overline{w})/(1 - \overline{x}_{in}^2), \qquad i = 1, \ldots, n,$$

we find using (14) and (9) that

$$\overline{\ell}_{in}(x; w) = \frac{k_{n-1}(\overline{w})}{k_n(\overline{w})} \overline{\mu}_{in} p_{n-1}(\overline{x}_{in}; \overline{w}) \frac{p_n(x; \overline{w})(1 - x^2)}{x - \overline{x}_{in}}.$$

Hence, by (10)

$$\overline{E}_n\left(\overline{\ell}_{in}(\overline{w}) p_{n+1}(w)\right) = \frac{k_{n-1}(\overline{w})}{k_n(\overline{w})^3} \overline{\mu}_{in} p_{n-1}(\overline{x}_{in}; \overline{w}) k_n(w) \tag{36}$$

so that

$$A_n = (K, p_{n+1}(w)) B_n \left(\sum_{i=1}^{n} \overline{\mu}_{in} |p_{n-1}(\overline{x}_{in}; \overline{w})|^q \right)^{1/q} \tag{37}$$

where

$$0 < \lim_{n \to \infty} B_n < \infty$$

inasmuch as

$$0 < \lim_{n \to \infty} 2^{-n} k_n(v) < \infty$$

for any $v \in GSJ$ [5 (21)].

Now, for the values of q satisfying (33), we have that

$$
\left.
\begin{array}{l}
1 < q < \infty, \quad \gamma_k < 0; \\[2mm]
2 - \dfrac{2}{\gamma_k + 2} < q < 2 + \dfrac{2}{\gamma_k}, \quad \gamma_k > 0, \quad k = 1, \ldots, \rho \\[2mm]
1 < q < \infty, \quad \gamma_k \leq -\dfrac{1}{2}; \\[2mm]
2 - \dfrac{2}{2\gamma_k + 3} < q < 2 + \dfrac{2}{2\gamma_k + 1}, \quad \gamma_k > -\dfrac{1}{2}, \quad k = 0, \rho + 1
\end{array}
\right\}. \tag{38}
$$

Since from [9]

$$
|p_{n-1}(\overline{x}_{in}; \overline{w})| \sim \overline{w}(\overline{x}_{in})^{-1/2}(1 - \overline{x}_{in}^2)^{1/4} \tag{39}
$$

it follows that

$$
|p_{n-1}(\overline{x}_{in}; \overline{w})|^q < C\widetilde{w}(x_{in})
$$

where

$$
\widetilde{w}(x) := \prod_{k=0}^{\rho+1} |x - t_k|^{\delta_k}
$$

and the δ_k satisfy

$$
\delta_k > -1 - \gamma_k, \quad k = 0, \ldots, \rho + 1.
$$

Hence, $\widetilde{w}$ is MI with respect to w and T so that from (28)

$$
\sum_{i \in \tau(n, \widetilde{w})} \overline{\mu}_{in} \widetilde{w}(\overline{x}_{in}) \to I(w\widetilde{w}) \quad \text{as} \quad n \to \infty.
$$

This implies that A_n^* is bounded where A_n^* has the same form as A_n in (37) except that the index of the sum ranges over the set $\tau(n, \widetilde{w})$, inasmuch as the Fourier coefficients $(K, p_{n+1}(w))$ are uniformly bounded for $K \in L_{q,w}$. We are essentially through since we have shown that

$$
\left| \sum_{i \in \tau(n, \widetilde{w})} w_{in}(k) g(\overline{x}_{in}) \right| \leq C_2 \|g\|_{p,w} \|K\|_{q,w}.
$$

Since

$$
\left| \sum_{i \notin \tau(n, \widetilde{w})} w_{in}(k) g(\overline{x}_{in}) \right| = o(1)
$$

because g is bounded in $[t_1 - \delta, t_\rho + \delta]$ and $w_{in}(k) = o(1)$, our theorem is proved.

Before we discuss our four examples, we state another theorem which corresponds to Theorem 4.3 in [12].

Theorem 3. Let $w \in GSJ$ be such that $v := w/w_{sr} \in GSJ$ and assume that $K := k/w \in L_{q,v}$ where q satisfies (35). If $|w_{sr}f|^p$ is MI with respect for v, then

$$I\big(kL_n(w; f)\big) \to I(kf) \quad \text{as} \quad n \to \infty. \tag{40}$$

Proof: As in [12], we have that $w_{sr}L_n(w; f) = \overline{L}_n(\overline{v}; w_{sr}f)$ since $f(\pm 1) = 0$. Hence

$$
\begin{aligned}
I\big(kL_n(w; f)\big) &= I\big((k/w_{sr})w_{sr}L_n(w; f)\big) \\
&= I\big((k/w_{sr})\overline{L}_n(\overline{v}; w_{sr}f)\big) \to I\big((k/w_{sr})w_{sr}f\big) = I(kf).
\end{aligned}
$$

Here, we have used Theorem 2 with k replaced by k/w_{sr} and w by v together with the fact that $\overline{v} = w$.

We note that Theorem 3 is meaningful only for the cases $r + s \geq 1$ and that (40) is a convergence result only for G points with respect to w. Thus, we have two convergence results for such points, one given in Theorem 2 and one in Theorem 3. However, Theorem 3 is only applicable for weights w such that w/w_{sr} is integrable.

We now apply Theorems 2 and 3 to our examples.

Example 1. $w(x) := (1 - x^2)^{-1/2}$. From (38), we see that q can take on any value in $(1, \infty)$. If we want f to be as singular as possible, we must choose p close to 1. Hence, if $f = (1 - x^2)^\beta$, β must be greater than $-1/2$ to insure that $|f|^p$ is MI with respect to w. For any $\beta > -1/2$, we find the appropriate p. Then the corresponding q places a restriction on k. This restriction is the same one as in required by the condition that $kf \in L_1(J)$, so that if $k = (1-x^2)^\gamma$, then $\beta+\gamma > -1$. Indeed, if $-1/2 + \beta p = \delta > -1$ and $\beta + \gamma = \epsilon > -1$, then if $q(\epsilon - \delta) + \delta > 1$, $-1/2 + (\gamma + 1/2)q > -1$ and $K \in L_{q,w}$. Thus, in contrast to Theorem 1, we see that product integration based on the Chebyshev zeros converges for all f such that $\beta > -1/2$. Theorem 3 is not applicable since $w/w_{sr} \notin GSJ$.

Example 2. $w(x) = 1$. From (38), $1 < q < 4$ so that $p > 4/3$. If $f = (1 - x^2)^\beta$, then β must be greater than $-3/4$ so that using Theorem 2, we can prove convergence of product integration based on the Gauss-Legendre points only for f with a singularity weaker than $(1 - x^2)^{-3/4}$ in contrast to Theorem 1 which allows singularities as strong as $(1 - x^2)^{-1}$. On the other hand, the conditions on k are weaker and not as rigid so that if the singularity on f is weaker than $-3/4$, we can allow a strong singularity on k. Essentially, all we need is that $kf \in L_1(J)$ as in Example 1. Again Theorem 3 is not applicable.

Example 3. $w(x) = (1-x)^{1/2}$. From (38), $1 < q < 3$ so that $p > 3/2$ and β must be greater than -1. Thus, the convergence result in this case is about as good as that in Theorem 1 with much more flexibility. If we apply Theorem 3, we get a stronger result that allows β to be greater than $-4/3$.

Example 4. $w(x) = (1-x)^{3/2}$. From (38), $1 < q < 5/2$ so that $p > 5/3$ and β must be greater than $-3/2$. As in Example 2, this result is not as good as that given by Theorem 1. If we apply Theorem 3, our condition on β is weakened to be $\beta > -19/10$ which is still not as good as the condition $\beta \geq -2$ of Theorem 1. Again, Theorems 2 and 3 impose weaker conditions on k.

4. Interior Singularities

As in [4] and [12], we shall assume that f has unbounded singularities at a set, Y, of ℓ points such that $Y \subset (-1,1)$ and we define $f(y_i) = 0$, $i = 1, \ldots, \ell$. These singularities are in addition to possible singularities at the endpoints. If $|f|^p$ is now MI with respect to w and Y for some $p > 1$, then we can state theorems corresponding to Theorems 2 and 3 except that here we must avoid the singularities, in contrast to the case in those theorems where we ignored the singularities. It is sufficient to state the theorems since the proofs are similar to those of Theorems 2 and 3.

Theorem 4. Let $w \in GSJ$ and $K \in L_{q,w}(J)$ where q is such that (33) is satisfied. Assume that $|f|^p$ is MI with respect to w and Y where $p^{-1} + q^{-1} = 1$. Then

$$\widehat{I}_n(f; k) := \sum_{i \in \tau(n,f)} w_{in}(k) f(\overline{x}_{in}) \to I(kf) \quad \text{as} \quad n \to \infty. \tag{41}$$

Theorem 5. Let $w \in GSJ$ be such that $v := w/w_{sr} \in GSJ$ and assume that $K \in L_{q,v}(J)$ where q satisfies (35). If $|w_{sr}f|^p$ is MI with respect to v, then

$$I\big(k\widehat{L}_n(w; f)\big) \to I(kf) \quad \text{as} \quad n \to \infty$$

where $\widehat{L}_n(w; f) := \sum_{i \in \tau(n,f)} \ell_{in}(x; w) f\big(x_{in}(w)\big)$.

Now, these theorems are the L_p analogues of similar L_2 theorems given in [12] and while they give some results about interior singularities, they impose global conditions on the functions involved and do not exploit the local properties of the weights $w_{in}(k)$ when k satisfies certain regularity properties in some neighborhoods of the singularities. We shall now deal with this case and as in previous treatments of singularities, e.g. in [8] and [10], we shall assume, for simplicity, that f has a single singularity at a point $\xi \in (-1,1)$, $\xi \neq t_k$, $k = 1, \ldots, \rho$. Further, we assume that $f \in M_d(\xi; k)$ where

$$M_d(\xi; k) := \{g : g \in C(\xi, 1], \exists F \geq 0 \text{ on } J \ni F = 0$$
$$\text{on } [-1, \xi], \ F \text{ is continuous and nonincreasing on } (\xi, 1],$$
$$|f| \leq F \text{ on } J \text{ and } kF \in L_1(J)\}.$$

We now recall the following convergence theorem.

Theorem A [10]. Let $\xi \in (-1, 1)$ and let $Q_n g$ be a PIR of the form

$$Q_n g := \sum_{i=1}^{n} w_{in} g(x_{in}).$$

Define the index κ such that $x_{\kappa-1,n} \leq \xi < x_{\kappa n}$ and the modified integration rule $\widetilde{Q}_n g$ by

$$\widetilde{Q}_n g := Q_n g - w_{\kappa n} g(x_{\kappa n}) \tag{42}$$

so that, $\widetilde{Q}_n g$ avoids the singularity. Let $\xi_0 := \overline{\lim}_{n \to \infty} x_{\kappa n}$. Assume that $Q_n g \to I(kg)$ for all piecewise continuous functions on J. Suppose that there exist a point $A \in (\xi_0, 1]$ and a positive constant C such that

$$|w_{in}| \leq C \int_{x_{i-1,n}}^{x_{in}} |k(x)| dx \tag{43}$$

for all n sufficiently large and for all i such that $x_{\kappa n} < x_{in} \leq A$. Then, for all $f \in M_d(\xi; k)$

$$\widetilde{Q}_n f \to I(kf) \quad \text{as} \quad n \to \infty.$$

Consequently, $Q_n f \to I(kf)$ for all $f \in M_d(\xi; k)$ if and only if $w_{\kappa n} f(x_{\kappa n}) \to 0$ as $n \to \infty$.

Theorem 6. Let $w \in GSJ$ and $\xi \in (-1, 1)$. Assume that $f \in M_d(\xi; k)$, $k(\xi) \neq 0$, k satisfies (26) for some $p > 1$ and k is either continuous and of bounded variation or satisfies a Lipschitz-Dini condition in $N_\delta(\xi) := [\xi, \xi + \delta]$ where δ is such that $N_\delta(\xi) \cap T = \phi$. Then $\widetilde{I}_n(f; k) \to I(kf)$ as $n \to \infty$ where $\widetilde{I}_n(f; k)$ is defined as in Theorem A.

Proof: Since k satisfies (26), $I_n(f; k) \to I(kf)$ for all $f \in R(J)$ and, a fortiori, for all piecewise continuous function. Hence, we need only show that $w_{in}(k) = O(n^{-1})$. For, then by (11) and by continuity of k in $N_\delta(\xi)$, (43) would hold. Since, from (23),

$$w_{in}(k) = \overline{\mu}_{in} S_{n+r+s-1}^K(w; \overline{x}_{in}) + \eta_{in}$$

and $\overline{\mu}_{in} \sim n^{-1}$ from (12), we need only show that $S_{n+r+s-1}^k(w; \overline{x}_{in})$ is uniformly bounded and that $\eta_{in} = O(n^{-1})$ for all x_{in} in $N_\delta(\xi)$. Now, the hypotheses of the theorem insure that the Lebesgue function of $\{\varphi_n(x; w)\}$ with respect to orthogonal expansions is $O(\log n)$ so that by classical results, $S_m^K(w; \overline{x}_{in})$ is uniformly bounded in $N_\delta(\xi)$ [6]. As for η_{in}, from (35) and (36),

$$\eta_{in} = O((K, p_{n+1}(w))\overline{\mu}_{in} p_{n-1}(\overline{x}_{in}; \overline{w})).$$

Since, by (39), $p_{n-1}(\overline{x}_{in}; \overline{w})$ is bounded in $N_\delta(\xi)$ and $(K, p_{n+1}(w))$ is bounded inasmuch as $K \in L_{1,w}$, it follows from (12) that $\eta_{in} = O(n^{-1})$ proving the theorem.

If w is a Jacobi weight, $w = w_{\alpha\beta}$, then we can do better and show that in certain cases, we can ignore the singularity. These results are similar to those in [8] and [15] for Gaussian integration and the reader is referred to these papers for the details.

5. Concluding Remarks

In a previous paper [12], the authors studied the L_2 theory of product integration in the presence of singularities. This theory is applicable for a more general family of weight functions than the L_p theory studied in this paper. Thus, in order to justify the present work, we should show that the convergence results achieved are stronger than those in [12]. If we consider several concrete cases, we find that this is indeed true but unfortunately, the results are not that much better.

We illustrate this in the case when $w(x) := (1 - x^2)^\alpha$, $f(x) := (1 - x^2)^\beta$ and $k(x) := (1 - x^2)^\gamma$ with $\beta + \gamma > -1$. In the L_2 theory, the result corresponding to Theorem 2 here requires that $f \in L_{2,w}$ and $K \in L_{2,w}$ which, in the case we are studying requires that $\beta > -(\alpha + 1)/2$ and $\gamma > (\alpha - 1)/2$. In the corresponding L_p case, we require that $f \in L_{p,w}$ and $K \in L_{q,w}$ $p^{-1} + q^{-1} = 1$, where $q \in (-1, \infty)$ when $\alpha \le -1/2$ and

$$2 - 2/(2\alpha + 3) < q < 2 + 2/(2\alpha + 1)$$

when $\alpha > -1/2$. The conditions on β and γ then become $\beta > -(2\alpha + 3)/4$ and $\gamma > \alpha/2 - 1/4$. We see that we can deal at best with a singularity stronger by $1/4$ using Theorem 2.

If we wish to apply Theorem 3 and the corresponding theorem in [12], then α must be restricted to be > 0. In the L_2 theory, the conditions on β and γ become $\beta > -(\alpha + 2)/2$ and $\gamma > \alpha/2$ while in the L_p theory, we get an improvement,

$\beta > -(\alpha+2)/2 - \alpha/(4\alpha+4)$ and $\gamma > \alpha/2 + \alpha/(4\alpha+4)$. Again, the improvement is not that significant.

The improvement in the L_p case is diminished further when we consider the more general weight function $(1-x)^\alpha(1+x)^{\alpha+\delta}$. In the L_2 case, the change in the exponent of $(1+x)$ only restricts the behavior of f near -1 whereas in the L_p case, this change restricts q as is shown in (38). Hence the best value of p is affected and, consequently, the behavior of f not only near -1 but also near 1. For example, if $\alpha = -1/2$ and $\delta = 1/2$, then q must satisfy $q < 4$ which implies the restriction $p > 4/3$. If we now consider $f(x) := (1-x)^\beta(1+x)^{\beta+\epsilon}$, we find that using the L_2 theory, $\beta > -1/4$ as before but $\epsilon > -1/4$ so that the condition on $(1+x)$ is completely independent of the condition on $(1-x)$. In the L_p case, we find that β must satisfy $\beta > -3/8$ instead of $\beta > -1/2$ and the improvement over the L_2 case drops from 1/4 to 1/8. However, the condition on ϵ is $\epsilon > -3/8$ so that the improvement in the exponent of $(1+x)$ is still 1/4. And if we consider weight functions with singularities in $(-1,1)$, namely when $\rho > 0$ and some $\gamma_\kappa \gg 0$, the situation can get worse since from (38), the restrictions on q can limit the value of p. But in any case, since the upper limit on q is greater than 2, the best value of p is less than 2 and the L_p result is always an improvement over the L_2 result, which is some consolation for the effort involved in deriving the L_p theory.

Acknowledgement

The first author wishes to thank Dr. E.L. Ortiz, Department of Mathematics, Imperial College, London and Prof. U. Ascher, Department of Computer Science, University of British Columbia, Vancouver for their hospitability during the period when this paper was written.

References

1. V. Badkov, Convergence in the mean and almost everywhere of Fourier series in polynomials orthogonal on an interval, *Math. USSR* – Sb. **24** (1974) 223–256.

2. P.J. Davis and P. Rabinowitz, *Methods of Numerical Integration*, Second Edition, Academic Press, New York, 1984.

3. D. Elliott and D.F. Paget, Product integration rules and their convergence, *BIT* **16** (1976) 32–40.

4. D.S. Lubinsky and A. Sidi, Convergence of product integration rules for func-

tions with interior and endpoint singularities over bounded and unbounded intervals, *Math. Comp.* **46** (1986) 229–245.

5. P. Nevai, Mean convergence of Lagrange interpolation. III, *Trans. Amer. Math. Soc.* **282** (1984) 669–698.

6. P. Nevai, Private communication.

7. P. Rabinowitz, The convergence of interpolatory product integration rules, *BIT* **26** (1986) 131–134.

8. P. Rabinowitz, Numerical integration in the presence of an interior singularity, *J. Comp. Appl. Math.* **17** (1987) 31–41.

9. P. Rabinowitz, Numerical evaluation of Cauchy principal value integrals with singular integrands, *Math. Comp.* **55** (1990) 265–276.

10. P. Rabinowitz and I.H. Sloan, Product integration in the presence of a singularity, *SIAM J. Numer. Anal.* **21** (1984) 149–166.

11. P. Rabinowitz and W.E. Smith, Interpolatory product integration for Riemann-integrable functions, *J. Austral. Math. Soc.* Ser. B **29** (1987) 195–202.

12. P. Rabinowitz and W.E. Smith, Interpolatory product integration in the presence of singularities: L_2 theory, to appear in *J. Comp. Appl. Math.* (1992).

13. I.H. Sloan and W.E. Smith, Properties of interpolatory product integration rules, *SIAM J. Numer. Anal.* **19** (1982) 427–442.

14. W.E. Smith and I.H. Sloan, Product-integration rules based on the zeros of Jacobi polynomials, *SIAM J. Numer. Anal.* **17** (1980) 1–13.

15. P. Vértesi, Remarks on convergence of Gaussian quadrature for singular integrals, *Acta Math. Hung.* **53** (1989) 399–405.

THE NUMERICAL EVALUATION OF DEFINITE INTEGRALS AFFECTED BY SINGULARITIES NEAR THE INTERVAL OF INTEGRATION

D. B. HUNTER
Department of Mathematics
University of Bradford
Bradford
Yorkshire
England

ABSTRACT. A method is described for the numerical evaluation of definite integrals over the complete real line where the integrand, as a function of a complex variable, has singularities near the real axis. The main method is concerned with the case when the singularities are poles of arbitrary order, but an extension to deal with branch singularities is briefly considered. A modification for the evaluation of Cauchy principal value integrals is also described.

1. Introduction.

It is well-known that most of the standard methods of evaluating numerically a definite integral

$$\int_a^b f(x)dx \qquad (1)$$

where f is a function of a complex variable, and a and b are real numbers with a<b, are adversely affected by the presence of singularities of f near the interval [a, b]. A number of ways of dealing with this problem have been proposed; see, e.g., Chiarella

111

T. O. Espelid and A. Genz (eds.), Numerical Integration, 111–120.
© 1992 *Kluwer Academic Publishers.*

and Reichel [2], Matta and Reichel [10], Hunter and Regan [7], Lether [8,9], Monegato [11].   In many of these papers the singularities are simple poles.   An extension to poles of arbitrary order was mentioned briefly in Hunter [4], but it is only recently that the problem has been considered in detail - see Bialecki [1], Hunter and Okecha [6].

The integral we shall consider here is

$$I(f) = \int_{-\infty}^{\infty} f(x)dx \qquad\qquad (2)$$

The more general form (1), with at least one of a and b finite, can be dealt with by transforming to the form (2).   We shall not consider this aspect of the problem here, but a thorough investigation is given in Bialecki [1].

We shall consider first the case in which f is analytic within some region containing the real axis, except at a finite number of poles of arbitrary order which may be near but not on the real axis.   Later, the method will be extended to allow some simple poles to lie on the real axis, leading to Cauchy principal value integrals.   Finally, some tentative suggestions will be made for the case of branch singularities near the real axis.

To simplify the analysis, we shall assume that f is a real function, i.e., that $f(\bar{z})=\overline{f(z)}$ for all complex z, though this restriction is not essential.

2.   The method

Our basic method is very similar to that of Bialecki [1], but there are a number of differences of detail.   Suppose f is a real function of a complex variable with the following properties:
(i) f is integrable along the entire real axis;
(ii) there is a real number d>0 such that f is analytic over the region D_d of the complex plane C defined by
$$D_d= \{z \in C: |Im(z)| \le d\},$$
except at a finite number of poles of arbitrary order within D_d;
(iii) as R→±∞, f(R+iy)→0 uniformly for −d≤y≤d, where d is as in

(ii).

The technique we shall use is based on a modification of the following "shifted" form of the trapezoidal rule:

$$T(\lambda,h) = h\sum_{k=-\infty}^{\infty} f((k+\lambda)h) \qquad (3)$$

where $h>0$ is a suitable interval and λ is a "shift" parameter satisfying $0\le\lambda<1$. It is based on the following result.

Theorem. Suppose f is a real function of a complex variable satisfying conditions (i), (ii) and (iii) above. Suppose further that within D_d f has poles z_1, z_2, ... , none of which lie on the real axis, where z_r is of order n_r, $(r=1,2,$ ... $)$, and that the principal part of the Laurent expansion of f about z_r is

$$\sum_{k=1}^{n_r} \rho_{rk}(z-z_r)^{-k}$$

Then

$$I(f) = T(\lambda,h) + R(\lambda,h,d) + E(\lambda,h,d) \qquad (4)$$

where

$$R(\lambda,h,d) = 4\pi \text{Im} \sum_r^+ \frac{\rho_{r1}}{\exp[2\pi i(z_r/h-\lambda)]-1}$$

$$+2\pi Rl \sum_r^+ \sum_{k=1}^{n_r} \rho_{rk}(\pi/h)^{k-1} G_k(\cot\pi(z_r/h-\lambda)) \qquad (5)$$

and

$$E(\lambda,h,d) = -2Rl \int_{-\infty}^{\infty} \frac{f(x+id)dx}{\exp[2\pi d/h-2\pi i(x/h-\lambda)]-1} \qquad (6)$$

Here, $\sum_r^+$ denotes summation over those poles z_r within D_d which lie above the real axis, and the functions G_k are defined by the equation

$$G_k(\cot\alpha) = \frac{d^{k-1}}{d\alpha^{k-1}} \cot\alpha \,/(k-1)! \qquad (7)$$

Proof.   The theorem can be proved by integrating $f(z)\cot\pi(z/h-\lambda)$
round a rectangular contour with vertices $\pm R \pm id$, and then letting
$R\to\infty$.   For the details see, e.g., McNamee [12], Bialecki [1],
Hunter [5].

   The above theorem suggests that we might evaluate $I(f)$ by
applying $R(\lambda,h,d)$ as a correction to the trapezoidal value $T(\lambda,h)$,
to obtain a modified trapezoidal rule

$$T^*(\lambda,h,d) = T(\lambda,h) + R(\lambda,d,h). \qquad (8)$$

The error in this approximation is thus $E(\lambda,h,d)$.

   There are three problems which must be considered here:
(i) the choice of d;
(ii) the choice of λ;
(iii) the evaluation of the functions $G_k(z)$.

We shall consider these in turn, and illustrate them by an
example.

   (i) The choice of d.   This is, at first sight, quite
straightforward; we choose a value large enough to ensure that all
the poles of f which are near enough to the real axis to have a
significant effect on $T(\lambda,h)$ are contained in D_d.   As d
increases, the values of $R(\lambda,h,d)$ and $E(\lambda,h,d)$ change only when
the horizontal edges of D_d pass through poles.   Thus the precise
value of d is not very important.   However, it follows from eqn.
(6) that

$$|E(\lambda,h,d)| \leq \frac{2}{\exp(2\pi d/h)-1} \int_{-\infty}^{\infty} |f(x+id)|\,dx \qquad (9)$$

This suggests that, to obtain an error bound, we should try to choose d so as to minimize the right-hand side of this inequality.

(ii) The choice of λ.   The parameter λ can be chosen so as to control the effect of rounding errors in the calculation. Suppose f has a pair of conjugate complex poles z_1 and $z_2 = \bar{z}_1$ close to the real axis.   Then intuitively we would expect that if $Rl(z_1)$ is close to one of the nodes $(k+\lambda)h$ then $T(\lambda,h)$ will tend to overestimate $I(f)$, so that $R(\lambda,h,d)$ will be opposite in sign to $T(\lambda,h)$, leading to a possible loss of significant digits.   This suggests that it may be better to choose λ so that $Rl(z_1)$ is, if possible, midway between two consecutive nodes, in the hope that then $T(\lambda,h)$ will underestimate $I(f)$, and so $R(\lambda,h,d)$ will have the same sign as $T(\lambda,h)$.   If there are several pairs of conjugate complex poles, it may be difficult to choose λ to suit all of them.   In particular, if f is an even function, the values $\lambda = 0$ or 0.5 should normally be used, to ensure that the nodes $(k+1)h$ are symmetrically distributed about 0.   It may then be difficult to ensure that $Rl(z_1)$ lies midway between consecutive nodes.

(iii) The evaluation of the functions $G_k(\cot\alpha)$.
    From (7),
$$G_1(\cot\alpha) = \cot\alpha,$$
$$G_2(\cot\alpha) = -\csc^2\alpha.$$
Expressions for the other functions $G_k(\cot\alpha)$ can then be obtained from eqn.(7).   Alternatively, we can use the recurrence formulae

$$(2m-1)G_{2m}(\cot\alpha) = -[G_m(\cot\alpha)]^2 - 2\sum_{j=1}^{m-1} G_j(\cot\alpha)G_{2m-j}(\cot\alpha), \quad (m\geq 2)$$

(10)

$$2mG_{2m+1}(\cot\alpha) = -2\sum_{j=1}^{m} G_j(\cot\alpha)G_{2m-j+1}(\cot\alpha), \quad (m\geq 1).$$

Example. $f(z) = \exp(-z^2)/(z^2+a^2)^3$, (a real and positive).
    Here there are two conjugate poles of order 3 at $z_1 = ia$ and $z_2 = -ia$.   The coefficients in the principal part of the

Laurent expansion about z_1 are

$$\rho_{11} = -i\exp(a^2)(3-4a^2+4a^4)/16a^5,$$
$$\rho_{12} = -\exp(a^2)(3-4a^2)/16a^4,$$
$$\rho_{13} = i\exp(a^2)/8a^3.$$

From (9), the error term $E(\lambda,h,d)$ satisfies

$$|E(\lambda,h,d)| \le \frac{2}{\exp(2\pi d/h)-1} \int_{-\infty}^{\infty} \left| \frac{e^{-(x+id)^2}}{\left[(x+id)^2 + a^2\right]^3} \right| dx$$

Provided $d>a$, this leads to

$$|E(\lambda,h,d)| \le 2\sqrt{\pi} e^{d^2}/(e^{2\pi d/h}-1)(d^2-a^2)^3.$$

The minimum value of the expression on the right occurs when d is close to the value π/h; this leads to

$$|E(\lambda,h,d)| \le \sqrt{\pi}\ \text{cosech}(\pi^2/h^2).(\pi^2/h^2-a^2)^3,$$

provided $a<\pi/h$.

As the integrand is an even function with its two poles on the imaginary axis, the criterion proposed in (ii) above suggests that we should choose $\lambda = 0.5$.   This choice is borne out by the following results for the case $a = 0.01$. Choosing, e.g., $h = 0.5$, we get $|E(1,0.5,2\pi)| < 4.1235\times10^{-22}$.

The following results were obtained:
When $\lambda = 0$, $T(0,0.5) = 0.5\times10^{12}$,

$$R(0,0.5,d) = -0.48821942\times10^{12}, \text{ giving}$$
$$T^* (0,0.5,d) = 1.17805786\times10^{10},$$

with an appreciable loss of significant digits.

On the other hand, when $\lambda = 0.5$, we get
$$T(0.5,0.5) = 3832.681561,$$

$$R(0.5, 0.5, d) = 1.17805760 \times 10^{10}, \text{ giving}$$

$$T^* (0.5, 0.5, d) = 1.17805798 \times 10^{10}.$$

3.   Application to Cauchy principal value integrals.

The techniques of Section 2 can be adapted to deal with Cauchy principal value integrals.   For the sake of simplicity, assume that the poles z_1, z_2, ... are all simple and situated on the real axis, though it is not difficult to deal with cases where some poles are on the real axis and others off it.   The arguments used in proving the theorem of Section 2 carry through with a few minor modifications.   Equations (4) and (6) still hold, but (5) becomes

$$R(\lambda, h, d) = \pi \sum_r \rho_{r1} \cot\pi(z_r/h - \lambda) \qquad (11)$$

provided none of the values $z_r/h - \lambda$ are integers.

If there is just one pole z_1, λ can be chosen so that $z_1/h - \lambda$ has fractional part 0.5, and consequently $R(\lambda, h, d) = 0$.   In this way we can try to avoid the possibility of a serious loss of accuracy which can occur if $T(\lambda, h)$ and $R(\lambda, h, d)$ are large and of opposite sign.   If there are several poles, however, it will usually be difficult to choose λ and h to suit all of them.   This will happen, in particular, if f is an even function, when λ will normally be assigned the value 0 or 0.5.

Example. $f(z) = \text{sech} z/(z-a)$,   (a real).   Here the integrand has a simple pole at $z_1 = a$, with residue $\rho_{11} = \text{sech} a$.   So

$$R(\lambda, h, d) = \pi \text{sech} a \cot\pi(a/h - \lambda).$$

If we set $\lambda = a/h - [a/h] \pm 0.5$, where the sign of the last term is chosen so as to ensure that $0 \leq \lambda < 1$, then $R(\lambda, h, d) = 0$, so that $T^*(\lambda, h, d) = T(\lambda, h)$.

For example, when $a = 0.1$ and $h = 0.5$, we choose $\lambda = 0.7$. Then $T^*(0.7, 0.5, d) = T(0.7, 0.5) = -0.2322322$.

The error $E(\lambda, h, d)$ satisfies

$$|E(\lambda,h,d)| \le 2\pi|secd|/d[exp(2\pi d/h)-1]$$

where, due to the poles at $\pm i\pi/2$, $0<d<\pi/2$. When $h = 0.5$, the minimum value of the expression on the right occurs around the value $d = 1.5$, giving $|E(\lambda,h,d)| \le 3.85640\times10^{-7}$. The accuracy here can be improved by taking account of some of the poles on the imaginary axis, at the points $z = (r-0.5)i\pi$, with residues $\rho_r = (-1)^{r-1}i/[a-(r-0.5)i\pi]$. In fact, if we set $d = n\pi$, and then let $n\to\infty$, it is easy to see that $E(\lambda,h,d)\to0$, whence

$$I(a) = T^*(\lambda,h) + S(\lambda,h)$$

where

$$S(\lambda,h) = 4\pi\sum_{r=1}^{\infty}Rl(-1)^{r-1}/[a-(r-.5)i\pi][exp((2r-1)\pi^2/h+2\lambda i\pi)-1]$$

The sum on the right converges very rapidly. For instance, when $a = 0.1, h = 0.5$ and $\lambda = 0.7$, its first two terms are -2.0691979×10^{-8} and 4.8873714×10^{-26}.

4.   Branch   singularities.
In this section, a few tentative suggestions are made on the extension of the techniques of Section 2 to deal with branch singularities. We shall deal with one special case only, when

$$f(z) = (z^2+a^2)^\alpha g(z),$$

where $a>0$ and α are real, $\alpha>-1$, α is not an integer, and g is analytic within a strip containing the real axis.

We proceed, as in the proof of the theorem of Section 2, by integrating $\pi(z^2+a^2)^\alpha g(z)cot\pi(z/h-\lambda)$ round a suitable contour, which, like the one used earlier, is rectangular with vertices $\pm R\pm id$, but, if $d>a$, slits must be made in it to avoid the branch points at $\pm ia$. Equation (4) still holds, but now the correction term $R(\lambda,h,d)$ is given by

$$R(\lambda,h,d) = 2\sin(\alpha\pi) \int_a^d (t^2-a^2)^\alpha \{g(it)/[\exp(2\pi(t/h+i\lambda))-1]$$
$$+g(-it)/[\exp(2\pi(t/h-i\lambda))-1]\}dt \quad (12)$$

The problem now becomes that of evaluating $R(\lambda,h,d)$. One possible way is to apply the substitution

$$t = (a+de^{2u})/(1+e^{2u}) \quad (13)$$

which transforms the interval $[a,d]$ to $(-\infty,\infty)$, and then use the trapezoidal rule; but more work remains to be done.

Example. $I(a) = \int_{-\infty}^\infty e^{-x^2}(x^2+a^2)^{-1/2}dx = \exp(a^2/2)K_0(a^2/2)$
where K_0 is a modified Bessel function of the second kind, (see Hunter [3]).   When a is small the branch points at $\pm ia$ cause problems.   For example, when a = 0.1, we get $T(1/2,1/2) = 4.43878372$.   On applying the trapezoidal rule with interval k = 0.25 to (12), after transforming by (13), we get $R(1/2,1/2,2\pi) = 1.00264448$.   These give $T^*(1/2,1/2,2\pi) = 5.44142820$, which is correct to 8 decimal places.

   The procedure has a number of snags.   The most obvious is the restriction $\alpha>-1$.   Also, the evaluation of $R(\lambda,h,d)$ may involve a large number of function-evaluations.   For instance, in the above example, 85 function-evaluations were required to obtain $R(1/2,1/2,2\pi)$.   On the other hand, to obtain comparable accuracy by using $T(1/2,h)$ alone, without $R(1/2,h,d)$, a value h<0.02 is needed, requiring over 200 function-evaluations.   Further work is in progress.

120

References.

[1] Bialecki, B. (1989) "A modified Sinc quadrature rule for functions with poles near the arc of integration", BIT 29, 464-476.

[2] Chiarella, C. and Reichel, A. (1968) "On the evaluation of integrals related to the error function", Math. Comp., 22, 137-143.

[3] Hunter, D.B. (1964) "The calculation of certain Bessel functions", Math. Comp., 18, 123-128.

[4] Hunter, D.B. (1971) "The evaluation of integrals of periodic analytic functions", BIT, 11, 175-180.

[5] Hunter, D.B.(1990) "The numerical evaluation of definite integrals affected by singularities near the interval of integration", University of Bradford, Report No. NA. 90-19.

[6] Hunter, D.B. and Okecha, G.E. (1986) "A modified Gaussian quadrature rule for integrals involving poles of any order", BIT, 26, 233-240.

[7] Hunter, D.B. and Regan, T (1972) "A note on the evaluation of the complementary error function", Math. Comp., 26, 539-541.

[8] Lether, F.G. (1977) "Modified quadrature formulas for functions with nearby poles", J. Comput. Appl. Math, 3, 3-6.

[9] Lether, F.G. (1977) "Subtracting out complex singularities in numerical integration", Math. Comp., 31, 223-229.

[10] Matta, F and Reichel, A (1971) "Uniform computation of the error function and other related functions", Math. Comp.,25, 339-344.

[11] Monegato, G. (1986) "Quadrature formulas for functions with poles near the interval of integration", Math. Comp., 47, 301-312.

[12]McNamee, J. (1964) "Error bounds for the evaluation of integrals by the Euler-Maclaurin formula and by Gauss-type formulae", Math. Comp., 18, 368-381.

APPLICATION OF COMPUTER ALGEBRA SOFTWARE TO THE DERIVATION OF NUMERICAL INTEGRATION RULES FOR SINGULAR AND HYPERSINGULAR INTEGRALS

NIKOLAOS I. IOAKIMIDIS
Division of Applied Mathematics and Mechanics
School of Engineering, University of Patras
P. O. Box 1120, GR-261.10 Patras, Greece

ABSTRACT. Computer algebra software has been used for the construction of numerical integration rules for Cauchy-type singular (principal value) integrals and the corresponding hypersingular (finite-part) integrals (like Mangler-type integrals). The modern and powerful computer algebra system MATHEMATICA was employed in this task. A case of the Gauss–Chebyshev quadrature rule as well as the Gauss–Legendre quadrature rule are used for the illustration of the whole approach both for singular and for hypersingular integrals and the corresponding elementary MATHEMATICA procedures are given together with some related results. The case when the singular point coincides with a node was also considered in some detail and related sample results are presented. Finally, the question of the closed-form evaluation of the corresponding fundamental principal value integral is also studied using MATHEMATICA. The obtained results were verified with the help of the already available formulae. The present results can be used in a variety of scientific, engineering and other applications or even in the case of ordinary quadrature rules.

1 Introduction

Cauchy-type principal value singular integrals and the related hypersingular integrals of the forms

$$I(y) = \int_a^b w(x)\,\frac{g(x)}{x-y}\,dx, \quad a < y < b, \tag{1.1}$$

and

$$J_p(y) = \int_a^b w(x)\,\frac{g(x)}{(x-y)^p}\,dx, \quad a < y < b, \quad p = 1,2,\dots \quad (\text{with } J_1(y) \equiv I(y)) \tag{1.2}$$

play a very important rôle in a variety of problems in science and engineering including the classical problems of fluid dynamics (e.g., airfoils) and elasticity (e.g., cracks) (see, e.g., Ashley and Landhahl (1965) and Ioakimidis (1976), respectively). Numerical integration rules and, especially, Gaussian rules (Davis and Rabinowitz, 1984) are very useful in practical situations. Related references include the results by Hunter (1972), Ioakimidis

T. O. Espelid and A. Genz (eds.), Numerical Integration, 121–131.
© *1992 Kluwer Academic Publishers.*

(1976, 1981, 1985), Ioakimidis and Theocaris (1977), Elliott and Paget (1979), Tsamasphyros and Theocaris (1983) (the latter with an obvious error in the convergence results for the Gauss rule), Rabinowitz (1984) and Lu (1984). An excellent review of the related applications to the corresponding singular integral equations was recently published by Golberg (1990).

On the other hand, computer algebra systems (or, equivalently, systems for symbolic manipulations and computations), having appeared at first in 1965–68, have reached a high efficiency and popularity as is the case with the computers (including personal computers and workstations) employed. The most efficient of these systems seem to be MAPLE (Char *et al.*, 1988) and MATHEMATICA (Wolfram, 1991), having originally appeared in 1980 (with recent version 5) and 1988 (with recent version 2), respectively. Although both of these systems (and other systems) were available to us and could be used in the present application, we preferred MATHEMATICA since it is the most modern, popular and powerful computer algebra system, but also because there is a large related literature (see, e.g., Maeder (1990), Crandall (1991), Gray and Glynn (1991)).

After describing the approach in Section 2, we will present concrete procedures and results for the Gauss–Chebyshev (in a special case) and the Gauss–Legendre quadrature rules in Sections 3 and 4, respectively, and will consider the problems of coincidence of the singular point y in (1.1) and (1.2) with a node and of the closed-form evaluation of the fundamental related principal value integral $q_0(y)$, where

$$q_n(y) = \int_a^b w(x) \frac{p_n(x)}{x - y} \, dx, \quad p_0(x) \equiv 1, \quad a < y < b, \quad n = 0, 1, \ldots, \tag{1.3}$$

(with $p_n(x)$ the polynomial with roots the nodes used) in Sections 5 and 6, respectively. Finally, our conclusions and related discussion will be included in Section 7.

2 The Approach

For the Cauchy-type principal value integral $I(y)$ in (1.1) we will use the standard method of extracting the singularity at $x = y$, rewriting it (with the help of (1.3)) as

$$I(y) = \int_a^b w(x) \frac{g(x) - g(y)}{x - y} \, dx + q_0(y)g(y), \tag{2.1}$$

where the main integral is a regular one under the assumption that $g(x)$ possesses a first derivative. Next, for the evaluation of the hypersingular integral $J_p(y)$, we will use the property that it can be written as an appropriate derivative of $I(y)$ (see, e.g., Ioakimidis (1981)), that is,

$$J_p(y) = \frac{1}{(p-1)!} \frac{d^{p-1}}{dy^{p-1}} I(y), \quad p = 1, 2, \ldots. \tag{2.2}$$

Now an appropriate (possibly interpolatory or Gaussian) rule of the general form

$$\int_a^b w(x) f(x) \, dx = \sum_{k=1}^n A_{k,n} f(x_{k,n}) + E_n, \tag{2.3}$$

where, as usual, $x_{k,n}$ denote the nodes, $A_{k,n}$ the corresponding weights and E_n the error term, can be directly used in (2.1) and, further, (2.2) with the help of $q_0(y)$ (assumed known in advance). We will proceed to two special cases in the following two sections.

3 A Gauss–Chebyshev Rule

We consider at first a special case of the classical Gauss–Chebyshev rule:

$$\int_{-1}^{1} \sqrt{1-x^2}\, f(x)\, dx = \sum_{k=1}^{n} A_{k,n} f(x_{k,n}) + E_n, \quad w(x) = \sqrt{1-x^2}, \qquad (3.1)$$

where $x_{k,n}$ and $A_{k,n}$ are given in closed form by (see, e.g, Lu (1984))

$$x_{k,n} = \cos\frac{(2k-1)\pi}{2n}, \quad A_{k,n} = \frac{(1-x_{k,n}^2)\pi}{n}, \quad k = 1, 2, \ldots, n, \quad q_0(y) = -\pi y. \qquad (3.2)$$

This case is of great interest in elasticity, e.g., in contact problems for $I(y) \equiv J_1(y)$ and in crack problems for $J_2(y)$. We display below the corresponding MATHEMATICA procedure, the GC procedure, based on the above formulae; a great part of this procedure is devoted to the simplification of the obtained results as can easily be observed.

```
GC[n_Integer?Positive,p_Integer?Positive]:=Block[{},
        (* nodes and weights/π *)
Table[{x[k,n]=Chop[N[Cos[(2*k-1)*Pi/(2*n)]]],
    A[k,n]=(1-x[k,n]^2)/n},{k,1,n}];
        (* application of (2.1) for I(y) *)
si=Collect[Sum[A[k,n]*(g[x[k,n]]-g[y])/(x[k,n]-y),{k,1,n}]-
    y*g[y],g[y]];
        (* simplification of the result for I(y) *)
si1:=ReplaceAll[si,g[y]->0]+Chop[ExpandDenominator[
    Together[Coefficient[si,g[y]]]]]*g[y];
        (* differentiation for Jp(y) *)
hsi=D[si1,{y,p-1}];
        (* simplification of the result for Jp(y) *)
t=Table[D[g[y],{y,j}]->0,{j,0,p-1}];
hsi1=hsi/.t;
Do[c[j]=Together[Coefficient[hsi,D[g[y],{y,j}]]],{j,0,p-1}];
        (* final function, Jp(y), output *)
HSI[y_]:=(Pi/(p-1)!)*(hsi1+Sum[c[j]*D[g[y],{y,j}],{j,0,p-1}]);
        (* end of the procedure *)
]
```

We present below the result for $n = 3$ and $p = 1$:

```
In[3]:= HSI[y]
```

124

```
  |           0.0833333 g[-0.866025]     g[0]     0.0833333 g[0.866025]
Out[3]= Pi (------------------------- - ---- + --------------------- +
  |               -0.866025 - y          3 y        0.866025 - y
  |
  |                      2      4
  |        (-0.75 + 3.75 y  - 3 y ) g[y]
  >        ----------------------------)
  |                      3
  |               -2.25 y + 3 y
```

the result for $n = 3$ and $p = 2$:

```
In[5]:= HSI[y]
```

```
  |           0.0833333 g[-0.866025]     g[0]     0.0833333 g[0.866025]
Out[5]= Pi (------------------------- + ---- + --------------------- +
  |                           2           2                         2
  |            (-0.866025 - y)          3 y        (0.866025 - y)
  |
  |                       2      4      6
  |        (-1.6875 - 1.6875 y  + 9 y  - 9 y ) g[y]
  >        ------------------------------------------ +
  |                     2           2 2
  |                    y  (-2.25 + 3 y )
  |
  |                      2      4
  |        (-0.75 + 3.75 y  - 3 y ) g'[y]
  >        -------------------------------)
  |                      3
  |               -2.25 y + 3 y
```

and, finally, the result for $n = 3$ and $p = 3$:

```
In[7]:= HSI[y]
```

```
  |           0.166667 g[-0.866025]     2 g[0]     0.166667 g[0.866025]
Out[7]= (Pi (----------------------- - ------ + --------------------- +
  |                            3           3                         3
  |            (-0.866025 - y)          3 y         (0.866025 - y)
  |
  |                       2      4       6
  |        (-7.59375 + 30.375 y  - 20.25 y  + 27 y ) g[y]
  >        ------------------------------------------------ +
  |                2           2 2              3
  |               y  (-2.25 + 3 y )  (-2.25 y + 3 y )
  |
  |                      2      4      6
  |        (-3.375 - 3.375 y  + 18 y  - 18 y ) g'[y]
  >        ------------------------------------------- +
  |                     2           2 2
  |                    y  (-2.25 + 3 y )
  |
  |                      2      4
  |        (-0.75 + 3.75 y  - 3 y ) g''[y]
  >        -------------------------------)) / 2
  |                      3
  |               -2.25 y + 3 y
```

These results show clearly the efficiency of the present approach during the construction of quadrature rules for Cauchy-type singular integrals and the related hypersingular integrals of any order. TEX output of these results is also possible.

4 The Gauss–Legendre Rule

This is the most classical quadrature rule of the form

$$\int_{-1}^{1} f(x)\,dx = \sum_{k=1}^{n} A_{k,n} f(x_{k,n}) + E_n, \quad w(x) = 1, \tag{4.1}$$

where the nodes $x_{k,n}$ are the roots of the Legendre polynomial $P_n(x)$ and the weights $A_{k,n}$ and the integral $q_0(y)$ are given by the classical formulae

$$A_{k,n} = \frac{2(1 - x_{k,n}^2)}{(n+1)^2 P_{n+1}^2(x_{k,n})}, \quad k = 1, 2, \ldots, n, \quad q_0(y) = \log\left(\frac{1-y}{1+y}\right). \tag{4.2}$$

The Gauss–Legendre rule is the most useful quadrature rule with a wide range of applications (e.g., in boundary element techniques it can be used along each separate element).

The corresponding MATHEMATICA procedure for the Gauss–Legendre rule, called GL, is given below:

```
GL[n_Integer?Positive,p_Integer?Positive]:=Block[{},
r=N[Roots[LegendreP[n,y]==0,y]];
q0[y_]:=Log[(1-y)/(1+y)];
Table[x[k,n]=Last[r[[k]]], A[k,n]=2*(1-x[k,n]^2)/
    ((n+1)*LegendreP[n+1,x[k,n]])^2,{k,1,n}];
si=Collect[Sum[A[k,n]*(g[x[k,n]]-g[y])/(x[k,n]-y),{k,1,n}]+
    q0[y]*g[y],g[y]];
si1=ReplaceAll[si,g[y]->0]+Chop[ExpandDenominator[
    Together[Coefficient[si,g[y]]]]]*g[y];
hsi=D[si1,{y,p-1}];
t=Table[D[g[y],{y,j}]->0,{j,0,p-1}];
hsi1=hsi/.t;
Do[c[j]=Chop[Together[Coefficient[hsi,D[g[y],{y,j}]]]],{j,0,p-1}];
Do[c[j]=Collect[Numerator[c[j]],q0[y]]/Denominator[c[j]],{j,0,p-1}];
HSI[y_]:=(1/(p-1)!)*(hsi1+Sum[c[j]*D[g[y],y,j],{j,0,p-1}]);
]
```

We observe the obvious similarity of the above GL procedure to the GC procedure of the previous section, but now further commands for the simplification of the result, HSI[y], have been added. We observe also that we have determined the nodes $x_{k,n}$ and the weights $A_{k,n}$ by appropriate very simple commands in MATHEMATICA and not through the available tables or another (external) computer program. As an example of our results we display below the results for $n = 4$ and $p = 1$, where:

```
In[3]:= HSI[y]

|        0.347855 g[-0.861136]     0.652145 g[-0.339981]
Out[3]= -------------------------  + -------------------------  +
|            -0.861136 - y              -0.339981 - y
```

```
|   0.652145 g[0.339981]    0.347855 g[0.861136]
>   --------------------- + --------------------- +
|      0.339981 - y             0.861136 - y
|
|                3                          2    4      1 - y
|   (-1.04762 y + 2 y  + (0.0857143 - 0.857143 y  + y ) Log[-----]) g[y]
|                                                           1 + y
>   -----------------------------------------------------------------
|
|                                         2    4
|                     0.0857143 - 0.857143 y  + y
```

as well as the result for $n = 4$ and $p = 2$:

```
In[5]:= HSI[y]
|         0.347855 g[-0.861136]    0.652145 g[-0.339981]
Out[5]= --------------------- + --------------------- +
|                        2                         2
|           (-0.861136 - y)           (-0.339981 - y)
|   0.652145 g[0.339981]    0.347855 g[0.861136]
>   --------------------- + --------------------- -
|                      2                       2
|      (0.339981 - y)          (0.861136 - y)
|                        0.10449 g[y]
>   --------------------------------------------- +
|                                      2     4 2
|   (1 - y) (1 + y) (0.0857143 - 0.857143 y  + y )
|                3                          2    4      1 - y
|   (-1.04762 y + 2 y  + (0.0857143 - 0.857143 y  + y ) Log[-----]) g'[y]
|                                                           1 + y
>   -----------------------------------------------------------------
|
|                                         2    4
|                     0.0857143 - 0.857143 y  + y
```

We have taken particular care, through appropriate MATHEMATICA commands, so that our results appear in the most simplified form (as has been the case in the previous section).

5   Coincidence of the Singular Point with a Node

This case is also of particular importance, since the derivative of $g(x)$ at this node appears even for $p = 1$. The related analytical results require sufficient effort and are somewhat complicated. For example, in the case of the Gauss–Legendre quadrature rule we have the following formula if $y = x_{m,n}$ $(m = 1, 2, \ldots, n)$ (Ioakimidis and Theocaris, 1977)

$$I(x_{m,n}) = \int_{-1}^{1} \frac{g(x)}{x - x_{m,n}} \, dx = \sum_{\substack{k=1 \\ k \neq m}}^{n} A_{k,n} \frac{g(x_{k,n})}{x_{k,n} - x_{m,n}} + A_{m,n} g'(x_{m,n}) + \Lambda_{m,n} g(x_{m,n}) + E_n \quad (5.1)$$

with the new "weights" $\Lambda_{m,n}$ given by

$$\Lambda_{m,n} = -2 \frac{Q_{n-1}(x_{m,n})}{P_{n-1}(x_{m,n})} - (n+1) A_{m,n} \frac{x_{m,n}}{1 - x_{m,n}^2}, \quad (5.2)$$

where $Q_n(x)$ denotes the classical Legendre function of the second kind. Much more complicated is the analytical formula for hypersingular integrals $(p > 1)$.

It is quite possible to find the form of the Gauss–Legendre quadrature rule for $y = x_{m,n}$ by using a limiting procedure for a particular value of m. In this case HSI[y] should be given directly by appropriate differentiation of si (in the GL procedure of the previous section) without the additional symbolic computations in the GL procedure, which have been used only for the simplification of the final results.

We display below the results obtained for $n = 4$ nodes and $p = 1$ (Cauchy-type singular integrals) in the case of the Gauss–Legendre quadrature rule for all the four nodes $x_{k,4}$. Moreover, we display the analogous results for the first node $x_{1,4}$ in the cases when $p = 2$ and $p = 3$ (hypersingular integrals). In the latter case, we do not have a closed-form formula like (5.2). In all cases, MATHEMATICA had no difficulty at all in finding the limits requested, which include appropriate derivatives of $g(x)$ (of order p). The GLD procedure is completely similar to the GL procedure of the previous section, but without the commands aiming to the simplification of the final results, not required now and dangerous in the sense that MATHEMATICA has difficulties in finding limits for the simplified forms of the results displayed in the previous section.

```
In[9]:=GLD[4,1]

In[10]:= Limit[HSI[y],y->x[1,4]]

Out[10]= -0.201974 g[-0.861136] - 0.542949 g[-0.339981] -
>      1.25135 g[0.339981] - 0.599181 g[0.861136] + 0.347855 g'[0.861136]

In[11]:= Limit[HSI[y],y->x[2,4]]

Out[11]= 0.599181 g[-0.861136] + 1.25135 g[-0.339981] +
>      0.542949 g[0.339981] + 0.201974 g[0.861136] + 0.347855 g'[-0.861136]

In[12]:= Limit[HSI[y],y->x[3,4]]

Out[12]= -0.289609 g[-0.861136] - 0.95909 g[-0.339981] -
>      0.126911 g[0.339981] + 0.667469 g[0.861136] + 0.652145 g'[0.339981]

In[13]:= Limit[HSI[y],y->x[4,4]]

Out[13]= -0.667469 g[-0.861136] + 0.126911 g[-0.339981] +
>      0.95909 g[0.339981] + 0.289609 g[0.861136] + 0.652145 g'[-0.339981]

In[14]:= GLD[4,2]

In[15]:= Limit[HSI[y],y->x[1,4]]

Out[15]= 0.117272 g[-0.861136] + 0.452036 g[-0.339981] +
>      2.4011 g[0.339981] - 10.709 g[0.861136] - 0.599181 g'[0.861136] +
>      0.173927 g''[0.861136]

In[16]:= GLD[4,3]

In[17]:= Limit[HSI[y],y->x[1,4]]

Out[17]= (-0.136183 g[-0.861136] - 0.752693 g[-0.339981] -
```

```
>       9.21452 g[0.339981] - 41.4667 g[0.861136] - 0.599181 g''[0.861136] +
|                                    (3)
>       2 (-10.709 g'[0.861136] -0.173927 g    [0.861136]) +
|              (3)
>       0.463806 g    [0.861136]) / 2
```

The above values of the coefficients $\Lambda_{m,n}$ of $g(x_{m,n})$, for $p = 1$ only, were also checked on the basis of (5.2) (again with MATHEMATICA) and were found to be correct.

6   Closed-Form Computation of the Fundamental Integral

In several cases, the fundamental integral $q_0(y)$, required in (2.1), is well known or easily computed (as is the case, e.g., in the Gauss–Chebyshev and the Gauss–Legendre quadrature rules). In general, it is possible to compute this integral by using the extensive libraries available in computer algebra software. Of course, these libraries do not concern Cauchy-type singular integrals as is the case in (1.3) and are generally unable to compute this class of integrals. Moreover, strangely enough, they insist also to assume the integral of $1/x$ to be equal to $\log x$ and not to $\log |x|$. In any case, we have succeeded in computing in closed form Cauchy-type singular integrals by using MATHEMATICA availing ourselves of the fundamental definition of this class of integrals, that is,

$$\fint_a^b \frac{g(x)}{x-y}\,dx = \lim_{\epsilon \to 0}\left[\int_a^{y-\epsilon}\frac{g(x)}{x-y}\,dx + \int_{y+\epsilon}^b\frac{g(x)}{x-y}\,dx\right], \quad \epsilon > 0, \quad a < y < b. \tag{6.1}$$

We have prepared a MATHEMATICA procedure, called **PVI**, where (6.1) is taken into account together with some rules to be followed in the computations of MATHEMATICA for the logarithmic function and the sign function (since $a < y < b$). By using this procedure, we evaluated in closed form $q_0(y)$ for the Gauss–Legendre rule and we found the result

$$q_0(y) = \fint_{-1}^1 \frac{1}{x-y}\,dx = \log\left(\frac{1-y}{1+y}\right), \quad -1 < y < 1, \tag{6.2}$$

and for several more complicated additional cases of Cauchy-type singular integrals, e.g., $g(x) = x^m$ or $g(x) = e^{-x}$ or even $g(x) = \cos x$ for $[a,b] \equiv [-1,1]$. In the latter two cases, we had, as was expected, the exponential integral $\text{Ei}(x)$ and both the cosine and sine integrals (together with the cosine and sine functions), respectively. The case of Mangler-type principal value hypersingular integrals ($p = 2$) has also been studied by an analogous approach (yet somewhat more complicated than (6.1)) based on their definition (see, e.g., Ashley and Landahl (1965)). We believe that it is possible to extend the integration capabilities of computer algebra software to include Cauchy-type singular integrals and hypersingular integrals in cases where these integrals really exist in closed form and, probably, the direct use of (6.1) may not be the best possibility. (E.g., the Sokhotski–Plemelj classical limiting formulae can also be used instead.)

7 Conclusions—Discussion

From the above results it is clear that modern computer algebra systems (like MATHE-MATICA and MAPLE) can easily be used for the construction of quadrature rules both for Cauchy-type singular (principal value) integrals and for the corresponding hypersingular (finite-part) integrals. The various calculus, simplification, graphics and many more additional commands available in these systems make them ideal in applications like the present one. Moreover, these systems can be used both for the closed-form evaluation of Cauchy-type integrals, Mangler-type integrals, etc. in special cases when this is possible (with the approach of the previous section) as well as for the construction of the corresponding interpolatory (frequently Gaussian) quadrature rules as was already illustrated in sufficient detail. In general, essentially any kind of symbolic or numerical or mixed computation can be performed inside the computer quickly and accurately (since manual, often error prone, computations are avoided) to the required arbitrarily high precision at an extremely low money and time cost. This situation is very interesting for applications to problems in science and engineering and quite helpful for the researcher, whose 'obligation' to look for closed-form formulae for the integrals that he meets or for the appropriate quadrature rules (in the lack of such explicit formulae) is sufficiently facilitated.

This year we have had the opportunity to use and appreciate the power of MATHEMATICA and we really believe that it will become the standard computer algebra system very soon with its strong possibilities and the variety of facilities that it offers. Of course, the other popular computer algebra systems, especially MACSYMA and MAPLE, will continue to be used as will be the case with the less strong (but equally popular) systems DERIVE and REDUCE (requiring only an 8086/88 personal computer).

Nevertheless, in order to avoid misleading impressions, we should mention that the above-generated symbolic results, which are essentially closed-form formulae, may suffer (if used immediately for numerical computations) from numerical instability problems for certain ranges of the parameter y (exactly as would be the case with manual computations). Thus in the Gauss–Chebyshev rule of Section 3 for Cauchy-type principal value integrals there is a region of instability near the origin $y = 0$. On the other hand, a more serious situation arises when we try to generate the formula for the Gauss–Jacobi rule for Cauchy-type principal value integrals (with the classical Jacobi weight function $w(x) = (1 - x)^{\alpha}(1 + x)^{\beta}$, $-1 < x < 1$, $\alpha > -1$, $\beta > -1$, not having been considered above) and related Gaussian rules. In this case, there exist computational problems just with respect to the evaluation of the auxiliary function (Cauchy-type principal value integral) $q_0(y)$ (although the present approach works in principle well even in this case). These problems (for the Jacobi weight function) have been studied in detail (both analytically and numerically) by Gautschi and Wimp (1987) and three explicit approaches are given there for the computation of $q_0(y)$ (and, further, $q_n(y)$). The efficient numerical evaluation of $q_0(y)$ is sufficiently complicated in this case as clearly indicated in the above paper.

On the other hand, we have tested MATHEMATICA with the Jacobi weight function, but it simply failed to evaluate $q_0(y)$ analytically. In the new version of MATHEMATICA (version 2, just released), there is a special external package (**CauchyPrincipalValue.m**) with the related explicit command **CauchyPrincipalValue** for Cauchy-type principal value

130

integrals, but this version (and the related package) are not available to us at this moment. We hope that several problems (although, of course, not all) with Cauchy-type integrals will have been solved with this particular package, which (to the best of our knowledge) is the first one in computer algebra systems. Alternatively, we can rather easily 'teach' MATHEMATICA about special formulae for integrals (like the analytical formula (2.1) for $q_0(y)$ in the paper by Gautschi and Wimp (1987) including the gamma function and a hypergeometric function). Such formulae (from classical tables and/or handbooks of formulae) can be directly added to MATHEMATICA by the user when required; other formulae (simpler or more useful ones) are simply already included in MATHEMATICA. On the other hand, MATHEMATICA can perform numerical computations very efficiently (to the required precision) even if these include special functions (like the aforementioned gamma and hypergeometric functions, which are standard ones in its kernel).

Furthermore, in spite of all these activities in the field of computer algebra and their influence on applications, we are pessimistic about the application of these systems to less applied mathematical computations. In fact, we have been essentially unable to prove theorems of interest in our research (although we are aware of some related efforts) and we feel that by no means can computer algebra systems and related software and computer languages (like LISP and PROLOG) substitute for, or even significantly help, the human mind in nontrivial cases, where we are not restricted to concrete computations well defined in advance by an appropriate procedure. Although there is a reasonable claim that computer algebra systems fall into the software of *artificial intelligence* and, really, some of these are able to 'learn' quite easily complicated formulae and algebraic rules and combine these, nevertheless, they seem not to be particularly useful in the derivation of really *new* concepts and results and this may be one of the reasons that they are still of limited use in research in mathematics.

In our opinion, in the case of *mathematical* research, we may not require the enormous list of commands of MACSYMA, MAPLE or MATHEMATICA, but it is much better to use simpler software like REDUCE or even LISP itself, which may permit us to attempt to imitate the way that we are thinking ourselves. In fact, we have been able to construct the Cauchy-type singular integral equations for crack problems, by using REDUCE, but we are not satisfied by the time devoted to this task, probably much more than that required with manual computations (contrary to the present case of the derivation of quadrature rules). Computer algebra systems tend to become very popular and most students, teachers, engineers and scientists have their favourite system(s) available and several universities are frequently proud of their modern '*Math*' laboratories. Time will show whether there is a chance that these systems become essentially more '*clever*' than they are today.

Acknowledgement—The author gratefully acknowledges several helpful and interesting comments by the referee.

References

Ashley, H. and Landahl, M. (1965) *Aerodynamics of Wings and Bodies*, 1st ed., Addison-Wesley, Reading, MA. [1st Dover ed. (1985): Dover, New York, NY.]

Char, B. W., Geddes, K. O., Gonnet, G. H., Monagan, M. B. and Watt, S. M. (1988) *MAPLE Reference Manual*, 5th ed., Symbolic Computation Group, Department of Computer Science, University of Waterloo, and Waterloo Maple Publishing, Waterloo, Ontario, Canada.

Crandall, R. E. (1991) *Mathematica for the Sciences*, 1st ed., Addison-Wesley, Redwood City, CA.

Davis, P. J. and Rabinowitz, P. (1984) *Methods of Numerical Integration*, 2nd ed., Academic Press, Orlando, FL.

Elliott, D. and Paget, D. F. (1979) 'Gauss type quadrature rules for Cauchy principal value integrals', *Math. Comp.* **33**, 301–309.

Gautschi, W. and Wimp, J. (1987) 'Computing the Hilbert transform of a Jacobi weight function', *BIT* **27**, 203–215.

Golberg, M. A. (1990) 'Introduction to the numerical solution of Cauchy singular integral equations', in M. A. Golberg (ed.), *Numerical Solution of Integral Equations*, 1st ed., Plenum Press, New York, NY, chap. 5, pp. 183–308.

Gray, T. W. and Glynn, J. (1991) *Exploring Mathematics with Mathematica*, 1st ed., Addison-Wesley, Redwood City, CA.

Hunter, D. B. (1972) 'Some Gauss-type formulae for the evaluation of Cauchy principal values of integrals', *Numer. Math.* **19**, 419–424.

Ioakimidis, N. I. (1976) *General Methods for the Solution of Crack Problems in the Theory of Plane Elasticity*, Doctoral Thesis at the National Technical University of Athens, Athens, Greece. [Available from University Microfilms, Ann Arbor, MI; order no. 76-21,056.]

Ioakimidis, N. I. (1981) 'On the numerical evaluation of derivatives of Cauchy principal value integrals', *Computing* **27**, 81–88.

Ioakimidis, N. I. (1985), 'On the uniform convergence of Gaussian quadrature rules for Cauchy principal value integrals and their derivatives', *Math. Comp.* **44**, 191–198.

Ioakimidis, N. I. and Theocaris, P. S. (1977) 'On the numerical evaluation of Cauchy principal value integrals', *Rev. Roumaine Sci. Tech. Sér. Méc. Appl.* **22**, 803–818.

Lu, Ch.-K., (1984) 'A class of quadrature formulas of Chebyshev type for singular integrals', *J. Math. Anal. Appl.* **100**, 416–435.

Maeder, R. E. (1990) *Programming in Mathematica*, 1st ed., Addison-Wesley, Redwood City, CA.

Rabinowitz, P. (1984) 'On the convergence and divergence of Hunter's method for Cauchy principal value integrals', in A. Gerasoulis and R. Vichnevetsky (eds.), *Numerical Solution of Singular Integral Equations*, IMACS, New Brunswick, NJ, pp. 86–88.

Tsamasphyros, G. and Theocaris, P. S. (1983) 'On the convergence of some quadrature rules for Cauchy principal-value and finite-part integrals', *Computing* **31**, 105–114.

Wolfram, S. (1991) *Mathematica: A System for Doing Mathematics by Computer*, 2nd ed., Addison-Wesley, Redwood City, CA.

REMAINDER ESTIMATES FOR ANALYTIC FUNCTIONS[1]

WALTER GAUTSCHI
Department of Computer Sciences
Purdue University
West Lafayette, IN 47907-1398 USA
Email: wxg@cs.purdue.edu

ABSTRACT. A survey is given on contour integration methods for estimating the remainder of quadrature rules involving analytic functions. In addition to historical remarks, recent results are summarized concerning the remainder of Gauss-type formulae.

1. Introduction

Remainder terms for quadrature rules (and other approximation processes) are traditionally expressed in terms of some high-order derivative of the function involved. The problematic nature of the resulting error estimates are too well known to be elaborated upon here. In an influential paper by Davis [12], which was soon followed by work of Davis and Rabinowitz [13] and Yanagiwara [42], derivative-free estimates were developed by considering the remainder term $R(f)$ as a bounded linear functional in some appropriate Hilbert space $\mathbb{H}$ of analytic functions f. Then the obvious inequality

$$|R(f)| \leq \sigma \parallel f \parallel_{\mathbb{H}}, \quad \sigma = \parallel R \parallel, \tag{1.1}$$

can be used to estimate the error. Here, $\parallel R \parallel$ is the norm of the functional R, and $\parallel f \parallel_{\mathbb{H}}$ the norm of f in the Hilbert space $\mathbb{H}$. Important features of the estimate (1.1), besides being derivative-free, are its sharpness (equality is attained for some $f \in \mathbb{H}$), its natural conduciveness to a comparison of different quadrature- (or approximation-) processes on a common basis, and the neat separation it expresses between the influence of the quadrature rule (given by the first factor σ on the right of (1.1)) and the function to which it is applied (given by the second factor). Nevertheless, the bound in (1.1) requires analysis of the function f in the complex plane $\mathbb{C}$, and is thus applicable only in relatively simple cases where, indeed, it is desirable to have a reliable error bound. Furthermore, the evaluation of the norm of R may be costly (depending on the choice of the space $\mathbb{H}$). The idea of functional-analytic estimates of the type (1.1) has been further developed in a series of

---

[1]This work was supported in part by the National Science Foundation under grant DMS - 9023403.

T. O. Espelid and A. Genz (eds.), Numerical Integration, 133–145.
© 1992 *Kluwer Academic Publishers.*

papers by Hämmerlin [24–26], with a view toward deriving upper bounds of $\parallel R \parallel$ for concrete quadrature processes and spaces $\mathbb{H}$. This early work inspired many others to further implement and refine the method. A theme that became popular was the problem of minimizing the norm σ in (1.1) over various classes of quadrature rules. For a list of references, see [14, §4.7], [17, §4.1.2], and for additional work, [1–10, 15, 31–35].

A more elementary, and older, technique of deriving error estimates of a type similar to (1.1) is based on Cauchy's theorem. The earliest relevant reference known to the author, in the case of Gaussian quadrature rules, is a 1932 Russian paper by Fock [16], but the same technique has already been used in 1878 by Hermite [29], and in 1881 by Heine [27, p.16], to derive error estimates for polynomial interpolation; these, when integrated, yield error estimates for interpolatory quadrature rules. This alternative approach enjoys some of the same features mentioned above, and has also led to a large literature, producing strict upper bounds, or asymptotic estimates, for quadrature errors. For an account of these, see [14, §4.6], [17, §4.1.1] and [11, 37–40]. Lower bounds, based on the theory of optimal recovery, have been obtained by Rivlin [36].

In the present paper, we review error estimates obtainable by this second approach – Cauchy's formula – paying special attention to Gauss-type quadrature rules.

2. Contour Integral Representation, and Estimates, for the Remainder Term

We consider only quadrature over a finite interval, taken to be $[-1,1]$, and quadrature rules of the simplest kind,

$$\int_{-1}^{1} f(t)w(t)dt = \sum_{\nu=1}^{n} w_\nu f(t_\nu) + R_n(f), \tag{2.1}$$

involving, however, a weight function w, assumed integrable over $[-1,1]$. The t_ν are certain distinct nodes contained in $[-1,1]$,

$$-1 \leq t_n < t_{n-1} < \cdots < t_1 \leq 1, \tag{2.2}$$

and w_ν appropriate coefficients ("weights").

2.1. ANALYTIC FUNCTIONS

We first assume f to be analytic in a simply connected domain D containing the interval $[-1,1]$,

$$f \in \mathcal{A}(D), \quad [-1,1] \subset D. \tag{2.3}$$

With Γ denoting any contour in D surrounding the interval $[-1,1]$, we then have for any $t \in [-1,1]$, by Cauchy's formula,

$$f(t) = \frac{1}{2\pi i} \oint_\Gamma \frac{f(z)}{z-t} \, dz, \quad t \in [-1,1]. \tag{2.4}$$

Applying the linear functional

$$R_n(f) = \int_{-1}^{1} f(t)w(t)dt - \sum_{\nu=1}^{n} w_\nu f(t_\nu) \tag{2.5}$$

to both sides of (2.4), on the right under the integral sign, yields

$$R_n(f) = \frac{1}{2\pi i} \oint_\Gamma R_n\left(\frac{1}{z-\cdot}\right) f(z)dz,$$

that is,

$$R_n(f) = \frac{1}{2\pi i} \oint_\Gamma K_n(z)f(z)dz. \tag{2.6}$$

The function

$$K_n(z) = R_n\left(\frac{1}{z-\cdot}\right) \tag{2.7}$$

in (2.6) is referred to as the *kernel* of the functional R_n. From (2.6) we obtain the remainder estimate

$$|R_n(f)| \leq \frac{1}{2\pi} \max_{z\in\Gamma} |K_n(z)| \cdot \int_\Gamma |f(z)|\, |dz|$$
$$\leq \frac{\ell(\Gamma)}{2\pi} \max_{z\in\Gamma} |K_n(z)| \cdot \max_{z\in\Gamma} |f(z)|, \tag{2.8}$$

where $\ell(\Gamma)$ is the length of Γ. While the contour Γ in (2.6) is arbitrary, subject to the conditions mentioned above, and the value of $R_n(f)$ is independent of Γ, the bounds in (2.8) do depend on the contour Γ. There is room, therefore, for optimizing these bounds over suitable families of contours Γ. Contours most frequently used are concentric *circles*,

$$C_r = \{z \in \mathbb{C} : |z| = r\}, \quad r > 1, \tag{2.9}$$

and confocal *ellipses* (having foci at ± 1 and the sum of semiaxes equal to ρ),

$$\mathcal{E}_\rho = \{z \in \mathbb{C} : z = \frac{1}{2}(\rho e^{i\vartheta} + \rho^{-1}e^{-i\vartheta}), \; 0 \leq \vartheta \leq 2\pi\}, \quad \rho > 1. \tag{2.10}$$

These, as $\rho \downarrow 1$, shrink to the interval $[-1,1]$ run back and forth, and become inflated and more circle-like as ρ increases.

2.2. MEROMORPHIC FUNCTIONS

Assume, more generally, that in each bounded domain $D \subset \mathbb{C}$, the function f has a finite number of poles, p_i, none of them in $[-1,1]$. For simplicity we assume all poles p_i to be simple; multiple poles, however, can be treated similarly [31]. Then, with $\Gamma = \partial D$ again surrounding $[-1,1]$, we can use the residue theorem,

$$\frac{1}{2\pi i} \oint_\Gamma \frac{f(z)}{z-t}\, dz = f(t) + \sum_i \frac{\mathrm{res}_{p_i} f}{p_i - t}, \tag{2.11}$$

where $\mathrm{res}_{p_i} f$ denotes the residue of f at p_i and the summation extends over all poles inside of Γ. One then obtains, similarly as before, the remainder formula

$$R_n(f) = \frac{1}{2\pi i} \oint_\Gamma R_n\left(\frac{1}{z-\cdot}\right) f(z)dz - \sum_i R_n\left(\frac{1}{p_i-\cdot}\right) \mathrm{res}_{p_i} f.$$

It is useful to interprete this as a "modified" quadrature rule with remainder,

$$\int_{-1}^1 f(t)w(t)dt = \sum_{\nu=1}^n w_\nu f(t_\nu) - \sum_i K_n(p_i)\mathrm{res}_{p_i} f$$
$$+ \frac{1}{2\pi i} \oint_\Gamma K_n(z)f(z)dz, \tag{2.12}$$

where the first two terms on the right constitute the modified quadrature rule, and the last term the respective remainder term. Note that in deriving (2.12), we have used (2.7).

We emphasize the dual role of the kernel K_n: on the one hand, it serves to express, and eventually estimate, the quadrature error, either in (2.6), (2.8), or in (2.12), and on the other hand, it allows us, as in (2.12), to modify, and thus improve, a quadrature rule by taking into account the influence of nearby poles. For both purposes it is important to be able to compute the kernel $K_n(z)$ for any $z \in \mathbb{C}\backslash[-1,1]$.

3. The Kernel K_n

By definition (2.7) of the kernel, we have

$$K_n(z) = \int_{-1}^1 \frac{w(t)}{z-t}dt - \sum_{\nu=1}^n \frac{w_\nu}{z-t_\nu}, \quad z \in \mathbb{C}\backslash[-1,1]. \tag{3.1}$$

This clearly is a one-valued analytic function in the complex plane cut along the interval $[-1,1]$. As z approaches the cut $[-1,1]$, the behavior of $K_n(z)$ is dictated by the two terms on the right of (3.1). The first is a Cauchy-type integral, known (according to the Sokhotskiy-Plemelj formula; cf. [28, p.94]) to approach two different (finite) values as z approaches an interior point of $[-1,1]$ from above and below, and to be continuous, or logarithmically singular, at the end points, according as $w = 0$ or $w \neq 0$ there. The second term, on the other hand, has poles at the nodes t_ν. Thus, near the cut, the kernel K_n behaves rather capriciously but, further away from it, settles down to a smooth and regular regimen. This is vividly illustrated by many contour maps of K_n in Takahasi and Mori [39–40] for a variety of standard quadrature rules.

3.1. THE KERNEL FOR INTERPOLATORY RULES

Interpolatory rules (2.1) are characterized by the weights w_ν being given by

$$w_\nu = \int_{-1}^1 \ell_\nu(t)w(t)dt, \quad \nu = 1, 2, \ldots, n, \tag{3.2}$$

where ℓ_ν is the νth elementary Lagrange interpolation polynomial associated with the nodes $t_1, t_2, \ldots, t_n$:

$$\ell_\nu \in \mathbb{P}_{n-1}, \quad \ell_\nu(t_\mu) = \delta_{\nu\mu} = \begin{cases} 1, & \nu = \mu, \\ \\ 0, & \nu \neq \mu. \end{cases} \tag{3.3}$$

Inserting (3.2) in (3.1), we can write the kernel in the form

$$K_n(z) = \int_{-1}^{1} \frac{1}{z-t} \left\{ 1 - \sum_{\nu=1}^{n} \ell_\nu(t) \frac{z-t}{z-t_\nu} \right\} w(t)dt.$$

Here, the function between braces, say $p(t)$, is a polynomial of degree n, which, by virtue of (3.3), vanishes at all the nodes t_μ and has the value 1 at $t = z$. Consequently,

$$p(t) = \frac{\pi_n(t)}{\pi_n(z)},$$

where π_n is the node polynomial

$$\pi_n(t) = \prod_{\nu=1}^{n} (t - t_\nu). \tag{3.4}$$

We therefore obtain

$$K_n(z) = \frac{\rho_n(z)}{\pi_n(z)}, \quad \rho_n(z) = \int_{-1}^{1} \frac{\pi_n(t)}{z-t} w(t)dt. \tag{3.5}$$

In (3.1), the kernel K_n is expressed as the difference of two quantities which are almost equal (K_n, after all, is typically small!). Evaluation of K_n based on (3.1) is therefore subject to severe cancellation errors. In contrast, the form (3.5) of K_n achieves the smallness of $|K_n|$, at least in part, through division by $\pi_n(z)$ – a large quantity if z is away from $[-1,1]$ – and is therefore more suitable for computation. This is particularly so for Gauss-type rules, as will be further discussed in the next subsection.

3.2. THE KERNEL FOR GAUSSIAN RULES

If (2.1) is a Gaussian rule relative to a nonnegative weight function w, i.e., $R_n(f) = 0$ for all $f \in \mathbb{P}_{2n-1}$, then the node polynomial π_n in (3.4) is the nth-*degree orthogonal polynomial* with respect to the weight function w,

$$\pi_n(t) = \pi_n(t; w). \tag{3.6}$$

The function ρ_n in (3.5),

$$\rho_n(z; w) = \int_{-1}^{1} \frac{\pi_n(t; w)}{z-t} w(t)dt, \tag{3.7}$$

is then referred to as the nth *associated function*, or the nth *function of the second kind*. The kernel

$$K_n(z; w) = \frac{\rho_n(z; w)}{\pi_n(z; w)} \qquad (3.8)$$

can now be computed using the basic three-term recurrence relation

$$y_{k+1} = (z - \alpha_k)y_k - \beta_k y_{k-1}, \qquad k = 0, 1, 2, \ldots, \qquad (3.9)$$

where $\alpha_k = \alpha_k(w)$, $\beta_k = \beta_k(w)$ are real, resp. positive, numbers uniquely determined by the weight function w, and

$$\beta_0 = \beta_0(w) = \int_{-1}^{1} w(t)dt. \qquad (3.10)$$

Indeed, the denominator in (3.8) is simply obtained by forward recurrence in (3.9), using the initial values

$$y_{-1} = 0, \quad y_0 = 1 \quad \text{(for } y_n = \pi_n). \qquad (3.11)$$

The numerator in (3.8), on the other hand, is a *minimal solution* of (3.9) for $z \in \mathbb{C}\backslash[-1, 1]$ (since $\rho_n/\pi_n \to 0$ as $n \to \infty$) and is uniquely determined by *one* initial value, namely

$$y_{-1} = 1 \quad \text{(for } y_n = \rho_n). \qquad (3.12)$$

There are well-known backward recursion algorithms for computing this minimal solution from the starting value $y_{-1} = 1$ (cf. [18, 41]).

In the next two sections, the behavior of the kernel (3.8) (for Gaussian formulae) on circular, resp. elliptic contours, is considered, and an attempt is made to determine where exactly on the contour the modulus of the kernel attains its maximum.

4. Estimation of $K_n(\cdot; w)$ on Circles

A detailed study of the kernel $K_n(z; w)$ on the circle $z \in C_r$ (cf. (2.9)) has been made in [22]. The following lemma (cf. [19]) turns out to be useful in this context and has found application elsewhere [2–4].

Lemma 4.1.(a) *If $w(t)/w(-t)$ is nondecreasing on $(-1,1)$, then $R_n(t^k) \geq 0$ for all $k \geq 2n$.*
(b) *If $w(t)/w(-t)$ is nonincreasing on $(-1,1)$, then $(-1)^k R_n(t^k) \geq 0$ for all $k \geq 2n$.*

Since we are dealing with Gaussian quadrature rules, we clearly have $R_n(t^k) = 0$ for all $k < 2n$. We also note that if the ratio $w(t)/w(-t)$ is constant, then the constant must be 1, and thus w is an even function, $R_n(t^k) = 0$ for k odd, and both statements (a), (b) of the lemma hold simultaneously. The second, incidentally, follows from the first by applying (a) to $w^*(t) = w(-t)$.

The proof of Lemma 4.1(a) uses an interesting result of Hunter [30], valid for any (monic) orthogonal polynomial $\pi_n(\cdot; w)$ whose weight function w is such that $w(t)/w(-t)$ is strictly increasing on $(-1,1)$, namely that all coefficients $c_k^{(n)}$ in the expansion

$$\frac{1}{z^n \pi_n(\frac{1}{z}; w)} = 1 + c_1^{(n)} z + c_2^{(n)} z^2 + \cdots$$

are strictly positive for each $n \geq 1$.

We can now easily prove the following theorem.

Theorem 4.2. *For each $r > 1$, there holds*

$$
\max_{z \in C_r} |K_n(z; w)| = \begin{cases} K_n(r), \\[2ex] |K_n(-r)|, \end{cases} \tag{4.1}
$$

the first relation being valid if the ratio $w(t)/w(-t)$ is nondecreasing on $(-1,1)$, the second if it is nonincreasing.

Proof. Expand

$$
\frac{1}{z-t} = \frac{1}{z} \frac{1}{1-t/z}
$$

in powers of t/z to get

$$
K_n(z; w) = R_n\left(\frac{1}{z-\cdot}\right) = \sum_{k=2n}^{\infty} \frac{R_n(t^k)}{z^{k+1}}, \qquad |z| > 1.
$$

Therefore,

$$
\max_{z \in C_r} |K_n(z; w)| \leq \sum_{k=2n}^{\infty} \frac{|R_n(t^k)|}{r^{k+1}}. \tag{4.2}
$$

Assume now that $w(t)/w(-t)$ is nondecreasing. By part (a) of Lemma 4.1 it then follows that

$$
K_n(r; w) = \sum_{k=2n}^{\infty} \frac{R_n(t^k)}{r^{k+1}} = \sum_{k=2n}^{\infty} \frac{|R_n(t^k)|}{r^{k+1}}.
$$

Comparison with (4.2) proves the first relation in (4.1). The second is proved similarly, using part (b) of the lemma. $\square$

As an example, for the Jacobi weight function $w(t) = (1-t)^\alpha (1+t)^\beta$, $\alpha, \beta > -1$, one has

$$
\frac{w(t)}{w(-t)} = \left(\frac{1+t}{1-t}\right)^{\beta-\alpha},
$$

which is increasing on $(-1,1)$ if $\alpha < \beta$, and decreasing if $\alpha > \beta$. Therefore, the first relation in (4.1) holds if $\alpha \leq \beta$, the second otherwise.

5. Estimation of $K_n(\,\cdot\,; w)$ on Ellipses

The behavior of $K_n(\,\cdot\,; w)$ on the ellipse $\mathcal{E}_\rho$ (cf. (2.10)) is considerably more difficult to analyze. Firm results are known only for the four Chebyshev weights

$$w_1(t) = (1 - t^2)^{-1/2}, \quad w_2(t) = (1 - t^2)^{1/2},$$

$$w_3(t) = (1 - t)^{-1/2}(1 + t)^{1/2}, \quad w_4(t) = (1 - t)^{1/2}(1 + t)^{-1/2}, \tag{5.1}$$

where the result for the last one is easily deducible from that for the preceding one by virtue of $w_4(t) = w_3(-t)$.

We discuss in some detail the case of the first Chebyshev weight $w = w_1$. Here one finds [22, §5.1], for $z \in \mathcal{E}_\rho$, the explicit formula

$$|K_n(z; w_1)| = \frac{2\pi}{\rho^n}[a_2(\rho) - \cos 2\vartheta]^{-1/2}[a_{2n}(\rho) + \cos 2n\vartheta]^{-1/2}, \quad 0 \le \vartheta \le 2\pi, \tag{5.2}$$

where

$$a_j(\rho) = \frac{1}{2}\left(\rho^j + \rho^{-j}\right), \quad j = 1, 2, 3, \ldots . \tag{5.3}$$

The quantities in (5.3), which appear in all the formulae for kernels belonging to Chebyshev weight functions, satisfy the following inequality, among many others (cf. [21, §3.1]).

Lemma 5.1. *For any $\rho > 1$, there holds*

$$\frac{a_{2n}(\rho) - 1}{a_2(\rho) - 1} \ge n^2, \quad n = 1, 2, 3, \ldots . \tag{5.4}$$

This is the key for proving that $|K_n(\,\cdot\,; w_1)|$ on $\mathcal{E}_\rho$ attains its maximum on the real axis:

Theorem 5.2. *For any $\rho > 1$ and each $n = 1, 2, 3, \ldots$, one has*

$$\max_{z \in \mathcal{E}_\rho} |K_n(z; w_1)| = K_n\left(\frac{1}{2}(\rho + \rho^{-1}); w_1\right) = \frac{4\pi}{\rho^n}\frac{1}{(\rho - \rho^{-1})(\rho^n + \rho^{-n})}.$$

Proof. According to (5.2) one must prove that

$$(a_2 - \cos 2\vartheta)(a_{2n} + \cos 2n\vartheta) \ge (a_2 - 1)(a_{2n} + 1), \quad 0 \le \vartheta \le \pi/2, \tag{5.5}$$

where $a_j = a_j(\rho)$ (there is symmetry with respect to both coordinate axes). Simple trigonometry will show that (5.5), if $\vartheta > 0$, is equivalent to

$$a_{2n} + 1 - (a_2 - 1)\frac{\sin^2 n\vartheta}{\sin^2 \vartheta} - 2\sin^2 n\vartheta \ge 0. \tag{5.6}$$

Since $|\sin n\vartheta/\sin \vartheta| \le n$, the left-hand side of (5.6) is greater than

$$a_{2n} + 1 - n^2(a_2 - 1) - 2 = (a_2 - 1)\left\{\frac{a_{2n} - 1}{a_2 - 1} - n^2\right\},$$

which is ≥ 0 by Lemma 5.1. $\square$

For the Chebyshev weight of the second kind, $w = w_2$, one finds for $z \in \mathcal{E}_\rho$ that

$$|K_n(z; w_2)| = \frac{\pi}{\rho^{n+1}} \left(\frac{a_2(\rho) - \cos 2\vartheta}{a_{2n+2}(\rho) - \cos 2(n+1)\vartheta} \right)^{1/2}.$$

From this it is rather straightforward to show that the maximum of $|K_n(\,\cdot\,; w_2)|$ on $\mathcal{E}_\rho$ is attained on the imaginary axis, i.e., for $\vartheta = \pi/2$, if n is odd. If n is even, however, this is true only for ρ sufficiently large. The precise result is as follows [23].

Theorem 5.3. *There holds*

$$\max_{z \in \mathcal{E}_\rho} |K_n(z; w_2)| = \left| K_n \left(\frac{i}{2}(\rho - \rho^{-1}); w_2 \right) \right| = \frac{\pi}{\rho^{n+1}} \frac{\rho + \rho^{-1}}{\rho^{n+1} + (-1)^n \rho^{-(n+1)}} \qquad (5.7)$$

for all $\rho > 1$, if n is odd, and for all $\rho \geq \rho_{n+1}$, if n is even, where ρ_{n+1} is the unique root of

$$\frac{\rho + \rho^{-1}}{\rho^{n+1} + \rho^{-(n+1)}} = \frac{1}{n+1}, \qquad \rho > 1. \qquad (5.8)$$

The following numerical values of ρ_{n+1} show that the ellipse $\mathcal{E}_\rho$ has to be quite slim for (5.7) to fail when n is even. It can be shown, indeed, that ρ_n tends monotonically to 1 as $n \to \infty$.

n	ρ_{n+1}
2	1.932
4	1.618
8	1.386
16	1.232
32	1.136

Table 5.1. *The roots ρ_{n+1}*

Finally, for the Chebyshev weight $w = w_3$, an argument similar to the one in the proof of Theorem 5.2 will establish:

Theorem 5.4. *For all $\rho > 1$ and each $n = 1, 2, 3, \ldots$, there holds*

$$\max_{z \in \mathcal{E}_\rho} |K_n(z; w_3)| = K_n \left(\frac{1}{2}(\rho + \rho^{-1}); w_3 \right) = \frac{2\pi}{\rho^{n+1/2}} \frac{\rho + 1}{\rho - 1} \frac{1}{\rho^{n+1/2} + \rho^{-(n+1/2)}}.$$

The case of general Jacobi weights, in contrast to circular contours (cf. §4), is presently beyond the reach of mathematical analysis. Based on numerical experimentation, it can be conjectured, however, that the result for $w = w_1$ (cf. Theorem 5.2) is typical for all Jacobi weights w with $-1 < \alpha \leq -1/2$, $\alpha \leq \beta$, as far as the location on $\mathcal{E}_\rho$ of the maximum of $|K_n|$ is concerned. There are, of course, no longer explicit formulae for $K_n(z; w)$, and its value at $z = \frac{1}{2}(\rho + \rho^{-1})$ has to be computed by the method indicated in §3.2. Likewise, the result of Theorem 5.3 seems to be representative, in the same sense, for all Jacobi weights w with $0 \leq \alpha = \beta$. In the remaining cases, $-\frac{1}{2} < \alpha = \beta < 0$ or $-\frac{1}{2} < \alpha < \beta$, the maximum is attained on the positive real axis only for sufficiently large ρ and n.

6. Gauss-Radau and Gauss-Lobatto Rules

The work described in §§4–5 has been extended to Gauss-Radau and Gauss-Lobatto rules with simple end points in [20], and with double end points in [21]. Here, we only quote two sample results for elliptic contours, both for Gauss-Lobatto rules corresponding to the Chebyshev weight function $w = w_1$, the first for simple, the other for double end points.

Theorem 6.1. (a) (Gauss-Lobatto rule GL_s with simple end points) *For the Gauss-Lobatto rule with n interior nodes and two simple nodes at ± 1, there holds for all $\rho > 1$ and each $n = 1, 2, 3, \ldots$ that*

$$\max_{z \in \mathcal{E}_\rho} |K_{n+2}^{GL_s}(z; w_1)| = K_{n+2}^{GL_s}\left(\frac{1}{2}(\rho + \rho^{-1}); w_1\right) = \frac{4\pi}{(\rho - \rho^{-1})(\rho^{2n+2} - 1)}. \tag{6.1}$$

(b) (Gauss-Lobatto rule GL_d with double end points) *For the Gauss-Lobatto rule with n interior nodes and two double nodes at ± 1, there holds for all $\rho > 1$ and each $n = 1, 2, 3, \ldots$ that*

$$\max_{z \in \mathcal{E}_\rho} |K_{n+2}^{GL_d}(z; w_1)| = K_{n+2}^{GL_d}\left(\frac{1}{2}(\rho + \rho^{-1}); w_1\right)$$
$$= \frac{4\pi}{(\rho^2 - 1)\rho^{n+2}} \frac{\rho^2 - \frac{n+1}{n+3}}{\frac{n+1}{n+3}[\rho^{n+3} - \rho^{-(n+3)}] - [\rho^{n+1} - \rho^{-(n+1)}]}. \tag{6.2}$$

Note that the maximum as $\rho \to \infty$ is $O(\rho^{-(2n+3)})$ in (a), and $O(\rho^{-(2n+5)})$ in (b), the increase of two orders being due to the increase by 1 of the multiplicity of each end node.

References

1. Akrivis, G., *Fehlerabschätzungen für Gauss-Quadraturformeln*, Numer. Math. 44 (1984), 261–278.

2. Akrivis, G., *Die Fehlernorm spezieller Gauss-Quadraturformeln*, in *Constructive Methods for the Practical Treatment of Integral Equations* (G. Hämmerlin and K.-H. Hoffmann, eds.), pp. 13–19. ISNM 73, Birkhäuser, Basel, 1985.

3. Akrivis, G., *The error norm of certain Gaussian quadrature formulae*, Math. Comp. 45 (1985), 513–519.

4. Akrivis, G. and Burgstaller, A., *Fehlerabschätzungen für nichtsymmetrische Gauss-Quadraturformeln*, Numer. Math. 47 (1985), 535–543.

5. Barnhill, R.E., *Complex quadratures with remainders of minimum norm*, Numer. Math. 7 (1965), 384–390.

6. Barnhill, R.E., *Optimal quadratures in $L^2(E_\rho)$*. I, II, SIAM J. Numer. Anal. 4 (1967), 390–397; *ibid.* 534–541.

7. Barnhill, R.E., *Asymptotic properties of minimum norm and optimal quadratures*, Numer. Math. 12 (1968), 384–393.

8. Barnhill, R.E. and Wixom, J.A., *Quadratures with remainders of minimum norm*. I, II, Math. Comp. 21 (1967), 66–75; *ibid.* 382–387.

9. Chawla, M.M. and Kaul, V., *Optimal rules for numerical integration round the unit circle*, BIT 13 (1973), 145–152.

10. Chawla, M.M. and Raina, B.L., *Optimal quadratures for analytic functions*, BIT 12 (1972), 489–502.

11. Chen, T.H.C., *An upper bound for the Gauss-Legendre quadrature error for analytic functions*, BIT 22 (1982), 530–532.

12. Davis, P.J., *Errors of numerical approximation for analytic functions*, J. Rational Mech. Anal. 2 (1953), 303–313.

13. Davis, P.J. and Rabinowitz, P., *On the estimation of quadrature errors for analytic functions*, Math. Tables Aids Comput. 8 (1954), 193–203.

14. Davis, P.J. and Rabinowitz, P., *Methods of Numerical Integration*, 2nd ed., Academic Press, Orlando, 1984.

15. Eckhardt, U., *Einige Eigenschaften Wilfscher Quadraturformeln*, Numer. Math. 12 (1968), 1–7.

16. Fock, V., *On the remainder term of certain quadrature formulae* (Russian), Bull. Acad. Sci. Leningrad (7) (1932), 419–448.

17. Gautschi, W., *A survey of Gauss-Christoffel quadrature formulae*, in *E.B. Christoffel* (P.L. Butzer and F. Fehér, eds.), pp. 72–147. Birkhäuser, Basel, 1981.

18. Gautschi, W., *Minimal solutions of three-term recurrence relations and orthogonal polynomials*, Math. Comp. 36 (1981), 547–554.

19. Gautschi, W., *On Padé approximants associated with Hamburger series*, Calcolo 20 (1983), 111–127.

20. Gautschi, W., *On the remainder term for analytic functions of Gauss-Lobatto and Gauss-Radau quadratures*, Rocky Mountain J. Math., 1991, to appear.

21. Gautschi, W. and Li, S., *The remainder term for analytic functions of Gauss-Radau and Gauss-Lobatto quadrature rules with multiple end points*, J. Comput. Appl. Math. 33 (1990), 315–329.

144

22. Gautschi, W. and Varga, R.S., *Error bounds for Gaussian quadrature of analytic functions*, SIAM J. Numer. Anal. 20 (1983), 1170–1186.

23. Gautschi, W., Tychopoulos, E. and Varga, R.S., *A note on the contour integral representation of the remainder term for a Gauss-Chebyshev quadrature rule*, SIAM J. Numer. Anal. 27 (1990), 219–224.

24. Hämmerlin, G., *Über ableitungsfreie Schranken für Quadraturfehler*, Numer. Math. 5 (1963), 226–233.

25. Hämmerlin, G., *Über ableitungsfreie Schranken für Quadraturfehler, II. Ergänzungen und Möglichkeiten zur Verbesserung*, Numer. Math. 7 (1965), 232–237.

26. Hämmerlin, G., *Zur Abschätzung von Quadraturfehlern für analytische Funktionen*, Numer. Math. 8 (1966), 334–344.

27. Heine, E., *Anwendungen der Kugelfunctionen und der verwandten Functionen*, 2nd ed., Reimer, Berlin, 1881. [I. Theil: *Mechanische Quadratur*, 1–31.]

28. Henrici, P., *Applied and Computational Complex Analysis*, vol. 3, Wiley, New York, 1986.

29. Hermite, C., *Sur la formule d'interpolation de Lagrange*, J. Reine Angew. Math. 84 (1878), 70–79. [Oeuvres III, 432–443.]

30. Hunter, D.B., *Some properties of orthogonal polynomials*, Math. Comp. 29 (1975), 559–565.

31. Hunter, D.B. and Okecha, G.E., *A modified Gaussian quadrature rule for integrals involving poles of any order*, BIT 26 (1986), 233-240.

32. Knauff, W. and Kress, R., *Optimale Approximation mit Nebenbedingungen an lineare Funktionale auf periodischen Funktionen*, Numer. Math. 25 (1975/76), 149–152.

33. Paulik, A., *On the optimal approximation of bounded linear functionals in Hilbert spaces of analytic functions*, BIT 16 (1976), 298–307.

34. Raina, B.L. and Kaul, N., *A class of optimal quadrature formulae*, IMA J. Numer. Anal. 3 (1983), 119–125.

35. Raina, B.L. and Kaul, N., *Estimating errors of certain Gauss quadrature formulae*, Calcolo 22 (1985), 229–240.

36. Rivlin, T.J., *Gauss quadrature for analytic functions*, in *Orthogonal Polynomials and their Applications* (M. Alfaro et al., eds.), pp. 178–192. Lecture Notes Math. 1329, Springer, Berlin, 1988.

37. Smith, H.V., *Global error bounds for Gauss-Gegenbauer quadrature*, BIT 21 (1981), 481–490.

38. Smith, H.V., *Global error bounds for the Clenshaw-Curtis quadrature formula*, BIT 22 (1982), 395–398.

39. Takahasi, H. and Mori, M., *Error estimation in the numerical integration of analytic functions*, Rep. Comput. Centre Univ. Tokyo 3 (1970), 41–108.

40. Takahasi, H. and Mori, M., *Estimation of errors in the numerical quadrature of analytic functions*, Applicable Anal. 1 (1971), 201-229.

41. Wimp, J., *Computation with Recurrence Relations*, Pitman, Boston, 1984.

42. Yanagiwara, H., *A new method of numerical integration of Gaussian type* (Japanese), Bull. Fukuoka Gakugei Univ. III 6 (1956), 19–25.

ERROR BOUNDS BASED ON APPROXIMATION THEORY

H. Brass
Institut für Angewandte Mathematik
Technische Universität Braunschweig
Pockelsstraße 14
W 3300 Braunschweig
Federal Republic of Germany

ABSTRACT. We give examples for the usefulness of approximation theory in the discussion of error bounds for quadrature rules. Our main point is that this method is not only simple and general, but that it leads to sharp estimates in many cases.

1. Introduction

1.1 Let R be a functional defined on the normed space X and let $U \subset X$ be a subspace with $R[U] = 0$. Then we have for any $f \in X$ and $u \in U$

$$\left| R[f] \right| = \left| R[f - u] \right| \le \|R\| \, \|f - u\|$$

and therefore

$$\left| R[f] \right| \le \|R\| \inf_{u \in U} \|f - u\| = : \|R\| \operatorname{dist}(f; U). \tag{1}$$

This inequality or some refinements of it are the source of a great number of error bounds in quadrature theory. This technique is by no means new (e.g. BERNSTEIN [2], STROUD [29], BAKER [1], LOCHER/ZELLER [21], HAUSSMANN/ZELLER [18], DAVIS/RABINOWITZ [11] p.332), but it seems to me that its generality, simplicity and flexibility deserve more attention.

It is the aim of this paper to give various applications of (1) and to discuss the efficiency of the generated bounds.

1.2 Quadrature theory deals with the approximation of a given functional I of the type

$$I[f] : = \int_{-1}^{1} f(x)k(x)dx, \qquad k \in L^1$$

defined on the normed space $C[-1, 1]$ (sup norm) by functionals Q ("quadrature rules") of the type

$$Q[f] : = \sum_{\nu=1}^{n} \alpha_\nu f(x_\nu), \qquad x_\nu \in [-1, 1].$$

T. O. Espelid and A. Genz (eds.), Numerical Integration, 147–163.
© 1992 Kluwer Academic Publishers.

The error is the functional $R := I - Q$. The degree of Q is the number $\deg[Q] := \sup\{m : R[\mathcal{P}_m] = 0\}$, where $\mathcal{P}_m$ denotes the space of polynomials of degree at most m. A rule is called interpolatory, if it has n evaluation points and degree at least $n - 1$. Special cases are the rule of FILIPPI Q_n^{Fi} ($k = 1$, evaluation points $x_{\nu,n} = -\cos(\nu(n+1)^{-1}\pi)$) and the rule of CLENSHAW/CURTIS Q_n^{CC} ($k = 1$, $x_{\nu,n} = -\cos((\nu - 1)(n - 1)^{-1}\pi)$). The most important interpolatory rules are the GAUSSian rules Q_n^{G}, which are defined only if $k \geq 0$. They are characterized as rules with n evaluation points and degree $2n - 1$.

In (1) any norm will do, but with the exception of sections 3.5 and 3.8 we will only use the sup norm on the fundamental interval $[-1, 1]$.

2. On the degree of approximation

We shall discuss (1) mainly in the special case $U = \mathcal{P}_m$. We shall use the following results from approximation theory:

$$\text{dist}(f; \mathcal{P}_m) \leq K_r \frac{(m + 1 - r)!}{(m + 1)!} \left\| f^{(r)} \right\| \qquad r = 1, 2, \ldots, \quad m = r - 1, r, \ldots \tag{2}$$

where the constants K_r are defined by

$$K_r := \frac{4}{\pi} \sum_{\nu=0}^{\infty} \frac{(-1)^{\nu(r+1)}}{(2\nu + 1)^{r+1}} \tag{3}$$

(SINWEL [27]). These bounds are asymptotically ($m \to \infty$) optimal if r is fixed. If m is near r, they are rather poor and should be replaced by the following ones:

$$\text{dist}(f, \mathcal{P}_m) \leq \frac{1}{2^m(m + 1)!} \left\| f^{(m+1)} \right\| \tag{4}$$

$$\text{dist}(f, \mathcal{P}_m) \leq \frac{1}{2^{m-s}(m - s + 1)!} \left(\frac{(2m - 2s + 2)! \, s!}{(2m - s + 2)!} \frac{2m - s + 2}{2m - 2s + 1} \right)^{\frac{1}{2}} \left\| f^{(m-s+1)} \right\|. \tag{5}$$

(4) is the classical inequality of BERNSTEIN, it has—in contrast to (2) and (5)—a simple proof. (5) is proven in BRASS/FÖRSTER [8].

The special case $r = 1$ of (2) reads

$$\text{dist}(f; \mathcal{P}_m) \leq \frac{\pi}{2(m + 1)} \| f' \|. \tag{6}$$

A localized version of this bound is the following theorem of NIKOLSKY [23]: The polynomial

$$p := \frac{a_0}{2} + \sum_{\nu=1}^{m-1} \eta_{\nu,m} a_\nu T_\nu \tag{7}$$

with

$$T_\nu := \cos(\nu \arccos x), \qquad \eta_{\nu,m} := \frac{\nu\pi}{2m} \text{ctg}\left(\frac{\nu\pi}{2m} \right),$$

$$a_\nu = a_\nu[f] := \frac{2}{\pi} \int_{-1}^{1} f(x) T_\nu(x)(1 - x^2)^{-\frac{1}{2}} dx \tag{8}$$

satisfies

$$\left|f(x) - p(x)\right| \leq \left(\frac{\pi}{2m}\sqrt{1-x^2} + c\frac{\ln m}{m^2}\right)\|f'\|, \tag{9}$$

where c denotes a constant. A more precise result is given by SINWEL [28]: There exists a $p \in \mathcal{P}_m$, which satisfies

$$\left|f(x) - p(x)\right| \leq \text{tg}\left(\frac{\pi}{2m}\right)\left(\sqrt{1-x^2} + \frac{3}{m}|x|\right)\|f'\|. \tag{10}$$

The construction of this polynomial is more complicated than that of the NIKOLSKY polynomial (7).

3. Applications of (1)

3.1 Let Q be a positive quadrature rule of degree m for the functional $I[f] := \int_{-1}^{1} f(x)dx$. In this situation $\|R\| = 4$ is easily checked and (1) gives

$$\left|R[f]\right| \leq 4\,\text{dist}(f; \mathcal{P}_m). \tag{11}$$

If we use now (4), we get immediately

$$\left|R[f]\right| \leq \frac{2^{-m+2}}{(m+1)!}\left\|f^{(m+1)}\right\|. \tag{12}$$

This result (LOCHER/ZELLER [21]) is very illustrative for the merits of the method (1): It has a simple proof and is of great generality (it holds for all positive quadrature rules of degree m). But the main point is: (12) is almost sharp, the best possible factor of $\|f^{(m+1)}\|$ is asymptotically $\frac{\pi}{8}\frac{2^{-m+2}}{(m+1)!}$ (BRASS/SCHMEISSER [6], BRASS [7]). But this last result has a complicated proof and uses specialized methods, whereas (11) and (12) are easily generalized to any positive I, one has only to replace 4 by $2\|I\|$.

3.2 Let $I[f] := \int_{-1}^{1} f(x)dx$ and let Q be a positive quadrature rule of degree m. Combining (11) and (6) leads to

$$\left|R[f]\right| \leq \frac{2\pi}{m+1}\|f'\|.$$

This is a reasonable bound, as it can be seen by a comparison with the following result of PETRAS [25]:

$$\left(1 - \frac{2}{n}\right)\frac{\pi^2}{8n} \leq \sup\left\{\left|R_n^G[f]\right| : \|f'\| \leq 1\right\} \leq \frac{\pi^2}{8n}.$$

But we can do better, if we use (10) instead of (6). This leads to

Theorem 1 Let $I[f] := \int_{-1}^{1} f(x)dx$ and let Q be a positive quadrature rule of degree $m - 1$. Then we have

$$\left|R[f]\right| \leq \frac{\pi^2}{4m}\left(1 + \frac{3.8}{m}\right)\|f'\|.$$

The special case $R = R_n^G$ shows that this result is best possible in the asymptotic sense.
The proof, which is a little bit tedious, is given in section 4.1 below.

3.3   Theorem 1 may be generalized to any positive weight function k, but I restrict myself here to an asymptotic result, which has a much simpler proof.

Theorem 2   Let Q_m $(m = 1, 2, \ldots)$ denote a sequence of positive quadrature rules for the positive functional I. Assume $\deg[Q_m] \geq m$ and $f_0(x) := \sqrt{1 - x^2}$. Then

$$\limsup_{m \to \infty} m \sup\left\{\left|R_m[f]\right| : \|f'\| \leq 1\right\} \leq \frac{\pi}{2} I[f_0].$$

Again this is unimprovable, for we have equality in the special case of the GAUSSian rules:

Theorem 3   If $k > 0$ a.e. holds, then we have (with $f_0 := \sqrt{1 - x^2}$)

$$\lim_{n \to \infty} n \sup\left\{\left|R_n^G[f]\right| : \|f'\| \leq 1\right\} = \frac{\pi}{4} I[f_0].$$

PETRAS [24] has proved this theorem for a more specialized class of weight functions with a method which is totally different from ours.

3.4   Although the bound (12) is almost unimprovable in the set of all positive quadrature rules of degree m, it may give poor results for a special rule. As an example may be mentioned

$$\left|R_m^{CC}[f]\right| \leq \frac{8}{(m+1)(m-1)(m-3)} \frac{\left\|f^{(m+1)}\right\|}{2^{m-2}(m+1)!}, \qquad m > 3, \text{ odd.}$$

(BRASS [3]). This result and many related results can be obtained with (1), if we apply (1) (not to the error R, but) to a modified functional $\hat{R}$.

If Q has degree m, we define $\hat{R}$ by

$$\hat{R}[f] := R[f] - \sum_{\nu=m+1}^{m+p} R[T_\nu] a_\nu[f]. \tag{13}$$

Obviously $\hat{R}[\mathcal{P}_{m+p}] = 0$ holds, hence we have

$$\left|\hat{R}[f]\right| \leq \left\|\hat{R}\right\| \operatorname{dist}(f; \mathcal{P}_{m+p})$$

$$\leq \left(\|R\| + \sum_{\nu=m+1}^{m+p} \|a_\nu\| \left|R[T_\nu]\right|\right) \operatorname{dist}(f; \mathcal{P}_{m+p})$$

$$\leq \left(\|R\| + \frac{4}{\pi} \sum_{\nu=m+1}^{m+p} \left|R[T_\nu]\right|\right) \operatorname{dist}(f; \mathcal{P}_{m+p}). \tag{14}$$

From (1) we have further

$$\left|a_\nu[f]\right| \leq \frac{4}{\pi} \operatorname{dist}(f; \mathcal{P}_{\nu-1}). \tag{15}$$

We construct a polynomial p by collocation with f at the zeros of T_{m+1}. Then we have, as is well known

$$\left| f(x) - p(x) \right| \leq \frac{|T_{m+1}(x)|}{2^m(m+1)!} \left\| f^{(m+1)} \right\| .$$

Using this, we obtain

$$
\left| a_{m+1}[f] \right| = \frac{2}{\pi} \left| \int_{-1}^{1} f(x) \frac{T_{m+1}(x)}{\sqrt{1-x^2}} dx \right| = \frac{2}{\pi} \left| \int_{-1}^{1} \left(f(x) - p(x) \right) \frac{T_{m+1}(x)}{\sqrt{1-x^2}} dx \right|
$$

$$
\leq \frac{2}{\pi} \frac{\left\| f^{(m+1)} \right\|}{2^m(m+1)!} \int_{-1}^{1} \frac{T_{m+1}^2(x)}{\sqrt{1-x^2}} dx = \frac{\left\| f^{(m+1)} \right\|}{2^m(m+1)!} . \tag{16}
$$

We introduce now (16), (15), (14) in (13) and get

$$
\left| R[f] \right| = \left| \sum_{\nu=m+1}^{m+p} R[T_\nu] a_\nu[f] + \hat{R}[f] \right|
$$

$$
\leq \frac{\left| R[T_{m+1}] \right|}{2^m(m+1)!} \left\| f^{(m+1)} \right\| + \frac{4}{\pi} \sum_{\nu=m+2}^{m+p} \left| R[T_\nu] \right| \operatorname{dist}(f; \mathcal{P}_{\nu-1})
$$

$$
+ \left(\| R \| + \frac{4}{\pi} \sum_{\nu=m+1}^{m+p} \left| R[T_\nu] \right| \right) \operatorname{dist}(f; \mathcal{P}_{m+p}) .
$$

In a last step we use (5) to construct bounds which are multiples of $\| f^{(m+1)} \|$ for all terms $\operatorname{dist}(f; \mathcal{P}_\mu)$ in this formula. We have proved

Theorem 4   Let R be the error of a quadrature rule of degree m and $p \in \{2, 3, \ldots\}$, then we have

$$
\left| R[f] \right| \leq \frac{\left| R[T_{m+1}] \right|}{2^m(m+1)!} (1 + \varepsilon_m) \left\| f^{(m+1)} \right\|
$$

with

$$
\varepsilon_m := \frac{4}{\pi} \sum_{\kappa=1}^{p-1} \left| \frac{R[T_{m+\kappa+1}]}{R[T_{m+1}]} \right| \left(\frac{(2m+2)! \kappa!}{(2m+\kappa+2)!} \frac{2m+\kappa+2}{2m+1} \right)^{\frac{1}{2}}
$$

$$
+ \frac{1}{\left| R[T_{m+1}] \right|} \left(\| R \| + \frac{4}{\pi} \sum_{\nu=m+1}^{m+p} \left| R[T_\nu] \right| \right) \left(\frac{(2m+2)! p!}{(2m+p+2)!} \frac{2m+p+2}{2m+1} \right)^{\frac{1}{2}} .
$$

This theorem gives us the asymptotically best value of c_n in the bound

$$
\left| R_n^{CC}[f] \right| \leq c_n \left\| f^{(m)} \right\| , \qquad m = 2 \left\lceil \frac{n}{2} \right\rceil
$$

(if we choose $p = 7$) and in the bound

$$
\left| R_n^{Fi}[f] \right| \leq c_n \left\| f^{(m)} \right\| , \qquad m = 2 \left\lceil \frac{n}{2} \right\rceil
$$

(if we choose $p = 3$) to name only two.

For these cases other methods have been given to compute or estimate the c_n. But look at the situation for R_n^{Fi}, n even. The best possible value for c_n is determined (BRASS/SCHMEISSER [5]), but the method is not simple and gives little hope of possible generalizations.

3.5 Let

$$f_\sigma(x) := |x - \xi|^\sigma, \qquad \xi \in\,]-1, 1[\,, \quad \sigma \in\,]0, 1]\,,$$

then

$$c_1(\sigma, \xi) m^{-\sigma} \le \mathrm{dist}(f_\sigma; \mathcal{P}_m) \le c_2(\sigma, \xi) m^{-\sigma}$$

(where $c_1, c_2 > 0$) holds (TIMAN [30] p.417). An application of (11) leads to

$$\left| R_n^{\mathrm{G}}[f_\sigma] \right| = O(n^{-\sigma})\,,$$

but the better estimation

$$\left| R_n^{\mathrm{G}}[f_\sigma] \right| = O(n^{-\sigma-1}) \tag{17}$$

is known and we have similar results for many other quadrature rules (DAVIS/RABINOWITZ [11] p.313, KÜTZ [19]). It is possible to prove (17) as a special case of (1). For this purpose we have to use another norm. We use in this section

$$\|f\|_1 := \int_{-1}^1 |f'(x)| dx\,, \tag{18}$$

where f' is interpreted as a weak derivative. Strictly speaking, this is a norm only in a suitable quotient space, but this is of no importance to us. If R is the error of a quadrature rule, we have the PEANO kernel theorem

$$R[f] = \int_{-1}^1 f'(x) K(x) dx\,, \qquad K(x) := R\left[(\cdot - x)_+^0 \right]$$

and therefore

$$\|R\|_1 = \|K\|\,.$$

If Q has degree m and is positive, we have the bounds

$$\|K\| \le \frac{\pi}{m - 3} \quad \text{if } k = 1 \qquad \text{(FÖRSTER/PETRAS [16])},$$

$$\|K\| \le \frac{4\pi}{m} \quad \text{if } 0 \le k(x) \le (1 - x^2)^{-\frac{1}{2}} \qquad \text{(FREUD [17])}.$$

The last bound is improved by FÖRSTER in his unpublished "Habilitationsschrift" to

$$\|K\| \le \frac{2\pi}{m} \quad \text{if } 0 \le k(x) \le (1 - x^2)^{-\frac{1}{2}}\,. \tag{19}$$

The change from $\|f\|$ to $\|f\|_1$ makes it necessary to replace $\mathrm{dist}(f; \mathcal{P}_m)$ by

$$\mathrm{dist}_1(f; \mathcal{P}_m) := \inf \left\{ \int_{-1}^1 |f'(x) - p'(x)|\, dx : p \in \mathcal{P}_m \right\}\,.$$

There are various bounds for these numbers in the literature. For our purpose is useful

$$\text{dist}_1(f; \mathcal{P}_m) \le \left(1 + \frac{\pi}{2}\right) \sup\left\{\int_{-1}^{1} |f'(x) - f'(x+h)| \, dx : |h| \le \frac{1}{m}\right\}$$

(where f' is arbitarily continued to $\left[-1 - \frac{1}{m}, 1 + \frac{1}{m}\right]$) which can easily be derived from the result of TIMAN [30] (p.292) concerning trigonometric approximation. Combining this bound with (19) yields

$$\left|R[f_\sigma]\right| \le 2\pi \left(1 + \frac{\pi}{2}\right) 4m^{-1-\sigma},$$

if Q is a positive quadrature rule of degree m and $0 \le k(x) \le (1 - x^2)^{-\frac{1}{2}}$ holds.

In a similar manner we may discuss all functions which are piecewise convex or concave, an interesting special case are functions with a singularity of the logarithmic type.

3.6 If we want to discuss the function

$$f_{\sigma,1}(x) := (1 - x)^\sigma \qquad \sigma \in]0,1[$$

with the help of (1), then neither the norm $\|\cdot\|$ nor the norm $\|\cdot\|_1$ leads to the correct asymptotics of $R_n^G[f_{\sigma,1}]$. In this case the introduction of a weight function in (18) is necessary, this is done in the paper of DEVORE/SCOTT [12].

f_σ and $f_{\sigma,1}$ are prototype functions with a singularity in the interior respective at the boundary of the fundamental interval. If the singularity is outside $[-1, 1]$, then the norm $\|\cdot\|$ leads to realistic bounds. As an example I cite the following result: For $k = 1$ and $f_a(x) := (a - x)^\sigma$ $(a > 1, \sigma \in]0,1[)$ we have

$$\lim_{n \to \infty} \frac{\left|R_n^G[f_a]\right|}{\text{dist}(f_a; \mathcal{P}_{2n-1})} = 2\pi \frac{a^2 - 1}{\left(a + \sqrt{a^2 - 1}\right)^2}. \tag{20}$$

We have the same limit relation for a wide variety of holomorphic functions with a singularity at $a > 1$ and no further singularity on the ellipse of holomorphy (BRASS [4]). We cannot expect that a relation like (20) holds for all holomorphic functions, a simple counterexample is given by odd functions (: $f(x) = -f(-x)$). Since every function f is the sum of an even function f_e and an odd function f_o and we have $R_n^G[f_o] = 0$ if $k = 1$, we may restrict ourselves to the class of even functions. Therefore the following theorem is a justification for the use of (11) if f is holomorphic.

Theorem 5   Let f be an even function which is holomorphic in a domain that includes $[-1, 1]$, but is not a polynomial. If $k = 1$, then we have

$$\limsup_{n \to \infty} \frac{\left|R_n^G[f]\right|}{\text{dist}(f; \mathcal{P}_{2n-1})} > 0.$$

3.7 Theorem 5 means that for a holomorphic function the bound

$$\left|R[f]\right| \le \|R\| \, \text{dist}(f; \mathcal{P}_m) \qquad \text{if } R[\mathcal{P}_m] = 0$$

is a realistic one, if applied to the GAUSSIAN rule. But this is not necessarily so for other rules. A striking example is

Theorem 6   If $f^{(n+1)} \geq 0$ and n is odd, then we have

$$\left|R_n^{\mathrm{Fi}}[f]\right| \leq \frac{4}{n+1}\,\mathrm{dist}(f;\mathcal{P}_n).$$

3.8   Until now we have used (1) only in the special situation $U = \mathcal{P}_m$, but there are a lot of other possibilities. As an example we discuss the numerical computation of FOURIER coefficients.

In this section all functions are 2π-periodic and the norm is $\|f\| := \sup_x |f(x)|$. We have

$$I_p[f] := \frac{1}{\pi}\int_0^{2\pi} f(x)\cos(px)\,dx$$

and the usual quadrature rule is

$$Q_{n,p}[f] := \frac{2}{n}\sum_{\nu=1}^{n} f(x_\nu)\cos(px_\nu), \quad x_\nu := \frac{\nu 2\pi}{n}, \quad n \equiv 0 (\mathrm{mod}\,4), \; n > 2p.$$

Here we may apply (1) with $U = \mathcal{T}_{\frac{n}{2}-1}$, where

$$\mathcal{T}_m := \left\{ t : t(x) = \sum_{\nu=0}^{m}(\gamma_\nu\cos(\nu x) + \delta_\nu\sin(\nu x)), \; \gamma_\nu, \delta_\nu \in \mathbb{R} \right\}$$

is the subspace of trigonometric polynomials of degree m. We use now

$$\|I_p\| = \frac{4}{\pi}, \qquad \|Q_{n,p}\| \leq \frac{4}{\pi} \qquad\qquad (\text{BRASS } [9])$$

and

$$\mathrm{dist}(f;\mathcal{T}_m) \leq \frac{\pi}{2(m+1)^r}\|f^{(r)}\| \qquad\qquad (\text{e.g. FISHER } [13])$$

and get the bound

$$\left|R_{n,p}[f]\right| \leq 4\frac{2^r}{n^r}\|f^{(r)}\|.$$

The best possible result of this type has 1.45 instead of 4 (BRASS [10]). We can repeat the commentary at the end of 3.1 and 3.4 .

4. The proofs

4.1   Proof of theorem 1.

4.1.1   We begin with $m = 1$. The operator H_0 with

$$H_0[f](x) := \frac{1}{2}\int_{-1}^{1} f(u)\,du$$

carries $C[-1,1]$ into $\mathcal{P}_0$. If we apply PEANO kernel theory, it is not difficult to prove

$$\left\|f - H_0[f]\right\| \leq \|f'\|. \qquad\qquad (21)$$

We have now

$$\left|I[f] - Q[f]\right| = \left|I[f - H_0[f]] - Q[f - H_0[f]]\right| = \left|Q[f - H_0[f]]\right|$$
$$\leq Q[1]\,\|f'\| = 2\|f'\|\,.$$

This is more than the assertion of theorem 1 in this special case. If we replace H_0 by

$$H_1[f](x) := \frac{1}{2}\int_{-1}^{1} f(u)du + x\int_{-1}^{1} uf(u)du$$

and (21) by

$$\left\|f - H_1[f]\right\| \leq \frac{1}{2}\|f'\|$$

then we obtain

$$\left|I[f] - Q[f]\right| \leq \|f'\|\,,$$

which is (more than) the special case $m = 2$.

We could proceed in the same manner, if we had suitable operators $H_2, \ldots$. We seek numbers $\varrho_{\nu,n}$ such that the operators

$$H_n[f] := \frac{1}{2}\int_{-1}^{1} f(u)du + \sum_{\nu=1}^{n} \varrho_{\nu,n} P_\nu \int_{-1}^{1} f(u)P_\nu(u)du$$

(where P_ν denote the LEGENDRE polynomials) gives a sufficiently good approximation for f. The form of the operator guarantees $I[f - H_n[f]] = 0$ and this makes the estimation so simple. A general result is not known, but Miss S.LANGE has given in her diploma thesis [20] operators H_n ($n = 1, 2, \ldots, 10$) which are sufficient for our purpose.

In the following we make the assumption $m \geq 12$.

4.1.2 We use now the polynomial $p \in \mathcal{P}_{m-1}$ of SINWEL as a tool for approximation. We have

$$\left|I[f] - Q[f]\right| \leq \left|I[f - p]\right| + \left|Q[f - p]\right|\,. \tag{22}$$

By use of (10) we obtain (with $f_0(x) := \sqrt{1 - x^2}$, $f_1(x) := |x|$)

$$\left|Q[f - p]\right| \leq \mathrm{tg}\,\frac{\pi}{2m}\left(Q[f_0] + \frac{3}{m}Q[f_1]\right)\|f'\|\,. \tag{23}$$

Since $f_0^{(2\nu)}(x) < 0$ holds for $x \in]-1, 1[$ and $\nu \geq 1$, the theorem of MARKOFF/KREIN (see e.g. BRASS/SCHMEISSER [6], Theorem 6) can be applied. This leads to

$$Q[f_0] \leq Q_n^G[f_0], \qquad n := \left\lfloor \frac{m}{2} \right\rfloor\,. \tag{24}$$

Let α_ν^G ($\nu = 1, \ldots, n$) denote the coefficients and x_ν^G the evaluation points of Q_n^G, then we have

$$\alpha_\nu^G \leq \frac{2\pi}{2n+1}\left(1 - \left(x_\nu^G\right)^2\right)^{\frac{1}{2}}$$

(FÖRSTER/PETRAS [15]) and therefore

$$Q_n^G[f_0] \le \frac{2\pi}{2n+1} \sum_{\nu=1}^{n} \left(1 - \left(x_\nu^G\right)^2\right) = \frac{2\pi}{2n+1} \left(n - \frac{n(n-1)}{2n-1}\right)$$

$$= \frac{\pi}{2} + \frac{\pi}{2(2n+1)(2n-1)}.$$

Using (24) we obtain

$$Q[f_0] \le \frac{\pi}{2} + \frac{\pi}{2m(m-2)}. \tag{25}$$

We now turn to $Q[f_1]$. FÖRSTER [14] proved $Q[f_1] \le Q_{2n}^G[f_1]$ if Q is any positive rule with degree $4n-1$. We deduce

$$Q[f_1] \le Q_6^G[f_1] < 1.02 \tag{26}$$

The function $h(x) := x^{-3}(\mathrm{tg}(x) - x)$ is increasing (power series!), therefore

$$\mathrm{tg}\,\frac{\pi}{2m} \le \frac{\pi}{2m} + \left(\frac{\pi}{24}\right)^{-3}\left(\mathrm{tg}\,\frac{\pi}{24} - \frac{\pi}{24}\right)\left(\frac{\pi}{2m}\right)^3 \le \frac{\pi}{2m} + \frac{1.302}{m^3} \tag{27}$$

holds if $m \ge 12$. Combining (23), (25), (26) and (27) and using $m \ge 12$, we find

$$\left|Q[f-p]\right| \le \left(\frac{\pi^2}{4m} + \frac{5.254}{m^2}\right) \|f'\|. \tag{28}$$

We have now to estimate

$$I[f-p] = \int_{-1}^{1}\left(f(x) - p(x)\right)dx = \int_0^\pi \left(f(\cos t) - p(\cos t)\right)\sin t\,dt \tag{29}$$

For this purpose more precise information on the structure of the SINWEL polynomial is necessary. Using SINWEL's method we may derive

$$f(\cos t) - p(\cos t) = \int_0^{2\pi}\left(-\sin t\,S_1(t-u) + \cos t\,S_2(t-u) + c\cos t\cos\left((m-1)(t-u)\right)\right)f'(\cos u)du$$

with

$$S_1(x) := \sum_{\nu=1}^{m-1}(\gamma_{\nu-1} + \gamma_{\nu+1})\sin \nu x + \frac{1}{\pi}\sum_{\nu=m}^{\infty}\frac{\nu}{\nu^2-1}\sin \nu x,$$

$$S_2(x) := \frac{1}{2}(\delta_1 - \delta_{-1}) + \sum_{\nu=1}^{m-1}(\delta_{\nu+1} - \delta_{\nu-1})\cos \nu x - \frac{1}{\pi}\sum_{\nu=m}^{\infty}\frac{1}{\nu^2-1}\cos \nu x,$$

$$\gamma_\nu := \frac{1}{4m}\left(\frac{1}{z_\nu} - \mathrm{ctg}\,z_\nu\right), \qquad \delta_\nu := \frac{1}{4m}\left(\frac{1}{z_\nu} - \mathrm{cosec}\,z_\nu\right),$$

$$z_\nu := \frac{\nu\pi}{2m}, \qquad c := \frac{1}{4m}\,\mathrm{tg}\,\frac{\pi}{m}\left(1 - \mathrm{tg}\,\frac{\pi}{2m}\right).$$

If we introduce this formula in (29) and use more or less obvious transformations and estimations, we finally get

$$\left| \int_{-1}^{1} \left(f(x) - p(x) \right) dx \right| \le 3.896 \frac{1}{m^2} \|f'\| . \tag{30}$$

The theorem now follows from (22), (28) and (30).

4.2 Proof of theorem 2

For every positive ϵ there exist numbers $\gamma_1, \dots, \gamma_r$ such that

$$\int_{0}^{\pi} \left| k(\cos t) \sin t - \frac{2}{\pi} \sum_{\nu=0}^{r} \gamma_\nu \cos \nu t \right| dt < \epsilon$$

holds. If we substitute $x = \cos t$, the same formula reads

$$\int_{-1}^{1} \left| k(x) - \frac{2}{\pi} \sum_{\nu=0}^{r} \gamma_\nu (1 - x^2)^{-\frac{1}{2}} T_\nu(x) \right| dx < \epsilon$$

which means, that the functional $I^*[f] := \sum_{\nu=0}^{r} \gamma_\nu a_\nu[f]$ approximates the given functional I, more precisely $\left| I[f] - I^*[f] \right| \le \epsilon \|f\|$.

If we denote the Nikolsky polynomial (7) by p_m, we have

$$
\begin{aligned}
\left| I[f] - Q_m[f] \right| &= \left| I[f - p_m] - Q_m[f - p_m] \right| \\
&\le \left| I^*[f - p_m] \right| + \left| Q_m[f - p_m] \right| + \left| (I - I^*)[f - p_m] \right| \\
&\le \left| I^*[f - p_m] \right| + \left| Q_m[f - p_m] \right| + \epsilon \|f - p_m\| .
\end{aligned} \tag{31}
$$

Observing that $|1 - \eta_{\nu,m}| \le \frac{\nu^2}{m^2}$, we deduce from (7), if $m \ge r$

$$
\begin{aligned}
\left| I^*[f - p_m] \right| &= \left| \sum_{\nu=1}^{r} \gamma_\nu a_\nu[f] (1 - \eta_{\nu,m}) \right| = O(m^{-2}) \sum_{\nu=1}^{r} |\gamma_\nu| \left| a_\nu[f] \right| \\
&= O\left(m^{-2} \|f'\| \right) .
\end{aligned} \tag{32}
$$

Application of (9) gives (with $f_0 := \sqrt{1 - x^2}$)

$$\left| Q_m[f - p_m] \right| \le \left(\frac{\pi}{2m} Q_m[f_0] + \frac{c \ln m}{m^2} Q_m[1] \right) \|f'\|$$

The relation $Q_m[f_0] \to I[f_0]$ follows from the hypotheses of positivity and degree. We get

$$\left| Q_m[f - p_m] \right| \le \frac{\pi}{2m} I[f_0] \left(1 + o(1) \right) \|f'\| . \tag{33}$$

The assertion now follows from (31), (32), (33) and (9).

4.3 Proof of theorem 3

4.3.1 If A is a set, we denote by $|A|$ its cardinality.

Lemma   Let $Q_n[f] = \sum_{\nu=1}^{n} \alpha_{\nu,n} f(x_{\nu,n})$ be a sequence of quadrature rules. Let there exist a function v which is continuous and positive on $]-1, 1[$ such that

$$\lim_{n\to\infty} \frac{1}{n}\left|\{x_{\nu,n} : x_{\nu,n} \leq t\}\right| = \int_{-1}^{t} v(x)\,dx$$

holds. If $k \in L^1$ is any nonnegative weight function and $\eta \in]0, 1[$ is any number, then we have

$$\liminf_{n\to\infty} n \sup \left\{ \left| \int_{-1}^{1} f(x)k(x)\,dx - Q_n[f] \right| : \|f'\| \leq 1 \right\} \geq \int_{-\eta}^{\eta} \frac{k(x)}{4v(x)}\,dx\,.$$

In the special case $Q_n = Q_n^G$ and with the additional hypothesis $k(x) > 0$ a.e. we have $v(x) = \frac{1}{\pi}\left(1 - x^2\right)^{-\frac{1}{2}}$ (FREUD [17] p.134). Combining the lower bound of the lemma with the upper bound which follows from theorem 2, we get theorem 3.

4.3.2 We turn now to the proof of the lemma. We begin with

$$\varrho\,(Q_n) := \sup \left\{ \left| \int_{-1}^{1} f(x)k(x)\,dx - Q_n[f] \right| : \|f'\| \leq 1 \right\} \geq \int_{-1}^{1} g_n^*(x)k(x)\,dx$$

$$g_n^*(x) := \min_{0 \leq \nu \leq n+1} |x - x_\nu|\,, \qquad x_\nu := x_{\nu,n}\,, \quad x_0 := -1\,, \quad x_{n+1} := 1\,.$$

A truncated version of g_n^* is $g_n(x) := \min\left(g_n^*(x), \frac{M}{n}\right)$, where $M := \left(\inf_{x \in [-\eta,\eta]} v(x)\right)^{-1}$.

For any $\epsilon > 0$ there exists a step function t with $\int_{-1}^{1}\left|k(x) - t(x)\right|dx < \epsilon$. Denote by $\xi_1, \ldots, \xi_m$ a partition of the interval $[-\eta, \eta]$ such that all RIEMANN sums of the integral $J := \int_{-\eta}^{\eta} \frac{t(x)}{v(x)}\,dx$ which are based on this partition are contained in the interval $[J - \epsilon, J + \epsilon]$. There are partitions of this type, we may even add the further hypothesis that the knots of t are a subset of $\{\xi_1, \ldots, \xi_m\}$.

We have until now

$$\varrho(Q_n) \geq \int_{-\eta}^{\eta} g_n^*(x)k(x)\,dx \geq \int_{-\eta}^{\eta} g_n(x)k(x)\,dx$$

$$\geq \int_{-\eta}^{\eta} g_n(x)t(x)\,dx - \epsilon\|g_n\| \geq \int_{-\eta}^{\eta} g_n(x)t(x)\,dx - \epsilon\frac{M}{n}$$

$$= \sum_{\nu=1}^{m-1} t(\xi_\nu + 0) \int_{\xi_\nu}^{\xi_{\nu+1}} g_n(x)\,dx - \epsilon\frac{M}{n}\,. \tag{34}$$

Let $[\xi_\nu, \xi_{\nu+1}]$ be a fixed interval and n sufficiently large. We define numbers τ and λ by

$$x_{\tau-1,n} < \xi_\nu \leq x_{\tau,n} < \cdots < x_{\lambda,n} \leq \xi_{\nu+1} < x_{\lambda+1,n}\,.$$

Using the definition of g_n and some elementary analysis we obtain

$$\int_{\xi_\nu}^{\xi_{\nu+1}} g_n(x)\,dx \geq \frac{1}{4(\lambda - \tau + 1)^2}\,(x_{\lambda,n} - x_{\tau,n})^2$$

and in the limit

$$\liminf_{n \to \infty} n \int_{\xi_\nu}^{\xi_{\nu+1}} g_n(x)\,dx \geq (\xi_{\nu+1} - \xi_\nu)^2 \left(4 \int_{\xi_\nu}^{\xi_{\nu+1}} v(x)\,dx\right)^{-1} \geq \frac{(\xi_{\nu+1} - \xi_\nu)}{4\,\max_{x \in [\xi_\nu, \xi_{\nu+1}]} v(x)}.$$

The last relation implies that for any $\epsilon > 0$ there exists a number n_0 which is independent of ν, such that

$$n \int_{\xi_\nu}^{\xi_{\nu+1}} g_n(x)\,dx > (1 - \epsilon)\frac{(\xi_{\nu+1} - \xi_\nu)}{4\,\max_{x \in [\xi_\nu, \xi_{\nu+1}]} v(x)}$$

if $n > n_0$.

Combining this inequality with (34), we get

$$n\varrho(Q_n) \geq (1 - \epsilon) \sum_{\nu=1}^{m-1} t(\xi_\nu + 0)\frac{(\xi_{\nu+1} - \xi_\nu)}{4\,\max_{x \in [\xi_\nu, \xi_{\nu+1}]} v(x)} - \epsilon M \geq (1 - \epsilon) \int_{-\eta}^{\eta} \frac{t(x)}{4v(x)}\,dx - \epsilon(M + 1)$$

$$\geq (1 - \epsilon) \int_{-\eta}^{\eta} \frac{k(x)}{4v(x)}\,dx - \epsilon(2M + 1),$$

if $n > n_0(\epsilon)$. The lemma follows.

4.4 Proof of theorem 5

We have

$$f = \frac{a_0}{2} + \sum_{\nu=1}^{\infty} a_\nu T_\nu$$

where

$$\limsup_{\nu \to \infty} \sqrt[\nu]{|a_\nu|} =: \sigma_1 < 1$$

holds (RIVLIN [26] p.145). Let $\sigma \in]\sigma_1, 1[$, and choose a k with $a_k > 0$. Since $\sigma > \limsup_{n \to \infty} \sqrt[n]{|a_n|}$, there are only finitely many l for which $a_{k+l} \geq a_k \sigma^l$ holds. Let a_m denote the last of these. Then for any $\nu > 0$ we have $a_{m+\nu} < a_k \sigma^{m+\nu-k} \leq a_m \sigma^\nu$, so there exist infinitely many numbers $m_1, m_2, \ldots$ such that

$$|a_{m_\kappa + \nu}| \leq \sigma^\nu |a_{m_\kappa}| \qquad \nu = 0, 1, \ldots \tag{35}$$

holds.

Let now $m = 2n$ be one of these numbers. If $\epsilon > 0$ is given, choose k such that $4\sigma^{2k}(1-\sigma^2) < \epsilon$ holds. We have

$$\left|R_n^G[f]\right| = \left|\sum_{\nu=m}^{\infty} a_\nu R_n^G[T_\nu]\right| \geq \left|a_m R_n^G[T_m]\right| - \sum_{\nu=1}^{\infty}\left|a_{m+\nu}R_n^G[T_{m+\nu}]\right|$$

$$\geq |a_m|\left(\left|R_n^G[T_m]\right| - \sum_{\nu=1}^{\infty}\sigma^\nu\left|R_n^G[T_{m+\nu}]\right|\right)$$

$$\geq |a_m|\left(\left|R_n^G[T_{2n}]\right| - \sum_{\nu=1}^{k-1}\sigma^{2\nu}\left|R_n^G[T_{2n+2\nu}]\right| - 4\sum_{\nu=k}^{\infty}\sigma^{2\nu}\right)$$

$$\geq |a_m|\left(\left|R_n^G[T_{2n}]\right| - \sum_{\nu=1}^{k-1}\sigma^{2\nu}\left|R_n^G[T_{2n+2\nu}]\right| - \epsilon\right).$$

On the other hand we have

$$\text{dist}(f;\mathcal{P}_{2n-1}) \leq \sum_{\nu=m+1}^{\infty}|a_\nu| \leq |a_m|\sum_{\nu=0}^{\infty}\sigma^\nu = \frac{|a_m|}{1-\sigma},$$

hence

$$\frac{R_n^G[f]}{\text{dist}(f;\mathcal{P}_{2n-1})} \geq (1-\sigma)\left(\left|R_n^G[T_{2n}]\right| - \sum_{\nu=1}^{k-1}\sigma^{2\nu}\left|R_n^G[T_{2n+2\nu}]\right| - \epsilon\right).$$

The theorem follows now immediately if we use the following result of NICHOLSON et al. [22]:

$$\lim_{n\to\infty} R_n^G[T_{2n}] = -\lim_{n\to\infty} R_n^G[T_{2n+2}] = \frac{\pi}{2},$$

$$\lim_{n\to\infty} R_n^G[T_{2n+2\mu}] = 0 \quad \text{if } \mu > 1.$$

4.5 Proof of theorem 6

We denote by $dvd(y_1,\ldots,y_n)[f]$ the divided difference of f with arguments $y_1,\ldots,y_n$. Let R be the error of the interpolatory rule with evaluation points $(-1 <) x_1 < \cdots < x_n (< 1)$ for the functional $I[f] := \int_{-1}^1 f(x)dx$. Using NEWTON's interpolation formula we obtain

$$R[f] = \int_{-1}^1 \prod_{\nu=1}^n (x - x_\nu)\, dvd(x, x_1, \ldots, x_n)[f]\, dx.$$

If $f^{(n+1)} \geq 0$ holds, then $dvd(\cdot, x_1, \ldots, x_n)[f]$ is increasing and we can apply the second mean value theorem to get

$$R[f] = dvd(-1, x_1, \ldots, x_n)[f]\int_{-1}^\xi \prod_{\nu=1}^n (x - x_\nu)\, dx + dvd(1, x_1, \ldots, x_n)[f]\int_\xi^1 \prod_{\nu=1}^n (x - x_\nu)\, dx.$$

where $\xi \in [-1,1]$ is an (unknown) number. If we now add the further hypothesis

$$\int_{-1}^1 \prod_{\nu=1}^n (x - x_\nu)\, dx = 0$$

we obtain

$$R[f] = \left(dvd(-1, x_1, \ldots, x_n)[f] - dvd(1, x_1, \ldots, x_n)[f] \right) \int_{-1}^{\xi} \prod_{\nu=1}^{n} (x - x_\nu)\, dx$$

$$= -2\, dvd(-1, x_1, \ldots, x_n, 1)[f] \int_{-1}^{\xi} \prod_{\nu=1}^{n} (x - x_\nu)\, dx .$$

We use now once more (1) (applied to dvd) to get the following bound

$$\left| R[f] \right| \leq 2 \left\| dvd(-1, x_1, \ldots, x_n, 1) \right\| \operatorname{dist}(f, \mathcal{P}_n) \sup_{\xi \in [-1,1]} \left| \int_{-1}^{\xi} \prod_{\nu=1}^{n} (x - x_\nu)\, dx \right| .$$

We finally specialize this to the situation of theorem 6, here we have

$$\prod_{\nu=1}^{n} (x - x_\nu) = 2^{-n} U_n(x)$$

where U_n denotes the CHEBYSHEV polynomial of the second kind. Using the well-known relation $T'_{n+1} = (n+1)U_n$, we obtain

$$\left| \int_{-1}^{\xi} \prod_{\nu=1}^{n} (x - x_\nu)\, dx \right| = 2^{-n} \left| \frac{T_{n+1}(\xi) - T_{n+1}(-1)}{n+1} \right| \leq \frac{2^{-n+1}}{n+1} .$$

The last step is the observation that

$$\left\| dvd(-1, x_1, \ldots, x_n, 1) \right\| = \left| dvd(-1, x_1, \ldots, x_n, 1)[T_{n+1}] \right| = 2^n$$

holds.

5. References

1. BAKER, CH. T. H. (1968) On the nature of certain quadrature formulas and their errors, *SIAM J. Numer. Anal.* 5, 783–804.

2. BERNSTEIN, S. (1918) Quelques remarques sur l'interpolation, *Math. Ann.* 79, 1–12.

3. BRASS, H. (1973) Eine Fehlerabschätzung zum Quadraturverfahren von Clenshaw und Curtis, *Numer. Math.* 21, 397–403.

4. BRASS, H. (1979) Zur Approximation einer Klasse holomorpher Funktionen im Reellen, in G. MEINARDUS (Ed.): Approximation in Theorie und Praxis, B.I. Wissenschaftsverlag, Mannheim, Wien, Zürich, 103–121.

5. BRASS, H. and SCHMEISSER, G. (1979) The definiteness of Filippi's quadrature formulae and related problems, in G. HÄMMERLIN (Ed.): Numerische Integration, Birkhäuser Verlag, Basel, 109–119.

162

6.   BRASS, H. and SCHMEISSER, G. (**1981**) Error estimates for interpolatory quadrature formulae, *Numer. Math.* 37, 371–386.

7.   BRASS, H. (**1985**) Eine Fehlerabschätzung für positive Quadraturformeln, *Numer. Math.* 47, 395–399

8.   BRASS, H. and FÖRSTER, K.-J. (**1987**) On the estimation of linear functionals, *Analysis* 7, 237-258.

9.   BRASS, H. (**1988**) Fast-optimale Formeln zur harmonischen Analyse, *ZAMM* 68, T484–T485.

10.   BRASS, H. (**1991**) Practical Fourier analysis—error bounds and complexity, *ZAMM* 71, 3–20.

11.   DAVIS, P. J. and RABINOWITZ, P. (**1984**) Methods of numerical integration (second edition), Academic Press, Orlando.

12.   DEVORE, R. A. and SCOTT, L. R. (**1984**) Error bounds for Gaussian quadrature and weighted-L^1 polynomial approximation, *SIAM J. Numer. Anal.* 21, 400–412.

13.   FISHER, S. D. (**1977**) Best approximation by polynomials, *J. Approx. Theory* 21, 43–59.

14.   FÖRSTER, K.-J. (**1982**) A comparison theorem for linear functionals and its application in quadrature, in G. HÄMMERLIN (ed.): Numerical Integration, Birkhäuser Verlag, Basel, 66–76.

15.   FÖRSTER, K.-J. and PETRAS, K. (**1990**) On estimates for weights in Gaussian quadrature in the ultraspherical case, *Math. of Comp.* 55, 243–264.

16.   FÖRSTER, K.-J. and PETRAS, K. (**1991**) Error estimates in Gaussian quadrature for functions of bounded variation, to appear in *SIAM J. Numer. Anal.* 28.

17.   FREUD, G. (**1969**) Orthogonale Polynome, Birkhäuser Verlag, Basel.

18.   HAUSSMANN, W. and ZELLER, K. (**1982**) Quadraturrest, Approximation und Chebyshev-Polynome, in G. HÄMMERLIN (ed.): Numerical Integration, Birkhäuser Verlag, Basel, 128–137.

19.   KÜTZ, M. (**1984**) Asymptotic error bounds for a class of interpolatory quadratures, *SIAM J. Numer. Anal.* 21, 167–175.

20.   LANGE, S. (**1987**) Experimente und Abschätzungen zur Polynomapproximation, Diplomarbeit, Institut für Angewandte Mathematik, TU Braunschweig.

21.   LOCHER, F. and ZELLER, K. (**1968**) Approximationsgüte und numerische Integration, *Math. Zeitschr.* 104, 249–251.

22.   NICHOLSON, D., RABINOWITZ, P., RICHTER, N. and ZEILBERGER, D. (**1971**) On the error in the numerical integration of Chebyshev polynomials, *Math. of Comp.* 25, 79–86.

23.   NIKOLSKY, S. (**1946**) On the best approximation of functions satisfying Lipschitz's conditions by polynomials (Russian), *Bulletin de l'Académie des sciences de l'URSS (Izvestija Akad. Nauk. SSSR) Série math.* 10, 295–322.

24. PETRAS, K. (1988) Asymptotic behaviour of Peano kernels of fixed order, in H. BRASS and G. HÄMMERLIN (eds.) Numerical Integration III, Birkhäuser Verlag, Basel, 186–198.

25. PETRAS, K. (1989) Normabschätzungen für die ersten Peanokerne der Gauß-Formeln, *ZAMM* 69, T81–T83.

26. RIVLIN, TH. J. (1974) The Chebyshev polynomials, Wiley, New York, London.

27. SINWEL, H. F. (1981) Uniform approximation of differentiable functions by algebraic polynomials, *J. Approx. Theory* 32, 1–8.

28. SINWEL, H. F. (1981) Konstanten in den Sätzen von Jackson und Timan, Dissertationen der Johannes-Kepler Universität Linz, Bd. 27, VWGÖ Wien.

29. STROUD, A. H. (1966) Estimating quadrature errors for functions with low continuity, *SIAM J. Numer. Anal.* 3, 420–424.

30. TIMAN, A. F. (1963) Theory of approximation of functions of a real variable, Pergamon Press, Oxford.

ONE SIDED L1-APPROXIMATION
AND BOUNDS FOR PEANO KERNELS

Knut Petras
Institut für Angewandte Mathematik
Technische Universität Braunschweig
Pockelsstr. 14
3300 Braunschweig
Germany

ABSTRACT. We investigate the Peano kernels of positive quadrature formulae. The kernels that we shall consider have a fixed argument, while the formulae have degree at least m, where m is some integer. We will show that there is a duality between these values and the error of polynomial, one sided, L1-approximation of maximal degree m. Using this duality, it will be possible to determine the set of quadrature formulae (extremal formulae) for which extreme values of Peano kernels are attained. Such results obviously yield error bounds for numerical integration; however, it is also possible to express the error constants for one sided L1-approximation in terms of quadrature errors. In some, special cases, the extremal formulae which generate an extremum of a Peano kernel at a certain point, are determined explicitly. In the final section, asymptotic bounds on Peano kernels, and hence asymptotic error constants in one sided approximation, are derived from the asymptotic estimates for the Peano kernels of the corresponding extremal formulae.

1. Introduction and notation

One sided approximation is closely related to the estimation of monotone functionals. If L is such a functional, and if p and P are one sided approximations of a function f, i.e., $p \leq f \leq P$, then we have $L[p] \leq L[f] \leq L[P]$.

In this paper, we investigate the case that L is a positive quadrature formula Q_n, i.e.,

$$(1.1) \qquad Q_n[f] = \sum_{\nu=1}^{n} a_\nu f(x_\nu),$$

with nodes x_ν, where $-1 \leq x_1 < x_2 < \ldots < x_n \leq 1$, and with positive weights a_ν. Such formulae should estimate the integral

$$(1.2) \qquad I[f] = \int_{-1}^{1} f(x)\, d\alpha(x)$$

165

T. O. Espelid and A. Genz (eds.), Numerical Integration, 165–174.
© 1992 Kluwer Academic Publishers.

with respect to a positive Borel measure $d\alpha$. We thus look for a best, one sided L1-approximation on the interval $[-1,1]$ from a subspace of polynomials of given degree for the measure $d\alpha$ in order to obtain bounds for the corresponding remainder,

$$(1.3) \qquad R_n[f] = I[f] - Q_n[f].$$

The quadrature formulae that have been given the most attention in literature are the positive quadrature formulae Q with a high algebraic degree

$$(1.4) \quad \deg(R) = m \qquad :\Leftrightarrow \begin{cases} R[u_\nu] = 0 & \text{for } \nu = 0, 1, \ldots, m \\ R[u_{m+1}] \neq 0; \end{cases} \qquad u_\nu(x) := x^\nu$$

of its remainder. These include the positive quadrature formulae of interpolatory type ($\deg(R_n) \geq n - 1$, e.g., Clenshaw-Curtis, Polya, Filippi formulae etc.) and, in particular, those with $\deg(R_n) \approx 2n$ (Gaussian, Radau, Lobatto formulae etc.). Generally, we define the class $\mathfrak{P}_m$ of quadrature formulae by

$$(1.5) \qquad \mathfrak{P}_m := \{Q \mid Q \text{ is a positive quadrature formula} \\ \text{of the form } (1.1) \text{ and } \deg(R) \geq m\}.$$

A useful representation of quadrature errors is given by the Peano kernel theory. Indeed, let $\deg(R) \geq s - 1$, then,

$$(1.6) \qquad R[f] = \int_{-1}^{1} K_s(R, z) f^{(s)}(z)\, dz,$$

for all functions f with an absolutely continuous $(s - 1)^{th}$ derivative, where

$$(1.7) \qquad K_s(R, z) := R[h_{z,s}]$$

is the sth Peano kernel of R at the point z, and the functions $h_{z,s}$ given by

$$(1.8) \qquad h_{z,s}(x) := \frac{(x - z)_+^{s-1}}{(s - 1)!} := \begin{cases} 0 & \text{for } x < z \text{ and} \\ \dfrac{(x - z)^{s-1}}{(s - 1)!} & \text{for } x \geq z, \end{cases}$$

are the truncated power functions.

Our main result (Theorem 2.1) is a duality between the one sided approximations of truncated power functions from Π_m, i.e., approximations with polynomials of degree m, and Peano kernels of quadrature formulae from $\mathfrak{P}_m$. It will be shown under certain restrictions on α and s that the maximal (minimal) value of the Peano kernels with a fixed argument is equal to the error of the best one sided approximation from below (resp. from above) for the corresponding truncated power function. We will therefore introduce the notation

$$(1.9) \qquad C_{s,m}(z) := \inf \left\{ \int_{-1}^{1} [h_{z,s}(x) - p(x)]\, d\alpha(x) \mid p \leq h_{z,s} \text{ on } [-1,1], p \in \Pi_m \right\}$$

and

$$(1.10) \qquad c_{s,m}(z) := \sup \left\{ \int_{-1}^{1} [h_{z,s}(x) - P(x)] \, d\alpha(x) \mid P \geq h_{z,s} \text{ on } [-1,1], \ P \in \Pi_m \right\}.$$

If we could determine the functions $C_{s,m}$ and $c_{s,m}$ which separate approximation errors and values of Peano kernels, we would have the sharpest possible bounds for all Peano kernels of quadrature formulae $Q \in \mathfrak{P}_m$. Moreover, we would also obtain error coefficients for one sided L1-approximation (cf. Section 4). We shall give a weaker result, namely that of characterizing the set of quadrature formulae for which values of the separation functions are attained. In particular, we shall determine those extremal formulae which generate, $C_{2,m}(z)$, $c_{2s,m}(0)$ and $C_{2s,m}(0)$, explicitly (cf. Section 5). Since all extremal formulae exhibit a certain structure, asymptotic estimates for the separation functions may be calculated for some measures $d\alpha$ enabling asymptotic estimates for error coefficients of one sided L1-approximation to computed (cf. Section 6).

ASSUMPTION: For much of the following it will be necessary, and therefore generally presupposed, that the carrier $C(\alpha)$ of α has sufficiently many points, $|C(\alpha)| \geq m+1$, when approximating with polynomials of degree m.

2. A duality relation

Our starting point is the following simple duality between one sided L1-approximation and values of Peano kernels:

THEOREM 2.1: *Let* $|C(\alpha)| \geq m+1$ *and* $s \leq m+1$, *then,*

$$(2.1) \qquad \sup_{Q \in \mathfrak{P}_m} K_s(R, z) = C_{s,m}(z)$$

and

$$(2.2) \qquad \inf_{Q \in \mathfrak{P}_m} K_s(R, z) = c_{s,m}(z).$$

Before proving this, we shall first exclude the following, trivial cases:

LEMMA 2.1: *Let* $n = [(m+2)/2]$, *and* x_ν *be a node of the* n *-point Gaussian formula* Q_n^G *for odd* m, *and a node of the* n *-point Radau formula* $Q_n^{Ra,-1}$ *with node at* -1 *for even* m, *and let* $z > -x_n$. *Then a best one sided L1-approximation of* $h_{z,s}$ *from* Π_m *may have infinitely many points in common with* $h_{z,s}$ *if and only if* $z \geq x_n$, *and the approximation is from below. In this latter case the approximation is identically zero.*

REMARK 2.1: By reflecting the measure $d\alpha$ and the approximations of truncated power functions, we readily obtain a corresponding result for $z \leq -x_n$ for odd m. Some distinctions of cases are necessary as indicated in the proof.

REMARK 2.2: It may easily be seen that the best approximation from below is not unique if $z = x_n$ and $s = 1$ or $s = 2$. (Recall that best approximations have the form, $\varepsilon(x - x_n)\prod_{\nu=1}^{n-1}(x-x_\nu)^2$, where $\varepsilon > 0$, and where m is odd, or $\varepsilon(x-x_n)(x-x_1)\prod_{\nu=2}^{n-1}(x-x_\nu)^2$, for $\varepsilon > 0$, when m is even.) Otherwise, for $z \geq x_n$, the zero function is uniquely determined as the best approximation from below.

PROOF OF THE LEMMA: Let m be odd, let $z \geq x_n$ and let $p \in \Pi_m$ be the best one sided approximation of $h_{z,s}$ from below. Then, $I[p] = Q_n^G[p] \leq 0$, and $\tilde{p}(x) \equiv 0$ also is a best one sided approximation.

Now let $z < x_n$. If x^* is chosen such that $z < x^* < x_n$, we can find an $\varepsilon > 0$ so that if $p \in \Pi_m$ is defined by the formula,

$$(2.3) \qquad p(x) = \varepsilon(x - x^*)\prod_{\nu=1}^{n-1}(x - x_\nu)^2,$$

then p is a one sided approximation of $h_{z,s}$ from below, and satisfies the condition, $I[p] = Q_n^G[p] = a_n p(x_n) > 0$. Hence $\tilde{p}(x) \equiv 0$ is not the best one sided approximation.

For even m, we use the Radau formula analogously, and define, $p(x) = (x - x^*)(x - x_1)\prod_{\nu=2}^{n-1}(x - x_\nu)^2$, instead of (2.3).

The approximation may thus be zero only in the cases given in the Lemma. Furthermore, $q_{z,s}(x) = (x - z)^{s-1}/(s - 1)!$ might be the best approximation from below if s is even and from above if s is odd. We would then subtract $q_{z,s}$ from $h_{z,s}$ and reflect the measure. If $q_{z,s}$ were the solution of the original approximation problem, the zero function would be the solution of the transformed problem, and this case has already been treated above. If the best approximation is not identically zero and not equal to $q_{z,s}$, it coincides with $h_{z,s}$ only at finitely many points. $\square$

PROOF OF THE THEOREM: Let $s \geq 2$, and let $p \in \Pi_m$ be a nontrivial best one sided L1-approximation of $h_{z,s}$ from below. If there is no zero of $h_{z,s} - p$ at z, then $h_{z,s} - p$ has at most $[(m + 1)/2]$ distinct zeros in each of the intervals $[-1, z)$ and $(z, 1]$, since $h_{z,s} - p$, when restricted to each of these intervals, is a polynomial of degree m and all zeros, except perhaps one, are multiple zeros. If there is a zero at z, then $h_{z,s} - p$ has at most $[(m + 2)/2]$ zeros on $[-1, z]$, and $[m/2]$ on $(z, 1]$. Therefore, $h_{z,s} - p$ has at most $m + 1$ zeros, which we shall denote by, $x_1, \ldots, x_n$. By Theorem 4.1 of DeVore [1969], there is a positive quadrature formula $Q_n^G \in \mathfrak{P}_m$ with nodes x_ν. For each quadrature formula $Q \in \mathfrak{P}_m$, we have $Q[h_{z,s}] \geq Q[p] = I[p] = Q_n^*[p] = Q_n^*[h_{z,s}]$, such that,

$$(2.4) \qquad K_s(R, z) = I[h_{z,s}] - Q[h_{z,s}] \leq I[h_{z,s}] - Q_n^*[h_{z,s}] = K_s(R_n^*, z).$$

An analogous argumentation yields the lower bound, $c_{s,m}(z)$. For $s = 1$, we can give a similar argument using extreme values instead of the zeros of $h_{z,1} - p$.

If the best approximation is identically zero, its error is attained, either by the Peano kernel of R_n^G, or else by the Peano kernel of $R_n^{Ra,-1}$. $\square$

3. The extremal formulae

Theorem 2.1 is only valuable if we have a prospect of being able to determine, either the best one sided L1-approximations of truncated power functions, or else a quadrature formula for which the supremum or infinum of the respective Peano kernel is attained. In the following, we shall call $Q \in \mathfrak{P}_m$ an *extremal formula (with respect to $\mathfrak{P}_m$)* if any nontrivial (cf. Lemma 2.1) value of $C_{s,m}$ or $c_{s,m}$ is attained by $K_s(R, \cdot)$ for any argument and any value of s. Since the first Peano kernels may be discontinuous, we must consider one sided limits.

We also say that a quadrature formula $Q \in \mathfrak{P}_m$ is a representation of the moment space,

$$(3.1) \qquad \mathfrak{M}_{m+1} = \left\{ c = (c_0, \ldots, c_m) \in \mathbb{R}^{m+1} \mid c_i = \int_{-1}^{1} x^i d\alpha(x) \right\}$$

(cf. Karlin and Studden [1966, p.42]). Let the index $\mathrm{Ind}(Q)$ be the number of nodes of Q, where nodes at -1 and 1 are each counted as half nodes. For each point $t \in [-1, 1]$, there is therefore a unique representation, $Q^{(t)}$, with $\mathrm{Ind}(Q^{(t)}) \leq (m + 2)/2$, which has a node at t. These formulae are called canonical representations of $\mathfrak{M}_{m+1}$.

THEOREM 3.1: *The set of extremal formulae is equal to the set of canonical representations.*

REMARK 3.1: This theorem is in some sense an extension of Bernstein's result ([1937, Theorem 3]) on sums of weights of positive quadrature formulae. He proved that $C_{1,m}(t) = \lim_{x \to t+0} K_1(R^{(t)}, x)$, and $c_{1,m}(t) = \lim_{x \to t-0} K_1(R^{(t)}, x)$, where the limits denote the limits on the right and left at t, respectively. The other important and known part of Theorem 3.1 is the case in which $s = m + 1 = 2n$ which corresponds to having a Peano kernel of even, highest order. In this case, only the $n + 1$-point Lobatto formula generates $c_{s,m}$, while only the n-point Gaussian formula generates $C_{s,m}$ (cf. Braß and Schmeißer [1981]). Furthermore, some points of $C_{2,m}$ and $c_{2,m}$ that are generated by the Gaussian or the Lobatto formulae are given in Förster [1982].

PROOF OF THE THEOREM: We must prove the statement for $s > 2$, or for $s = 2$ with the approximation from above. The case $s = 2$ is proved in Theorem 5.1. For $s = 1$, see Remark 3.1.

Suppose that the best approximation is not one of the trivial polynomials mentioned in the proof of Lemma 2.1. Furthermore, let $p \in \Pi_m$ be a best, one sided L1-approximation of $h_{z,s}$, and suppose that $p - h_{z,s}$ has l zeros, where zeros are counted according to their multiplicities. Hence $(p - h_{z,s})^{(s-2)}$ has at least $l - s + 2$ zeros. Differentiating the restrictions twice on the intervals $[-1, z]$ and $[z, 1]$, we see that $p^{(s)} \in \Pi_{m-s}$ has at least $l - s - 2$ zeros such that $l - s - 2 \leq m - s$; i.e., $l - 2 \leq m$. Since each inner zero of $h_{z,s} - p$ is a multiple zero, the index of the quadrature formula corresponding to the one sided approximation p is less than or equal to $l/2 \leq (m + 2)/2$. □

4. Error coefficients for one sided L1-approximation

Let

(4.1)
$$V_s := \{f \mid \operatorname{Var} f^{(s-1)} \leq 1\}$$

and

(4.2)
$$V_{s,m}(d\alpha) := \sup_{f \in V_s} \inf_{\substack{P \in \Pi_m \\ P \geq f}} \int_{-1}^{1} [P(x) - f(x)] \, d\alpha(x)$$

$$= \sup_{f \in V_s} \inf_{\substack{p \in \Pi_m \\ p \leq f}} \int_{-1}^{1} [f(x) - p(x)] \, d\alpha(x).$$

It is obvious that equality in (4.2) holds as V_s is symmetric; i.e., $f \in V_s$ is equivalent to $-f \in V_s$. The error constants $V_{s,m}(d\alpha)$ are the least possible constants satisfying the inequality,

(4.3)
$$\inf_{\substack{P \in \Pi_m \\ P \geq f}} \int_{-1}^{1} [P(x) - f(x)] \, d\alpha(x) \leq V_{s,m}(d\alpha) \cdot \operatorname{Var} f^{(s-1)},$$

for all functions which have an $(s-1)^{th}$ derivative of bounded variation. We will show that these coefficients may be given explicitly as follows.

THEOREM 4.1:

(4.4)
$$V_{s,m}(d\alpha) = \max \{\|C_{s,m}\|_\infty, \|c_{s,m}\|_\infty\}$$

Bounds on the error constants in one sided L1-approximation now readily result from bounds on Peano kernels. For example, from the results in Förster and Petras [1991], we obtain:

COROLLARY 4.1: *Let* $d\alpha(x) = dx$ *and* $n = [(m+1)/2]$, *then,*

(4.5)
$$\frac{\pi}{2n+1}\left\{1 - \frac{2}{2n+1}\right\} \leq V_{1,m}(d\alpha) \leq \frac{\pi}{2n+1}\left\{1 - \sin\frac{\pi}{2n+1}\right\}^{-1}$$

By investigating only truncated power functions, it is easily seen that the left-hand side in Theorem 4.1 is greater than or equal to the right-hand side.

Freud [1955] has determined one sided approximations $g_{z,s}$ of truncated power functions $h_{z,s}$, and has used them to estimate the representation,

(4.6)
$$f(x) = p(x) + \frac{1}{s!} \int_{-1}^{1} h_{z,s}(x) \, df^{(s-1)}(z), \qquad p \in \Pi_{s-1}.$$

His method requires the integrability of the approximations $g_{z,s}$ with respect to the parameter z. This might not be the case for the best one sided approximations. The Theorem is, however, proved if we can show that we loose arbitrarily little if we replace the best approximations by the same approximation in a piecewise manner. This will be done in the following two lemmata.

LEMMA 4.1: For $z \in (-1, 1)$, the derivatives of all best one sided approximations of the functions $h_{z,s}$ are uniformly bounded on each compact set.

PROOF: By Theorem 3.1, for each $x \in (-1, 1)$, there is an extremal formula $Q \in \mathfrak{P}_m$ with a node at x. Since the inner weights of all extremal formula can be expressed as reciprocals of only two, different, nonvanishing polynomials (cf. Karlin and Studden [1966, Theorem IV.3.1]), they are bounded below by a fixed number, $\underline{a} > 0$. Let p be a best approximation from below of a truncated power function with $p(x) \leq 0$. From $\|h_{z,s}\|_\infty \leq 2^{s-1}/(s-1)!$, and $\sum_{\nu=1}^n a_\nu = \|I\|$ for extremal formulae, it follows that $I[p] = Q[p] \leq \underline{a}\, p(x) + \|I\| 2^{s-1}/(s-1)!$. The zero function is also an approximation from below, such that $I[p] \geq I[0] = 0$, and hence,

$$(4.7) \qquad p(x) \geq -\frac{2^{s-1}\|I\|}{(s-1)!\,\underline{a}}.$$

Analogously, we can show that the best approximations from above are bounded above simultaneously by a fixed number. By Markoff's inequality, the derivatives are also uniformly bounded. $\qquad\square$

LEMMA 4.2: Let $p_{z,s}$ and $P_{z,s}$ be the best one sided approximations of $h_{z,s}$ from below and above, respectively. Then, for each $\varepsilon > 0$, there is a $\delta > 0$, such that, for each interval $[a - \delta, a] \subset [-1, 1]$, the one sided approximations $P_{a-\delta,s}$ and $p_{a,s}$ satisfy the relations,

$$(4.8) \qquad \int_{-1}^1 \left\{ P_{a-\delta,s}(x) - h_{z,s}(x) \right\} d\alpha(x) \leq \int_{-1}^1 \left\{ P_{z,s}(x) - h_{z,s}(x) \right\} d\alpha(x) + \varepsilon$$

and

$$(4.9) \qquad \int_{-1}^1 \left\{ h_{z,s}(x) - p_{a,s}(x) \right\} d\alpha(x) \leq \int_{-1}^1 \left\{ h_{z,s}(x) - p_{z,s}(x) \right\} d\alpha(x) + \varepsilon,$$

for all $z \in [a - \delta, a]$.

PROOF: Let $\delta > 0$, let $z \in [a - \delta, a]$ and let

$$(4.10) \qquad M := \sup_{z \in [-1,1]} \|p'_{z,s}\|_*,$$

172

where $\|\cdot\|_*$ denotes the supremum norm on the interval $[-3,3]$. If we Define $p(x) := p_{z,s}(x-(a-z))$, we obtain $\|p-p_{z,s}\|_* \leq M\delta$. Hence, $p-M\delta$ is a one sided approximation of $h_{a,s}$ from below such that

$$(4.11) \qquad \int_{-1}^{1} p_{a,s}(x)\,d\alpha(x) \geq \int_{-1}^{1} p_{z,s}(x)\,d\alpha(x) - 2M\delta \int_{-1}^{1} d\alpha(x).$$

Choosing

$$(4.12) \qquad \delta = \frac{\varepsilon}{2M \int_{-1}^{1} d\alpha(x)},$$

we have the assertion (4.9). Formula (4.8) is proved analogously thereby proving Theorem 4.1. $\qquad\square$

5. Examples

In this section, we will give some examples where the extremal formulae corresponding to a certain truncated power function may be determined explicitly.

A. We first generalize a result of Bojanic and DeVore [1966] on the best one sided approximation of $f(x) = |x|$ which, by Theorem 2.1, is equivalent to Förster's result (cf. [1982]) on the values of $K_2(R,0)$

THEOREM 5.1: *Let $d\alpha$ be symmetric, and let Q_n^{Lo} be the n-point Lobatto formula, then,*
i) $C_{2s,m}(0) = K_{2s}(R_n^G,0)$ and $c_{2s,m}(0) = K_{2s}(R_{n+1}^{Lo},0)$ for
  a) $m = 4k$ or $m = 4k+1$, $n = 2k+1$ and odd s, or
  b) $m = 4k+2$ or $m = 4k+3$, $n = 2k+2$ and even s.
ii) $C_{2s,m}(0) = K_{2s}(R_{n+1}^{Lo},0)$ and $c_{2s,m}(0) = K_{2s}(R_n^G,0)$ for
  a) $m = 4k$ or $m = 4k+1$, $n = 2k+1$ even s, or
  b) $m = 4k+2$ or $m = 4k+3$, $n = 2k+2$ and odd s.

PROOF: The quadrature formulae for this problem are exact for $g(x) = x^{2s-1}/(2s-1)!$, whence the corresponding approximation problem is equivalent to the approximation problem of the symmetric function $f = (h_{0,2s} - g)/2$. By the uniqueness of the best approximation (DeVore [1968, Theorem 3.3]), the corresponding quadrature formula is symmetric and hence either the Gaussian or the Lobatto formula. Förster [1982] has shown that the second Peano kernels of these formulae are positive at 0 if and only if there is a node at the origin. By repeated integration and by symmetry arguments, the statement follows. $\square$

B. Förster [1982] has proved that Gauß and Lobatto formulae are extremal formulae for the second Peano kernel at their nodes. A description of the extremal formulae which generate $C_{2,m}$ at prescribed arguments is given in the following theorem.

THEOREM 5.2: *Let* $Q^{(z)} \in \mathfrak{P}_m$ *be the canonical representation involving the point* z, *then,*

$$(5.1) \qquad C_{2,m}(z) = K_2(R^{(z)}, z).$$

PROOF: Between any two points, $z_1, z_2 \in (-1, 1)$, where $z_1 \neq z_2$, and $z_i \neq z$, for which $h_{z,2}$ is tangent to its best one sided L1-approximation p, there are at least two points of inflection of p. Furthermore, there is a point of inflection between a point of contact at the boundary and a point of tangency different from z in $(-1, 1)$. Let p and $h_{z,s}$ have n points of tangency in $(-1, 1)$ that are distinct from z. Then, p is at least of degree $2n + j \leq m$, where j denotes the number of points of p in common with $h_{z,2}$ at the boundary. If z were a point of contact with $h_{z,2}$, i.e., a node of Q, we would obtain a quadrature formula of algebraic degree greater than or equal to $2n + j$ with $n + j$ nodes, which is impossible. $\qquad \square$

6. Asymptotic representation of the separation function

Let w be a weight function, i.e., $w(x)dx = d\alpha(x)$. The asymptotic behaviour of Peano kernels when the order s is fixed has been investigated for certain weight functions and quadrature rules, (cf. Petras [1988]). In particular, these results can be applied to formulae Q_n with $\deg(R_n) \geq 2n - 2$. Since the extremal formulae are closely related to such quadrature formulae, (cf. Karlin and Studden [1966, Theorem IV.3.1]), we conclude that the same asymptotic behaviour holds for extremal formulae. Thus we obtain:

THEOREM 6.1: *Let* w *be a weight function, let* $M \in \mathbb{R}^+$, *let* q *a polynomial with the zero set* $N = \{-1, \xi_1, \ldots, \xi_l, 1\} \subset [-1, 1]$, *such that*

$$(6.1) \qquad q^2(x) \leq w(x) \leq \frac{M}{\sqrt{1 - x^2}},$$

for all $x \in E := [-1, 1] \setminus N$, *and let* w *be continuous in* $[a, b] \subset E$. *Then, we have uniformly for all* $x \in [c, d] \subset (a, b)$, *that*

$$(6.2) \qquad \lim_{m \to \infty} m^s c_{s,m}(x) = \left(2\pi \sqrt{1 - x^2}\right)^s w(x) \min_{t \in [0,1]} B_s(t)$$

and

$$(6.3) \qquad \lim_{m \to \infty} m^s C_{s,m}(x) = \left(2\pi \sqrt{1 - x^2}\right)^s w(x) \max_{t \in [0,1]} B_s(t),$$

where B_s *is the* s^{th} *Bernoulli-polynomial (cf. Braß [1977, pp.52 ff.]).*

If the values of Peano kernels cannot only be estimated in such fixed subintervals as those described in Theorem 6.1, but also for all $x \in [-1, 1]$, the best error constant for one sided

174

L1-approximation may be estimated according to Theorem 4.1. For the ultraspherical case $d\alpha(x) = (1 - x^2)^{\lambda - 1/2} dx$, a result of Nevai [1972] shows that $C_{s,m}$ and $|c_{s,m}|$ decrease sufficiently fast to zero when the argument tends to the boundary. Hence, Theorem 6.1 yields:

THEOREM 6.2: *Let* $d\alpha(x) = (1 - x^2)^{\lambda - 1/2} dx$, *where* $\lambda > 0$, *and let* s *be arbitrary, or* $d\alpha(x) = (1 - x^2)^{-1/2} dx$ *and* $s \geq 2$, *then,*

$$(6.4) \qquad \lim_{m \to \infty} m^s V_{s,m}(d\alpha) = (2\pi)^s \|B_s\|_\infty.$$

REMARK 6.1: Comparing the asymptotic behaviour given in Theorem 6.2 with Freud's bounds [1955], we see that his bounds may be improved asymptotically by a factor of about $\sqrt{(2s-1)/4} \cdot e^{s+1} \cdot \pi^{s-1/2}$ for odd s and $5\sqrt{s/8} \cdot e^s \cdot \pi^{s-1/2}$ for even s.

References

1. S. N. BERNSTEIN, On quadrature formulae with positive coefficients, *Izv. Akad. Nauk SSSR, Ser. Mat.*, 4 (1937), 479-503. (Russian)

2. R. BOJANIC AND R. A. DEVORE, On polynomials of best one sided approximation, *L'Enseignement Math.* 12 (1966), 139-164.

3. H. BRASS, *Quadraturverfahren*, Vandenhoeck & Ruprecht, Göttingen, 1977.

4. H. BRASS AND G. SCHMEISSER, Error estimates for interpolatory quadrature formulae, *Num. Math.* 37 (1981), 371-386.

5. R. A. DEVORE, One-sided approximation of functions, *J. Approx. Th.* 1 (1968), 11-25.

6. K.-J. FÖRSTER, A comparison theorem for linear functionals and its application in quadrature, *Numerical Integration (G. Hämmerlin, ed.)*, ISNM 57, Birkhäuser, Basel 1982, 66-76.

7. K.-J. FÖRSTER AND K. PETRAS, Error estimates in Gaussian quadrature for functions of bounded variation, *SIAM J. Numer. Anal.* 18 (1991). (to appear)

8. G. FREUD, Über einseitige Approximation durch Polynome I, *Acta Sci. Math. (Szeged)* 16 (1955), 12-28.

9. KARLIN AND STUDDEN, *Tchebycheff Systems: with Applications in Analysis and Statistics*, Interscience Publishers, Wiley & Sons, 1966.

10. G. P. NEVAI, Einseitige Approximation durch Polynome, mit Anwendungen, *Acta Math. Acad. Sci. Hungaricae* 23 (1972), 495-506.

11. K. PETRAS, Asymptotic behaviour of Peano kernels of fixed order. *Numerical Integration III (G. Hämmerlin and H. Braß, ed.)*, ISNM 85, Birkhäuser, Basel 1988, 186-198.

AN ALGEBRAIC STUDY OF THE LEVIN TRANSFORMATION IN NUMERICAL INTEGRATION

RICOLINDO CARIÑO*
La Trobe University
Bundoora VIC 3083
Australia

IAN ROBINSON
La Trobe University
Bundoora VIC 3083
Australia

ELISE DE DONCKER
Western Michigan Univ.
Kalamazoo MI 49008
USA

ABSTRACT. The asymptotic error expansion for the m^N-copy quadrature rule approximation $Q^{(m)}f$ to $If = \int f(\bar{x})d\bar{x}$ over the hypercube H_N is known for some integrand functions $f(\bar{x})$. In this paper, we use Macsyma to investigate the outcome of applying the Levin transformation to a sequence of quadrature approximations $Q^{(m)}f$, $m = n + 1, n + 2, \ldots$ for which these known forms of error expansion are valid. The resulting expansions suggest that in certain cases, iterated extrapolation by a low-order Levin transformation is capable of obtaining approximations more accurate than those obtained by the straightforward use of the transformation. Numerical results are presented to illustrate the effectiveness of the iterated transformation in these cases.

1. Introduction

Let S_m, $m = 1, 2, \ldots$ be an infinite convergent sequence with limit S. Assume that S_m has the form

$$S_m = S + R_m p(m), \quad m = 1, 2, \ldots, \tag{1}$$

where R_m is an estimate of the error in S_m, and $p(x)$, considered as a function of the continuous variable x, is continuous for all $x \geq 1$, including $x = \infty$, and as $x \to \infty$, $p(x)$ has a Poincaré-type asymptotic expansion in inverse powers of x, given by

$$p(x) \sim \sum_{i=0}^{\infty} \beta_i / x^i, \text{ as } x \to \infty, \ \beta_0 \neq 0. \tag{2}$$

Then the Levin transformation is suitable for accelerating convergence to S [5]. Choosing $R_m = m a_m$ where $a_1 = S_1$, and $a_m = S_m - S_{m-1}, m \geq 2$, leads to the Levin u-transformation (see [3]).

From (1) and (2), the limit S and the first k coefficients β_i, $i = 0, 1, \ldots, k - 1$ can be approximated by T_{kn} and the constants γ_i, $i = 0, 1, \ldots, k - 1$, respectively, by solving the $k + 1$ linear equations

$$S_m = T_{kn} + R_m \sum_{i=0}^{k-1} \gamma_i / m^i, \quad m = n, n + 1, \ldots, n + k.$$

*On leave from the University of the Philippines at Los Baños, Laguna, Philippines.

T. O. Espelid and A. Genz (eds.), Numerical Integration, 175–186.
© 1992 Kluwer Academic Publishers.

A closed-form solution for T_{kn} is given in [3] by

$$T_{kn} = \frac{\sum_{j=0}^{k}(-1)^j \binom{k}{j}(n+j)^{k-1}(S_{n+j}/R_{n+j})}{\sum_{j=0}^{k}(-1)^j \binom{k}{j}(n+j)^{k-1}(1/R_{n+j})}, \tag{3}$$

with error

$$\mathcal{E}_{kn} = S - T_{kn} = \frac{\sum_{j=0}^{k}(-1)^j \binom{k}{j}(n+j)^{k-1}(E_{n+j}/R_{n+j})}{\sum_{j=0}^{k}(-1)^j \binom{k}{j}(n+j)^{k-1}(1/R_{n+j})}, \tag{4}$$

where $E_{n+j} = S - S_{n+j}$, and $R_{n+j} \neq 0$, $j = 0, 1, \ldots, k$.

In this paper, we are interested in sequences of quadrature approximations $\mathcal{Q}^{(m)}f$ to an integral $If = \int_{H_N} f(\tilde{x})d\tilde{x}$ where $f(\tilde{x}) = w(\tilde{x})g(\tilde{x})$, and $g(\tilde{x})$ is regular over the unit hypercube H_N. If $\mathcal{Q}f = \sum_{i=1}^{\nu} \delta_i f(\tilde{x}_i)$ is a quadrature rule approximation to If, then $\mathcal{Q}^{(m)}f$ is the m^N-copy version which is obtained by equally subdividing H_N into m^N smaller hypercubes and applying a properly scaled version of $\mathcal{Q}f$ to each. For the remainder of the paper, we assume the correspondence $S_m = \mathcal{Q}^{(m)}f$, $S = If$ and $E_m = If - \mathcal{Q}^{(m)}f$.

Of the known asymptotic error expansions for E_m, only a few can be written in the form given by (1) and (2). Instances include the cases in which

- $w(\tilde{x}) = 1$, and

- $w(x) = x^\lambda(1-x)^\lambda$, $\lambda > -1$.

However, if

- $w(\tilde{x}) = r^\alpha h(\tilde{x})\phi(\theta)$, where $r^2 = x_1^2 + x_2^2 + \cdots + x_N^2$, (r, θ) is the hyperspherical coordinate of $\tilde{x}$, and $h(\tilde{x})$ and $\phi(\theta)$ are analytic in their respective variables, or

- $w(\tilde{x}) = x_k^\lambda$, $1 \leq k \leq N$, $\lambda > -1$,

then the resulting error expansion for $\mathcal{Q}^{(m)}f$ can be written in the form $E_m = R_m p(m) + Z^{(d)}(m)$, where the remainder term $Z^{(d)}(m) \to 0$ as m increases, depending on the degree of precision d of the quadrature rule used. In these cases, if d is sufficiently high, then the Levin u-transformation becomes a useful and efficient extrapolation method (see [1]).

A more general and common form of error expansion than (1) which arises in numerical quadrature is

$$E_m = If - \mathcal{Q}^{(m)}f = G_m p(m) + H_m q(m), \tag{5}$$

where $q(x)$, considered as a function of the continuous variable x, has an expansion of the form (2). This "mixed" form of E_m is associated with an integrand function $f(\tilde{x})$ which has an algebraic and/or logarithmic singularity on a boundary of the region of integration. In these cases, G_m is of the form $1/m^s$ and H_m is of the form $1/m^t$ or $\log m/m^t$, respectively. We investigate the effect of applying the Levin u-transformation (3) in such cases.

2. Symbolic Computation of the Levin Transformation

We use the computer algebra system Macsyma [4] to symbolically compute $\mathcal{E}_{kn}$ based on (4) for the following forms of error expansion, all of which arise in numerical integration:

$$E_m = \frac{a_0}{m^s} + \frac{a_1}{m^{s+1}} + \frac{a_2}{m^{s+2}} + \cdots, \quad s > 0, \tag{6}$$

$$
\begin{aligned}
E_m = {} & \frac{a_0}{m^s} + \frac{a_1}{m^{s+1}} + \frac{a_2}{m^{s+2}} + \cdots \\
& + \frac{b_0}{m^t} + \frac{b_1}{m^{t+1}} + \frac{b_2}{m^{t+2}} + \cdots, \quad t > s > 0,
\end{aligned}
\tag{7}
$$

$$E_m = \frac{a_0 + b_0 \log m}{m^s} + \frac{a_1 + b_1 \log m}{m^{s+1}} + \frac{a_2 + b_2 \log m}{m^{s+2}} + \cdots, \quad s > 0, \tag{8}$$

$$
\begin{aligned}
E_m = {} & \frac{a_0}{m^s} + \frac{a_1}{m^{s+1}} + \frac{a_2}{m^{s+2}} + \cdots \\
& + \frac{b_0 \log m}{m^t} + \frac{b_1 \log m}{m^{t+1}} + \frac{b_2 \log m}{m^{t+2}} + \cdots, \quad t > s > 0.
\end{aligned}
\tag{9}
$$

The first of these expansions can be expressed in the form (1) and thus the Levin u-transformation is effective in successively eliminating the terms in the expansion. The other three expansions can be expressed in the form (5).

As operations on infinite series are restricted in Macsyma, we use truncated versions of the above expansions when substituting them into (4) (with $m = n, n+1, \ldots, n+k$) for the computation of $\mathcal{E}_{kn}$. In addition, we use the Macsyma function taylor() to generate truncated Taylor series expansions for each of $E_{n+1}, E_{n+2}, \ldots, E_{n+k}$ in terms of n. We include only the necessary number of terms in the these truncated Taylor series expansions to ensure the correctness of the leading terms in $\mathcal{E}_{kn}$. If j is the number of terms in the truncated version of E_n and l is the level of truncation in the Taylor series expansions for $E_{n+1}, E_{n+2}, \ldots, E_{n+k}$, then the terms in $\mathcal{E}_{kn}$ which remain constant after systematically increasing j and l during the computations are presumed to be correct. (See Appendix for a sample Macsyma script).

3. Results

When the form of E_n is given by (6), application of the Levin u-transformation produces the following expansions for $\mathcal{E}_{1n}$ and $\mathcal{E}_{2n}$:

$$\mathcal{E}_{1n} = \frac{a_0 s(s+1) + 2a_1}{2s^2 n^{s+1}} + \frac{a_2'}{n^{s+2}} + \cdots \tag{10}$$

and

$$\mathcal{E}_{2n} = \frac{a_0^2 s^2 (s^2 - 1) - 24a_0 a_2 s + 12a_1^2(s+1)}{6a_0 s^3 (s+1) n^{s+2}} + \cdots, \tag{11}$$

where a_2' depends on s, a_0, a_1, a_2. Thus, T_{1n} eliminates the first term in E_n and T_{2n} eliminates the second term. This is consistent with the known result that the u-transformation successively eliminates terms in an expansion of this form.

When the form of E_n is given by (7), we obtain the following expansion for $\mathcal{E}_{1n}$:

$$\mathcal{E}_{1n} = \frac{a_0 b_0 (t-s)^2}{a_0 s^2 n^t + b_0 t^2 n^s} + \frac{\sum_{i=0}^{3} c_{1i} n^{it+(3-i)s}}{2n^{t+s+1}(a_0 s^2 n^t + b_0 t^2 n^s)^2}$$

$$+ \frac{\sum_{i=0}^{4} c_{2i} n^{it+(4-i)s}}{12 n^{t+s+2}(a_0 s^2 n^t + b_0 t^2 n^s)^3} + \cdots, \tag{12}$$

where the c_{lj} depend on $s, t, a_k, b_k, \ k \le l$. Next, for n sufficiently large, we can write

$$\frac{1}{a_0 s^2 n^t + b_0 t^2 n^s} = \frac{1}{a_0 s^2 n^t} \left(1 + \frac{b_0 t^2}{a_0 s^2 n^{t-s}} \right)^{-1}$$

$$= \frac{1}{n^t} \left(\alpha_0 + \frac{\alpha_1}{n^{t-s}} + \frac{\alpha_2}{n^{2(t-s)}} + \cdots \right).$$

Expanding each term in (12) using power series in this way, $\mathcal{E}_{1n}$ simplifies to

$$\mathcal{E}_{1n} = \frac{a_1'}{n^{s+1}} + \frac{a_2'}{n^{s+2}} + \cdots \qquad + \frac{b_0'}{n^t} + \frac{b_1'}{n^{t+1}} + \frac{b_2'}{n^{t+2}} + \cdots$$

$$+ \sum_{i \ge 0} \left(\frac{c_{1i}'}{n^{2t-s+i}} + \frac{c_{2i}'}{n^{3t-2s+i}} + \frac{c_{3i}'}{n^{4t-3s+i}} + \cdots \right).$$

The simplified expansion obtained for $\mathcal{E}_{2n}$ in the same way is:

$$\mathcal{E}_{2n} = \frac{a_2''}{n^{s+2}} + \frac{a_3''}{n^{s+3}} + \cdots \qquad + \frac{b_0''}{n^t} + \frac{b_1''}{n^{t+1}} + \frac{b_2''}{n^{t+2}} + \cdots$$

$$+ \sum_{i \ge 0} \left(\frac{c_{1i}''}{n^{2t-s+i}} + \frac{c_{2i}''}{n^{3t-2s+i}} + \frac{c_{3i}''}{n^{4t-3s+i}} + \cdots \right).$$

Thus, for n sufficiently large, the first two steps of the u-transformation eliminate the terms in $1/n^s$ and $1/n^{s+1}$, successively, while the terms in $1/n^{t+i}$, $i \ge 0$ remain. At the same time however, terms of the form $1/n^{(l+1)t-ls+i}$, $l > 0$, $i \ge 0$ are introduced. These results suggest that T_{kn} eliminates the terms in $1/n^{s+i}$, $i = 0, 1, \ldots, k-1$, with

$$\mathcal{E}_{kn} = \frac{a_k^{(k)}}{n^{s+k}} + \frac{a_{k+1}^{(k)}}{n^{s+k+1}} + \cdots \qquad + \frac{b_0^{(k)}}{n^t} + \frac{b_1^{(k)}}{n^{t+1}} + \frac{b_2^{(k)}}{n^{t+2}} + \cdots$$

$$+ \sum_{i \ge 0} \left(\frac{c_{1i}^{(k)}}{n^{2t-s+i}} + \frac{c_{2i}^{(k)}}{n^{3t-2s+i}} + \frac{c_{3i}^{(k)}}{n^{4t-3s+i}} + \cdots \right). \tag{13}$$

As an aside, if we put $s = t$ in the above, the expansion (12) reduces to the same form as (10) and the expansion for $\mathcal{E}_{2n}$ reduces to (11).

When the form of E_n is given by (8), we obtain the following simplified expansions for $\mathcal{E}_{1n}$ and $\mathcal{E}_{2n}$:

$$\mathcal{E}_{1n} = \frac{c_0'}{n^s(\log n + d_0')} + \frac{a_1' + b_1' \log n}{n^{s+1}} + \frac{a_2' + b_2' \log n}{n^{s+2}} + \cdots$$

$$\mathcal{E}_{2n} = \frac{c_0''}{n^s(\log n + d_0''')} + \frac{c_1''}{n^{s+1}(\log n + d_1''')}$$

$$+ \frac{a_2'' + b_2'' \log n}{n^{s+2}} + \frac{a_3'' + b_3'' \log n}{n^{s+3}} + \cdots.$$

Here, the first two steps of the u-transformation modify the terms $(a_i + b_i \log n)/n^{s+i}$, $i = 0, 1$ into terms of the form $c_i/(n^{s+i}(\log n + d_i))$, $i = 0, 1$. We conjecture that $\mathcal{E}_{kn}$ is of the form

$$\mathcal{E}_{kn} = \frac{c_0^{(k)}}{n^s(\log n + d_0^{(k)})} + \frac{c_1^{(k)}}{n^{s+1}(\log n + d_1^{(k)})} +$$

$$\cdots + \frac{c_{s+k-1}^{(k)}}{n^{s+k-1}(\log n + d_{s+k-1}^{(k)})}$$

$$+ \frac{a_k^{(k)} + b_k^{(k)} \log n}{n^{s+k}} + \frac{a_{k+1}^{(k)} + b_{k+1}^{(k)} \log n}{n^{s+k+1}} + \cdots.$$

When the form of E_n is given by (9), the simplified expansions for $\mathcal{E}_{1n}$ and $\mathcal{E}_{2n}$ are:

$$\mathcal{E}_{1n} = \frac{a_1'}{n^{s+1}} + \frac{a_2'}{n^{s+2}} + \cdots$$

$$+ \frac{b_0' \log n + d_0}{n^t} + \frac{b_1' \log n + d_1}{n^{t+1}} + \frac{b_2' \log n + d_2}{n^{t+2}} + \cdots$$

$$+ \sum_{i \geq 0} \left(\frac{P_2(\log n)}{n^{2t-s+i}} + \frac{P_3(\log n)}{n^{3t-2s+i}} + \frac{P_4(\log n)}{n^{4t-3s+i}} + \cdots \right),$$

and

$$\mathcal{E}_{2n} = \frac{a_2''}{n^{s+2}} + \frac{a_3''}{n^{s+3}} + \cdots$$

$$+ \frac{b_0'' \log n + d_0'}{n^t} + \frac{b_1'' \log n + d_1'}{n^{t+1}} + \frac{b_2'' \log n + d_2'}{n^{t+2}} + \cdots$$

$$+ \sum_{i \geq 0} \left(\frac{P_2'(\log n)}{n^{2t-s+i}} + \frac{P_3'(\log n)}{n^{3t-2s+i}} + \frac{P_4'(\log n)}{n^{4t-3s+i}} + \cdots \right),$$

where $P_j(\log n)$ and $P_j'(\log n)$ are degree j polynomials in $\log n$. In this case, the first two steps of the u-transformation eliminate the terms in $1/n^s$ and $1/n^{s+1}$ while the terms in $\log n/n^{t+i}$, $i \geq 0$ remain. Further, terms of the forms $1/n^{t+i}$, $i = 0, 1, 2, \ldots$ and $\log^k n/n^{(l+1)t-ls+i}$, $k = 0, 1, \ldots, l+1$, $l = 1, 2, 3, \ldots$, and $i \geq 0$, are introduced. We conjecture that T_{kn} introduces terms of these forms as it eliminates the terms in $1/n^{s+i}$, $i = 0, 1, \ldots, k-1$.

4. Discussion

When E_n can be represented by (6), the results of the previous section indicate that $\mathcal{E}_{kn}$ has the same form as E_n but with the first k terms eliminated. It follows that the Levin u-transformation could be applied to the sequence $\{T_{k,n+j}\}$, $j = 0, 1, 2, \ldots$ with a similar effect. Thus, iterating the first k steps of the u-transformation on a sequence with error expansion given by (6) is a valid convergence acceleration procedure. However, for these cases, since the straightforward use of the transformation achieves the same goal more efficiently, such iterative use appears superfluous.

There may, however, be a benefit in iterating the transformation in the other cases considered. For the following discussion note that the results for $k = 1$ and $k = 2$ have been proven. We conjecture that they also hold for general $k > 2$.

When the form of E_n is given by (7), the form of $\mathcal{E}_{kn}$ is essentially the same as (7) but with the terms in $1/n^{s+i}$, $i < k$ eliminated and new terms $\sim O(1/n^{2t-s})$ introduced. Thus, provided $t \geq s + k$, the dominant term in $\mathcal{E}_{kn}$ (i.e., $O(1/n^{s+k})$) is eliminated when $\mathcal{E}_{k+1,n-1}$ is computed. On the other hand, if $t < s + k$, then $\mathcal{E}_{kn} = O(n^{-t})$, so continuing to increase k in the computation of the u-transformation leaves the order of the dominant error term unaffected. In this case, renaming $\{T_{k,n+j}\}$, $j = 0, 1, 2, \ldots$ as $\{S_m\}$, $m = 1, 2, \ldots$ and applying the first k steps of the Levin transformation (3) to this sequence produces, say, T_{kn}^* with an error of the form

$$\mathcal{E}_{kn}^* = \frac{a_k^*}{n^{s+k}} + \frac{a_{k+1}^*}{n^{s+k+1}} + \cdots + \frac{b_k^*}{n^{t+k}} + \frac{b_{k+1}^*}{n^{t+k+1}} + \cdots$$
$$+ \sum_{j \geq 0} \sum_{i \geq 0} \frac{c_{ij}^*}{n^{x_{ij}s + y_{ij}t + z_{ij}k + i}}.$$

The terms in $1/n^{s+i}$, $0 \leq i < k$ are eliminated while at the same time further terms in $1/n^l$, $l > s$ are introduced. $\mathcal{E}_{kn}^*$ is again of a form similar to (7), so the process can be repeated. On each iteration, if the term in $1/n^s$ is the dominant error term, it will be eliminated along with the associated terms in $1/n^{s+1}, 1/n^{s+2}, \ldots, 1/n^{s+k-1}$, but a large number of terms of order $1/n^{s+\gamma}$, $\gamma > 0$ are introduced.

From this we conclude that iterating the first k steps of the u-transformation on a sequence with error expansion given by (7) is effective, at least until the introduced terms become so numerous that further elimination of the most dominant terms no longer has a significant overall effect. This is numerically illustrated by Figure 1 in the next section.

The effect of the first k steps of the Levin u-transformation when applied to a a sequence with associated error expansion E_n of the form given by (8) is to reduce each of the first k terms in the expansion by a factor of order $\log^2 n$, while otherwise retaining the form of the expansion. To determine whether iteration of the transformation is worthwhile in this case, we need to investigate the effect of the transformation on an error expansion of the form

$$E_m = \frac{a_0}{m^s(\log m + b_0)} + \frac{a_1}{m^{s+1}(\log m + b_1)} + \frac{a_2}{m^{s+2}(\log m + b_2)} + \cdots. \tag{14}$$

Doing this, we obtain the simplified expansions

$$\mathcal{E}_{1n} = \frac{a_0'}{n^s P_3(\log n)} + \frac{a_1'}{n^{s+1}(\log n + b_1')} + \frac{a_2'}{n^{s+2}(\log n + b_2')} + \cdots,$$

$$\mathcal{E}_{2n} = \frac{a_0''}{n^s P_3'(\log n)} + \frac{a_1''}{n^{s+1} P_3''(\log n)}$$

$$+ \frac{a_2''}{n^{s+2}(\log n + b_2'')} + \frac{a_3''}{n^{s+3}(\log n + b_3'')} + \cdots.$$

Again, it appears that the first k terms in the expansion (14) are reduced by a factor of order $\log^2 m$, the form of the remaining terms being unchanged. We conclude that iteration of the u-transformation produces an improvement over its straightforward use in this case. This is numerically illustrated in Figure 2 in the next section.

We also note that when E_n has the form given by (9), the first k steps of the u-transformation eliminate the terms in $1/n^{s+i}$, $i < k$ and introduce terms in $\log^j n/n^{(j+1)t-js+i}$, $i, j \geq 0$. Thus, provided $t \geq s + k$, the dominant term in $\mathcal{E}_{kn}$ (i.e., $O(1/n^{s+k})$) is eliminated when $\mathcal{E}_{k+1,n-1}$ is computed. On the other hand, if $t < s + k$ then $\mathcal{E}_{kn} = O(\log n/n^t)$, so continuing to increase k in the computation of the u-transformation leaves the order of the dominant error term unaffected. In this case, $\mathcal{E}_{kn}$ takes on the form of (8), and the arguments given in the previous paragraph are applicable. A numerical illustration is given in Figure 3 in the next section.

5. Numerical Illustrations

In the following figures, we compare the accuracy attained by the straightforward use and the iterated use of the Levin u-transformation on quadrature approximations which have error expansions given by the forms (7)–(9). We denote the true relative error of the approximation $T_{0n} = \mathcal{Q}^{(n)}f$ by $\mathcal{E}_{0n}$. μ_n denotes the true relative error after application of the u-transformation to $T_{00}, T_{01}, \ldots, T_{0n}$, and ν_n denotes the error after iterating the u-transformation on T_{4n}. We plot the logarithms of the magnitudes of these errors against the cumulative number of function evaluations.

Typically, the approximations T_{kn} (and iterations on T_{4n}) are arranged in a triangular table with the original elements of the sequence forming the first column, and the rows generated one at a time. Note that 6 elements of the original sequence are required to obtain a single T_{4n} entry (5 elements, if $n = 1$). All computations are performed on a VAX 8800 using Fortran quadruple precision (approximately 32 digits). We use such a high accuracy in order to illustrate the effect of the transformation and its iterated use.

For the integral $If = \int_0^1 x^{-3/4} \cos x \, dx$, the error expansion for the m-copy midpoint rule approximation $\mathcal{Q}^{(m)}f$ can be expressed in the form

$$E_m = \mathcal{Q}^{(m)}f - If = \frac{a_0}{m^{1/4}} + \frac{a_1}{m^{5/4}} + \cdots \quad + \frac{b_2}{m^2} + \frac{b_4}{m^4} + \frac{b_6}{m^6} + \cdots.$$

This is of the form (7) and will be transformed into the form (13) by the Levin u-transformation. The results shown in Figure 1 indicate that for this example, iteration of the transformation on T_{4n} significantly increases the accuracy attainable by the straightforward application of the transformation. We note also that the lower order error terms introduced by the transformation limit the effectiveness of the iterated use to approximately 13 correct digits.

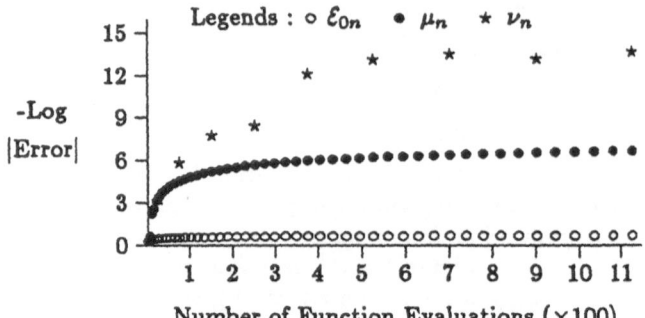

Figure 1: Results of iterated T_{4n} from midpoint rule approximations to $\int_0^1 x^{-3/4} \cos x \, dx$.

A second example is the integral $If = \int_0^1 x^{-1/2} \log x \, dx$, which we approximate by the m-copy degree 7 Gauss-Legendre rule. In this case, the error expansion is

$$E_m = \frac{a_0 + b_0 \log m}{m^{1/2}} + \frac{c_0}{m^8} + \frac{c_2}{m^{10}} + \cdots .$$

The dominant part of this expansion is of the form (8) and after the first step of the u-transformation, the leading term is reduced by a factor of order $\log^2 m$ and new terms of the form $(a_i + b_i \log m)/m^{i+1/2}$, $i = 1, 2, \ldots$ are introduced. Figure 2 indicates that for this integral, only the first few steps of the u-transformation are effective in improving the accuracy of the quadrature approximations, but iterated use of the transformation on T_{4n} can increase the accuracy attained from 3 digits to 7 digits. Again, introduced error terms prevent further improvement.

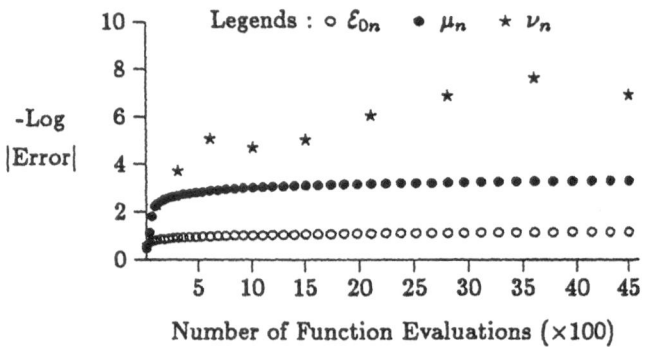

Figure 2: Results of iterated T_{4n} from degree 7 Gauss-Legendre rule approximations to $\int_0^1 x^{-1/2} \log x \, dx$

In Figure 3, we approximate the integral $If = \int_0^1 \int_0^1 x\sqrt{x^2 + y^2} \, dy \, dx$ by the m^2-copy midpoint rule. We use a rule with a low degree of precision here in order to illustrate the effect of iterating the u-transformation. For this example, the error expansion can be written as

$$E_m = \frac{a_0}{m^2} + \frac{a_2}{m^4} + \frac{b_0 \log m}{m^4} + \frac{a_4}{m^6} + \cdots$$

which is in the form (9) with $b_i = 0$, $i > 0$. Figure 3 indicates that iteration of the u-transformation on T_{4n} can increase the accuracy attainable by the straightforward application from about 12 digits to about 17 digits before being limited by the introduced error terms.

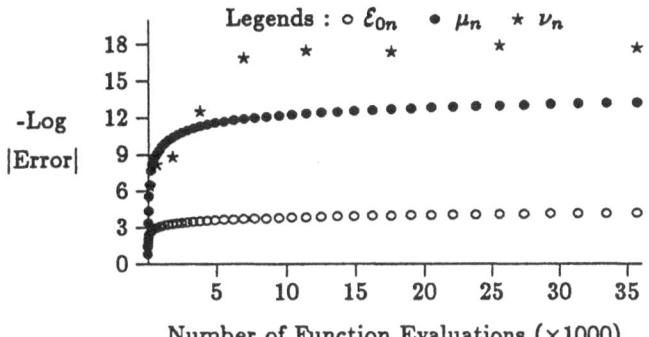

Figure 3: Results of iterated T_{4n} from product midpoint rule approximations to $\int_0^1 \int_0^1 x\sqrt{x^2 + y^2}\,dy\,dx$

The practical value of using the iterated u-transformation to evaluate integrals will depend on both the integrand function and the accuracy required in the final estimate. First, for regular functions and for other functions for which there is an associated expansion of the form

$$\varrho^{(m)}f = If + R_m p(m), \quad m = 1, 2, 3, \ldots,$$

where R_m and $p(m)$ are as defined in (1) and (2), straightforward use of the u-transformation is sufficient. However, for integrals of associated error expansions of the "mixed" form (5), the iterated u-transformation can be very efficient and effective, provided the accuracy requirement is moderate. (Accumulated round-off is frequently a limiting factor in obtainable precision with methods based on the harmonic mesh sequence, unless extra precautions are taken.)

A comparison of the performance of the iterated u-transformation and the ε-algorithm is given in Figure 4 for the integral $\int_0^1 x^{-3/4} \cos x\,dx$. The iterated u-transformation achieves approximately 12-figure accuracy using 378 function evaluations, whereas after 511 function evaluations, the estimate obtained by the ε-algorithm is accurate to only about 6 figures. To obtain a 12-figure accuracy using the ε-algorithm would require 4095 function evaluations.

The superiority of the iterated u-transformation is even more pronounced when one takes into account the need to use a practical error estimate to recognize when the desired accuracy has been achieved. Because of the dependence of the iterated u-transformation on the harmonic mesh sequence, generating an additional estimate of the integral in order to confirm the accuracy of a previous estimate is not very costly. This is not the case with the ε-algorithm which is based on a geometric mesh sequence. For this example, to confirm 12-figure accuracy would require 528 function evaluations using the iterated u-transformation compared with 8191 using the ε-algorithm.

More extensive testing is needed to better delineate the comparative merits of the iterated u-transformation and the ε-algorithm.

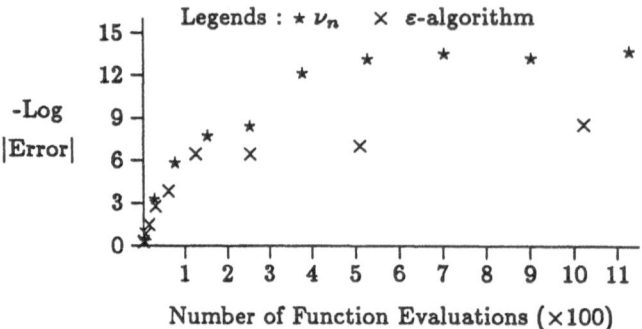

Figure 4: Errors of iterated T_{4n} and ε-algorithm for $\int_0^1 x^{-3/4} \cos x \, dx$.

6. Concluding Remarks

This algebraic study, supported by limited numerical examples, suggests that, to an extent, iteration of the Levin u-transformation is effective as an extrapolation procedure for quadrature approximations to integrals with asymptotic error expansions of "mixed" type. Such error expansions typically arise when the integrand function has an algebraic or logarithmic singularity at a vertex or along a boundary of the region of integration. The method is also efficient due to its use of the harmonic subdivision sequence. Further numerical examples of the application of the iterated Levin u-transformation in quadrature have been given in [2], where comparison is made with the ε-algorithm.

References

[1] R.L. Cariño, I. Robinson and E. de Doncker (1990), "Approximate integration by the Levin transformation," *Technical Report 9/90*, Dept. of Computer Science and Computer Engineering, La Trobe University, Australia (submitted for publication).

[2] R.L. Cariño, E. de Doncker and I. Robinson (1991), "Approximate integration using iterated Levin transformations," *in N. A. Sherwani, E. de Doncker and J. Kapenga (eds.), Lecture Notes in Computer Science Vol. 507*, Springer-Verlag, pp. 293–299.

[3] D. Levin (1973), "Development of nonlinear transformations for improving convergence of sequences," *Intern. J. Computer Math.*, v. B3, pp. 371–388.

[4] MATHLAB GROUP (1977), "Macsyma Reference Manual," Laboratory for Computer Science, MIT, Cambridge, MA.

[5] A. Sidi (1979), "Convergence properties of some nonlinear sequence transformations," *Math. Comp.*, v. 33, pp.315-326.

Appendix

```
;
; Sample Macsyma Script   (Lines starting with ; are comments)
;
; Macsyma batch file for computing the first K steps of the
;   Levin transformation of a sequence S
;
; Input variables:
;
;   S - the sequence under consideration
;   K - last column of Levin table to compute
;   NT - truncation level of Taylor series expansions
;
;
; Write results into file "results.mac"
      writefile(results.mac)$
;
; define the sequence, variable, constants and some properties
      s: (a0+b0*log(n))/n^r + (a1+b1*log(n))/n^(s+1)
         + (a2+b2*log(n))/n^(s+2);
      mainvar(n)$
      put(n,'inf,'limit)$
      declare([a0,a1,a2,b0,b1,b2,s],constant)$
      assume(s>0)$
;
; define NT and K
      nt:3; k:2;
      taylordepth:nt$
      kp1:k+1$
;
; form S(n+j) and R(n+j),  j=1,K+1    (NQ, DQ, respectively)
      for j:1 step 1 thru kp1 do (
         nq[j]:taylor(subst(n=n+j,s),n,'inf,nt),
         dq[j]:1/((n+j)*(nq[j]- if j>1 then nq[j-1] else s))  )$
;
; form S(n+j)/R(n+j),  j=1,K+1
      for j:1 step 1 thru kp1 do nq[j]:nq[j]*dq[j]$
;
; compute and display T(1), T(2), ..., T(k)
      for l:1 step 1 thru k do (
        snq:0, sdq:0,
        for j:0 step 1 thru l do (
          c:(-1)^j*binomial(l,j)*(n+j+1)^(l-1),
          snq:snq+c*nq[j+1],
          sdq:sdq+c*dq[j+1]  ),
```

```
    t[1]:snq/sdq,
    ldisplay(t[1])  )$
closefile()$
```

Note: The above script computes only T_{1n}, T_{2n}, ... and produces a long listing. Further Macsyma operations are required to formulate T_{kn} as expressions involving powers of N and LOG(N). To ensure the correctness of the leading terms, we systematically increase the number of terms in s and in taylor() (by using successively higher values for nt) and recalculating t. The terms that remain constant are presumed to be correct.

DEVELOPMENTS IN SOLVING INTEGRAL EQUATIONS NUMERICALLY

GÜNTHER HÄMMERLIN
Mathematisches Institut der
Ludwig-Maximilians-Universität
Theresienstr. 39
8000 München 2
Germany

ABSTRACT. After a short survey over the main methods for treating Fredholm integral equations practically, this article gives some information on the degenerate substitution kernel method and some recent results. In particular, the extension to Hammerstein integral equations, an investigation of Bateman's method as applied to eigenvalue problems and its convergence behaviour for Green's kernels are considered.

1. Preliminaries

The subject of this workshop as well as the place in which it is held form a framework very well suited to a talk about integral equations. To begin with the place: In Findø near Stavanger, not very far from here, Niels Henrik Abel was born in 1802. His name is one of the first names which occur to us, when we are engaged in integral equations. Erik Ivar Fredholm and Evert Johannes Nyström who will also have to be mentioned later were Scandinavians. Thus, we are found in the right vicinity. As to the subject, it is certainly in the practical treatment of integral equations where we recognise most directly that the problems of numerical integration extend beyond the mere computation of definite integrals. Integral equations together with differential equations are obviously the most important instruments required to model physical facts by the means of mathematical analysis. The term "integral equations" was introduced by Paul du Bois–Reymond in Berlin, known by his work on partial differential equations, who in 1887 expressed the following opinion: "From that time (since 1852) integral equations occured to me so often in the theory of partial differential equations, that I am convinced, that the progresses in this theory are connected with the treatment of integral equations, about which practically nothing is known".

To finish the introduction to this presentation of some aspects of the numerical treatment of integral equations, let us remember another Scandinavian mathematician, the famous Gösta Mittag–Leffler. He was not a numerical analyst; nevertheless, the following inscription over the entrance of the institute, founded by him in Djursholm, could also serve as a motto of this present workshop:

> The number is beginning and end of thinking,
> with the thought the number is born,
> the thought does not reach beyond the number.

T. O. Espelid and A. Genz (eds.), Numerical Integration, 187–201.
© 1992 *Kluwer Academic Publishers.*

2. Nyström, Ritz–Galerkin and collocation methods

In this overview, I shall mainly concentrate on some developments concerning the substitution kernel method. First of all, however, I want to mention several methods to solve Fredholm integral equations approximately which are well established in the arsenal of numerical devices.

Fredholm himself had already had the idea, to substitute a continuous kernel k in the integral equation

$$(2.1) \qquad \kappa\varphi(s) - \int_0^1 k(s,t)\varphi(t)dt = f(s)$$

by a step function, that is by a meshwise constant kernel over an equidistant quadratic grid, in order to produce a neighbouring integral equation with a neighbouring solution. He communicated this idea in 1899 in a letter to Mittag–Leffler, and from this idea he developed his theory. Quadrature methods as well as substitution kernel methods can be traced back to Fredholm's proposal, although these possibilities were followed up only much later with the aim of the numerical computation of a solution.

Indeed, it took thirty years, until Nyström in 1928 established the quadrature method in his paper [19]. He applies a quadrature formula to discretise the integral in the equation, such that in a first step we receive the functional equation

$$\kappa\varphi(s) - \sum_{\nu=1}^n w_\nu k(s,t_\nu)\varphi(t_\nu) = f(s)$$

with weights w_ν and nodes $0 \le t_1 < \ldots < t_n \le 1$. From this we get by choosing $s_\mu := t_\mu, 1 \le \mu \le n$, and collocation the fully discretised linear system of equations

$$\kappa\tilde\varphi(s_\mu) - \sum_{\nu=1}^n w_\nu k(s_\mu,t_\nu)\tilde\varphi(t_\nu) = f(s_\mu), \quad (\mu = 1,\ldots,n).$$

Solving this system, we receive the values $\tilde\varphi_\mu$ of the approximate solution in the nodes, and finally by

$$\kappa\tilde\varphi(s) = \sum_{\nu=1}^n w_\nu k(s,t_\nu)\tilde\varphi_\nu + f(s)$$

the complete numerical solution.

Nyström himself originally considered only Gaussian quadrature formulae and in a later paper also Chebyshev formulae. As to the mathematical analysis of the quadrature method, it lasted again thirty years, until Brakhage in 1960 [5] published his error bounds for the inhomogeneous and in 1961 [6] for the homogeneous Fredholm integral equation. This short article cannot adequately describe the great number of papers which followed. These papers contain extensions to singular and to nonlinear integral equations, results on overconvergence and other refinements. Good general references covering the topic of these remarks are the monographs by Atkinson [1], Baker [2], Delves–Mohamed [7] and Fenyö–Stolle [8].

After these remarks on the quadrature method we cast a glance at that group of methods which are based on variational principles, namely the Ritz–Galerkin methods. In order to give arguments in support of the key role which the problem of maximizing or minimizing a function plays in mathematics and its applications, we can appeal to one of our greatest predecessors as mathematicians, to Leonhard Euler (1707–1783). He expressed his belief – freely translated from an article in Commentationes Mechanicae – with the following sentences: "Since everything in the entire world is best possible, and since it is all the work of the wisest of creators, there is nothing in this world which is not blessed with either a maximum or minimum property. Thus, there can be no doubt that all of our worldly processes can as easily be derived by the method of maxima and minima as from their basic properties themselves."

The variational principle which applies to our case, can be described as follows: The solution of the linear operator equation $L\varphi = f$ in a suitable Hilbert space H, where L is a symmetric and positive operator, is equivalent to the search for a minimum of the functional

$$Fv := \langle Lv, v \rangle - 2\langle v, f \rangle.$$

Thus, instead of solving the integral equation

$$(\kappa I - K)\varphi = f,$$

we can determine the minimum of the functional

$$Fv = \langle (\kappa I - K)v, v \rangle - 2\langle v, f \rangle$$

or at least find an approximation to this minimum in a finite– dimensional subspace $\text{span}(u_1, \ldots, u_n) \subset H$ by setting $v = \sum_{\nu=1}^{n} \alpha_\nu u_\nu$ and solving the necessary conditions for an extremum $\frac{\partial F}{\partial \alpha_\nu} = 0$, $1 \leq \nu \leq n$.

These are, in modern terms, the essentials of the idea of Walter Ritz (1878–1909). The thus established linear system of equations for the coefficients $(\alpha_1, \ldots, \alpha_n)$ is determined by the matrix $(\langle K u_\mu, u_\nu \rangle)_{\mu,\nu=1}^{n}$. The same matrix results from the proposal, published in 1915 by the Russian engineer B.G. Galerkin, to compute an approximation $v \in H$ by minimizing the defect of the expression $Lv - f$ by orthogonalization with respect of $\text{span}(u_1, \ldots, u_n)$.

On the other hand, we also know that any Ritz–Galerkin procedure is equivalent to a substitution kernel method. We can construct a degenerate substitution kernel using the functions $u_1, \ldots, u_n$ of the Ritz–Galerkin process; we are led to the same system of equations determining the approximate solution. There is a link, therefore, between the Ritz–Galerkin and the degenerate substitution kernel method which we shall study subsequently in more detail.

As a last remark, collocation methods should be mentioned in this listing. As in the Ritz–Galerkin procedure, an approximate solution is sought in a finite–dimensional subspace, spanned by the n elements $u_1, \ldots, u_n$. But instead of orthogonality conditions, we postulate pointwise fulfillment of the original integral equation and in this way we are led to determining equations for the coefficients of the linear combination of the elements $u_1, \ldots, u_n$.

3. Degenerate kernels

The only type of Fredholm integral equations, which can generally be solved exactly, are the equations with degenerate kernels.

Let $k(s,t)$ be given in the finite representation

$$(3.1) \qquad k(s,t) = \sum_{\rho,\sigma=1}^{r} c_{\rho\sigma} x_\rho(s) y_\sigma(t);$$

inserting this kernel in (2.1) yields

(3.2)
$$\kappa\varphi(s) = \sum_{\rho=1}^{r} x_\rho(s)[\sum_{\sigma=1}^{r} c_{\rho\sigma} \int_a^b y_\sigma(t)\varphi(t)dt] + f(s)$$

or

$$\kappa\varphi(s) = \sum_{\rho=1}^{r} d_\rho x_\rho(s) + f(s).$$

Inserting this representation of the solution with unknown coefficients d_ρ again in (3.2), results in

$$\kappa\sum_{\rho=1}^{r} d_\rho x_\rho(s) = \sum_{\rho=1}^{r} x_\rho(s)[\sum_{\sigma=1}^{r} c_{\rho\sigma} \int_a^b y_\sigma(t)(\sum_{\tau=1}^{r} d_\tau x_\tau(t) + f(t))dt];$$

comparing the coefficients leads us to the linear system

(3.2.1)
$$\kappa\underline{d} = CY\underline{d} + C\underline{b},$$

where $\underline{d} := (d_1, \ldots, d_r)^T$, $C := (c_{\rho\sigma})_{\rho,\sigma=1}^r$, $Y := (\langle y_\sigma, x_\tau \rangle)_{\sigma,\tau=1}^r$, $\underline{b} := (\langle y_1, f \rangle, \ldots, \langle y_r, f \rangle)^T$ or, in the homogeneous case,

(3.2.2)
$$\kappa\underline{d} = CY\underline{d}.$$

In a slightly modified manner, we can also treat Hammerstein integral equations with degenerate kernels. For this purpose, let us regard the Hammerstein integral equation in the form

(3.3)
$$\varphi(s) = \int_a^b k(s,t)g(t,\varphi(t))dt.$$

Any Hammerstein equation is comprised in this representation although sometimes the inhomogeneous form

$$\varphi(s) = \int_a^b k(s,t)g(t,\varphi(t))dt + f(t),$$

f being a known function, may be useful. Repeating the consideration above, we are now led to the nonlinear system of equations determining the coefficients d_ρ:

(3.4)
$$\underline{d} = \underline{g}(\underline{d}), \quad \underline{g} := (g_1, \ldots, g_r),$$

where
$$g_\rho(\underline{d}) := \sum_{\sigma=1}^{r} c_{\rho\sigma}[\int_a^b y_\sigma(t)g(t, \sum_{\rho=1}^{r} d_\rho x_\rho(t))dt].$$

A solution of this nonlinear system is equivalent to a corresponding solution of the Hammerstein equation with degenerate kernel.

In essential, one finds this consideration already in Hammerstein's original paper [15] which appeared in 1930, dedicated to the memory of Fredholm, who had died three years earlier. In this paper, Hammerstein developed a rather complete theory of the type (3.3) integral equations, concerning existence, uniqueness and branching of solutions. His theorems refer to symmetric and positive kernels, square–integrable in the Lebesgue sense; weakly singular kernels are thus contained in this theory. Hammerstein himself uses the term of a "workably discontinuous" kernel (in German: brauchbar unstetig) with the meaning, that the theory of linear integral equations is supposed to hold. Under certain conditions concerning the function $g(t, u)$ and mainly its behaviour in u, Hammerstein proved, among other statements, the following

(3.5) THEOREM: Let $k(s, t)$ be workably discontinuous, symmetric and positiv. Assume that the continuous function $g(t, v)$ fulfills the conditions $(0 \leq t \leq 1, -\infty < v < +\infty)$

$$\int_0^u g(t, v)dv \geq -\frac{\lambda}{2}u^2 - C_1 \text{ and } |g(t, v)| \leq C_2|v| + C_3,$$

where C_1, C_2, C_3 are constants, independent of t and $u, 0 < \lambda < \lambda_1, \lambda_1$ the smallest eigenvalue of the operator K as a solution of the linear equation $\varphi = \lambda K\varphi$; K is defined by the kernel k. Then there exists a solution of the equation (3.3).

This theorem also illustrates the fact, that Hammerstein integral equations are still very near to linear integral equations. Hammerstein showed, in addition, that a change in the number of solutions can only occur, when $|\lambda|$ is equal to an eigenvalue of a linear integral equation, which coincides with the linear equation that appears in Newton's method as applied to find an iterative solution of the Hammerstein equation.

4. Degenerate substitution kernels

The obvious idea, to approximate the kernel of a Fredholm integral equation by a sequence of degenerate kernels has been utilized after Fredholm also by Hilbert and by E. Schmidt for theoretical considerations. In order to make use of it in practical respect, we apply two–dimensional splines for the construction of approximating kernels. Let

$$S_l(\Omega_n) := \{\sigma \in C_{l-1}(I) \mid \sigma \in \mathbb{P}_l \text{ in } s \in [s_\nu, s_{\nu+1}), \ 0 \leq \nu \leq n - 1\}$$

be the space of polynomial splines of degree l and $\{B_{l\nu}\}, -l \leq \nu \leq n - 1$, the local basis of B-splines with support $[s_\nu, s_{\nu+l+1}]$. Then any polynomial spline of degree l can be written as

$$\sigma(s) = \sum_{\nu=-l}^{n-1} \alpha_\nu B_{l\nu}(s), \ \alpha_\nu \in \mathbb{R}.$$

Here $I := [a, b]$, and the equidistant partition Ω_n is determined by the knots $s_\nu = a + \nu h, -l \leq \nu \leq n.$

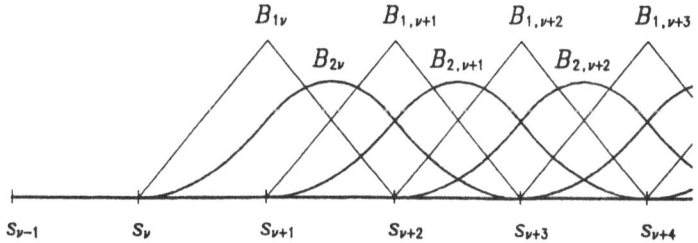

Conditions to determine the coefficients α_ν are for instance

a) Interpolation conditions : Let $\mathcal{P}\varphi$ be the projection of φ into the $(n+l)$–dimensional spline space $S_l(\Omega_n)$. Interpolation means

$$(\varphi - \mathcal{P}\varphi)(s_\nu) = 0, \ 0 \le \nu \le n,$$

and $(l+1)$ further conditions, which in the case of l odd (linear and cubic splines in particular) can be posed as symmetric end conditions; Hermitian and natural end conditions belong to this class. Or, as another example,

b) Orthogonal projection: The demand

$$\int_a^b [\varphi(s) - (\mathcal{P}\varphi)(s)]B_{l\nu}(s)ds = 0, \ -l \le \nu \le n-1,$$

also gives rise to a uniquely determined approximating spline of particularly high approximation order.

The direct two–dimensional generalisation is the tensor product. Let $G:=[a,b]\times[a,b]$, F a mapping $G \to \mathbb{R}, \mathcal{P}_s$ and $\mathcal{P}_t$ the projections in s- and t-direction, resp. Then

$$\mathcal{P}_s\mathcal{P}_t F := \sum_{\nu,\kappa=-l}^{n-1} \alpha_{\nu\kappa} B_{l\nu}(s)B_{l\kappa}(t)$$

is a two–dimensional spline over the quadratic grid with mesh–points $(s_\nu, t_\kappa), \ 0 \le \nu, \kappa \le n$.

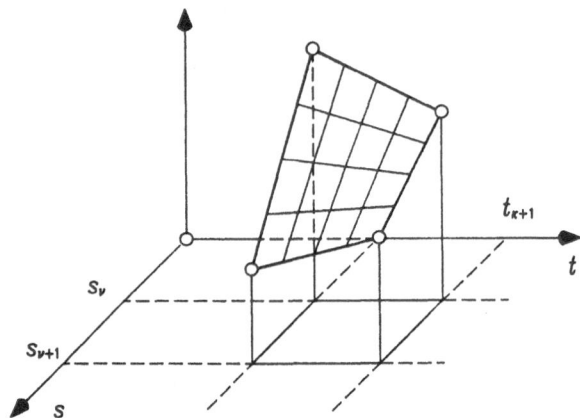

Interpolating bilinear tensor–product spline

Evidently, $\tilde{k} = \mathcal{P}_s\mathcal{P}_t k$ is a degenerate approximation of k ([10],[11],[14]). In the linear system (3.2) we have $C := (\alpha_{\nu\kappa})_{\nu,\kappa=1}^n$, $Y := (\langle B_{l\kappa}, B_{l\lambda}\rangle)_{\kappa,\lambda=1}^n$ so that we can establish the solution of the approximate integral equation with degenerate substitution kernel. The matrix Y can be calculated once and for all; in the bilinear case, e.g., we find the scheme

$$(\langle B_{l\kappa}, B_{l\lambda} \rangle) = \frac{h}{6} \begin{bmatrix} 2 & 1 & & & \\ 1 & 4 & 1 & & 0 \\ & \ddots & \ddots & \ddots & \\ 0 & & 1 & 4 & 1 \\ & & & 1 & 2 \end{bmatrix}.$$

The idea of spline–blended functions as introduced by Coons and Gordon [9] promises an even better approximation of k. Let $\mathcal{P}$ symbolize the projection which generates an interpolating spline. Then the spline–blended projection is defined by the Boolean sum

$$(\mathcal{P}_s \oplus \mathcal{P}_t)k := \mathcal{P}_s k + \mathcal{P}_t k - \mathcal{P}_s \mathcal{P}_t k;$$

it can easily be verified, that this approximation coincides with k not only in the mesh-points like the interpolating tensor–product spline, but even along the complete mesh–lines of the grid:

$$\begin{aligned} ((\mathcal{P}_s \oplus \mathcal{P}_t)k)(s, t_\kappa) &= k(s, t_\kappa), \ 0 \le \kappa \le n; \\ ((\mathcal{P}_s \oplus \mathcal{P}_t)k)(s_\nu, t) &= k(s_\nu, t), \ 0 \le \nu \le n. \end{aligned}$$

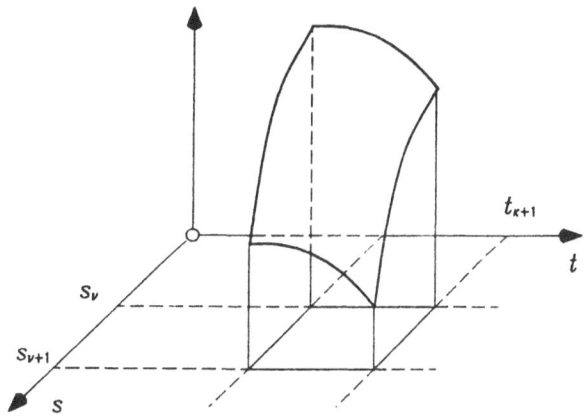

Linear spline blended approximation

The result

$$\tilde{k}(s,t) = \sum_{\nu=-l}^{n-1} \beta_\nu(t) B_{l\nu}(s) + \sum_{\kappa=-l}^{n-1} \gamma_\kappa(s) B_{l\kappa}(t) - \sum_{\nu,\kappa=-l}^{n-1} \alpha_{\nu\kappa} B_{l\nu}(s) B_{l\kappa}(t)$$

does, at a first glance, not appear as a degenerate kernel; by appropriate definition of the generating functions, however, we arrive at a degenerate representation. We are led to the matrices

$$
C := \left[\begin{array}{c|c} (-\alpha_{\nu\kappa}) & I_{n+l} \\ \hline I_{n+l} & 0 \end{array}\right], \qquad Y := \left[\begin{array}{c|c} (\langle B_{l\kappa}, B_{l\lambda}\rangle) & (\langle B_{l\kappa}, \beta_{\nu}\rangle) \\ \hline (\langle \gamma_{\kappa}, B_{l\nu}\rangle) & (\langle \gamma_{\kappa}, \beta_{\nu}\rangle) \end{array}\right]
$$

in (3.2), so that we can calculate the exact solutions and, in the homogeneous case, the eigenvalues of the approximating equation ([12]).

To apply one of the spline approximations to a Hammerstein integral equation, we have to treat (3.4) with $r = n + l$ or $r = 2(n + l)$ in the tensor–product or spline–blended case, respectively. Of course, it remains to solve additionally the nonlinear system (3.4).

5. Approximation quality for inhomogeneous integral equations

Let us consider the two neighbouring equations

(5.1)
$$
\kappa\varphi = K\varphi + f
$$
$$
\kappa\tilde{\varphi} = \tilde{K}\tilde{\varphi} + \tilde{f}
$$

Provided that $(I - K)^{-1}$ exists and $\|(I - K)^1\| \cdot \|K - \tilde{K}\| < 1$, we can easily estimate the deviation of $\tilde{\varphi}$ from φ by the bound

(5.2)
$$
\|\varphi - \tilde{\varphi}\| \leq \frac{\kappa^{-1} \cdot \|(I - K)^{-1}\|(\|K - \tilde{K}\| \cdot \|\varphi\| + \|f - \tilde{f}\|)}{1 - \|(I - K)^{-1}\| \cdot \|K - \tilde{K}\|}.
$$

In order to get a concrete estimation when applying different types of approximation operators K, we only need to know the approximation errors $\|K - \tilde{K}\|$ and $\|f - \tilde{f}\|$. Therefore, we can refer to the well–known error bounds for splines; from these bounds, we recognise the convergence orders $O(h^2)$ or $O(h^4)$ for bilinear or bicubic tensor–product splines as well as $O(h^4)$ or $O(h^8)$ for linear or cubic spline–blended approximations.

In a recent paper Kaneko and Xu [18] treat the problem of applying degenerate substitution kernels to Hammerstein equations under conditions which are partly stronger – insofar as they regard only continuous kernels – and partly weaker than Hammerstein's propositions, since definiteness is no longer required. Under the restrictions, that $(A, B, C$ appropriate constants)

$g(t, v)$ is continuous and $[\int_0^1 |g(t, \varphi(t))|^2 dt]^{\frac{1}{2}} \leq A\|\varphi\|_2$,

$g(t, v)$ is Lipschitz–bounded $|g(t, v) - g(t, \hat{v})| \leq B|v - \hat{v}|$,

$|k(s, t)| < C$ with $BC < 1$,

they are able to prove by a more or less standard application of the Banach fixed point theorem, that the Hammerstein equation (3.3) possesses a unique solution and that for the solution $\tilde{\varphi}$ of an approximate equation with kernel $\tilde{k}$ the error and convergence bound

(5.3) $$\|\varphi - \tilde{\varphi}\|_2 \le \frac{A\|\varphi\|_2}{1 - CB}\|k - \tilde{k}\|_2$$

holds. The estimation shows, that φ, relative to $\|\cdot\|_2$, is approximated by $\tilde{\varphi}$ of the same order as k by $\tilde{k}$. Of course, tensor–product splines and spline–blended functions fit in this estimate. This fact is affirmed by several numerical examples which were carried through in Munich by A. Nikolis. To the examples belong the Hammerstein equations

and
$$\varphi(s) = \int_0^1 e^{st} exp(-\varphi^2(t))dt + f_1(s)$$
$$\varphi(s) = \int_0^1 G(s,t)\varphi^2(t)dt + f_2(s)$$

with the Green's kernel

(5.4) $$G(s,t) := \begin{cases} s(1-t) & 0 \le s \le t \le 1 \\ t(1-s) & 0 \le t \le s \le 1 \end{cases}.$$

In both cases the equations were treated in the non–canonical form given above with f_1 and f_2 chosen such that the exact solutions $\varphi(s) = \sqrt{s}$ and, for the Green's kernel, $\varphi(s) = -\ln 2 + 2\ln(\csc \frac{c(s-0.5)}{2})$, $c = 1.3360556949$, , occurred. The first example was taken from [18], where the truncated power series expansion of the kernel had been used as a degenerate approximation. In the second example, that property of Green's kernels has again consequences which had already been proved earlier in 1979 [13]: The singularity in the derivative of the Green's kernel does not influence the convergence order of the spline approximation. Thus we have, for instance, $O(h^2)$-convergence using a bilinear tensor–product spline substitution kernel. In 1987, the second example had been treated by Kumar and Sloan [17], who introduced a new collocation–type method for Hammerstein equations. They found quadratic convergence, using a piecewise linear approximate solution. This is also a natural consequence of the above– mentioned property of Green's kernels, since the convergence bound in the new collocation–type method basically reduces the convergence rate to the order of approximation of a solution by a linear spline.

6. The eigenvalue problem

Let us compare the two neighbouring eigenvalue problems

$$\kappa\varphi = K\varphi \quad \text{and} \quad \tilde{\kappa}\tilde{\varphi} = \tilde{K}\tilde{\varphi}.$$

From a theorem of Weyl [23], published in 1913, we obtain for symmetric operators K and $\tilde{K}$ the following estimate for the distance of corresponding eigenvalues of the two operators: Let $\tilde{\kappa}_i$ be a simple eigenvalue of the operator $\tilde{K}$; then there exists an eigenvalue κ_i of K, such that

(6.1) $$|\kappa_i - \tilde{\kappa}_i| = O(\|K - \tilde{K}\|).$$

The same order of exactness can be shown to hold for the distance $\|\varphi_i - \tilde{\varphi}_i\|$ of the corresponding eigenfunctions [10]. As a consequence of the relation $\|K - \tilde{K}\| \le \|k - \tilde{k}\|$ which holds in $(C[a, b], \| \cdot \|_\infty)$ and in $L^p[a, b]$, any of the known bounds for $\|k - \tilde{k}\|$ produces an estimate for the error of solutions of the integral equations.

If we apply splines of odd degree to originate degenerate substitution kernels, then these estimates, based on the approximation error $\|k - \tilde{k}\|$, are optimal. For splines of even degree, however, the situation is different. The order of the quadrature error in the case, when the integrand is approximated by a polynomial of even degree, exceeds the approximation order of the integrand by one. This fact is familiar to every numerical analyst and is, for instance, the reason for the high quality of Simpson's rule.

Studying this phenomenon which had been corroborated by several examples, Schäfer ([21],[22]) could derive the following bound for the distance of the eigenvalues of the original kernel operator K from the eigenvalues of approximating operators K_n, built up on a grid of mesh–size $h = \frac{b-a}{n}$:Let H be a Banach space $L^p(I)$, $1 \le p < \infty$, or $(C(I), \| \cdot \|_\infty)$. Let $K, K_n : H \to H$ be compact integral operators and $\kappa \neq 0$ an eigenvalue of K with algebraic multiplicity μ, and $\lim_{n \to \infty} \|K - K_n\| = 0$. Then there exist exactly μ eigenvalues $\tilde{\kappa}, \ldots, \tilde{\kappa}_\mu$ of K_n, provided that h is small enough, which converge to κ for $n \to \infty$. In detail, we have the bound

$$(6.2) \qquad |\kappa - \frac{1}{\mu} \sum_{i=1}^{\mu} \tilde{\kappa}_i| \le c(\|K(K - K_n)K\| + \|K - K_n\|^2).$$

Let m be the approximation order of a polynomial spline and $\nu \ge m$ the convergence order of the quadrature formula which is generated by approximation of a sufficiently often differentiable integrand by the spline. Then we can derive from (6.2) the convergence order for the eigenvalues of a tensor–product substitution kernel

$$(6.3) \qquad |\kappa - \frac{1}{\mu} \sum_{i=1}^{\mu} \tilde{\kappa}_i| = O(h^{\min(2m, \nu)}).$$

For the bilinear ($m = 2$) and for the bicubic ($m = 4$) tensor–product spline we have $\nu = m$, such that we get again the convergence orders $O(h^2)$ and $O(h^4)$, resp. which may be deduced from (6.1), too. In the biquadratic ($m = 3$) case, however, $\nu = m+1$; therefore the resultant order is $O(h^4)$, the same as in the bicubic case.

The situation for spline–blended approximation is similar. Here we find from (6.2) the order

$$|\kappa - \frac{1}{\mu} \sum_{i=1}^{\mu} \tilde{\kappa}_i| = O(h^{\min(4m, 2\nu)})$$

which means $O(h^8)$ for quadratic ($m = 3, \nu = m + 1$) and also $O(h^8)$ for cubic spline–blending [3]. In [3] also those splines were examined which are generated by orthogonal projection. In this case, we have convergence of the order $O(h^{4m})$ which exceeds the above mentioned convergence speeds by far.

Needless to say, that all these bounds were examined and corroborated in numerous numerical examples which can be found in the cited papers.

7. Bateman's method

Based on results which go back to 1908, Bateman published in 1922 [4] a kernel approximation method which also amounts to the construction of a degenerate substitution kernel. He defines the substitution kernel k_n by the equation

$$
\begin{vmatrix}
k_n(s,t) & k(s,t_1) & \cdots & k(s,t_n) \\
k(s_1,t) & k(s_1,t_1) & \cdots & k(s_1,t_n) \\
\vdots & \vdots & & \vdots \\
k(s_n,t) & k(s_n,t_1) & \cdots & k(s_n,t_n)
\end{vmatrix} = 0 ,
$$

where a grid with meshpoints (s_μ, t_ν),

$$
a \leq s_1 < s_2 < \ldots < s_n \leq b, \quad a \leq t_1 < t_2 < \ldots < t_n \leq b,
$$

is taken as a basis. One recognises immediately by inspection of this determinant that $k_n(s_\mu, t) = k(s_\mu, t)$, $1 \leq \mu \leq n$, as well as $k_n(s,t_\nu) = k(s,t_\nu)$, $1 \leq \nu \leq n$, holds for all s and t. The Bateman-kernel k_n has this property in common with the spline–blended approximation of k; this fact points out that the Bateman approximation is of a particularly high quality.

Developing the determinant, we recognise the degenerate representation

$$
k_n(s,t) = \sum_{\mu,\nu=1}^{n} c_{\mu\nu} k(s,t_\nu) k(s_\mu, t).
$$

The first rigorous mathematical analysis of this substitution kernel method is due to Joe and Sloan [16]. Assuming a symmetric, non–negative and non–degenerate operator K, they examined the inhomogeneous case in general and succeeded in giving a convergence proof; for one–dimensional integral equations, order bounds were ascertained. In particular, the following asymptotic order relation has been proved: Let k have essentially–bounded generalised derivatives of order $\leq m$ on $[a,b] \times [a,b]$, in other words $k \in W_\infty^m([a,b] \times [a,b])$ as the element of a Sobolev–space. Let $h := \max_{\nu=2,\ldots,n} (|t_1 - a|, |t_\nu - t_{\nu-1}|, |b - t_n|)$ be the maximum step–width in the definition; then the sequence of approximate solutions (φ_n) converges to the true solution φ, the convergence order being given by the bound $\|\varphi - \varphi_n\| \leq ch^m \|\varphi\|$, where c depends on m but is independent of f and h.

This result of Joe and Sloan has been a motive for the investigation of the homogeneous case. In his original paper, Bateman himself had already communicated numerical results to illustrate his method when applied to the eigenvalue problem given by the Green's kernel (5.4). In his master's thesis [20] Prock examined the convergence problem for eigenvalues and eigenfunctions, hereby also using the ideas of Joe and Sloan. In particular, we are interested in the case of non–smooth kernels, typically represented by Green's kernels.

For the estimation of eigenvalues and eigenfunctions, we can again apply the bounds (6.1) of Weyl. These bounds turn out to be optimal for Bateman kernels and are simpler to evaluate than (6.2). That is to say, the bounds for the eigenvalues are determined by bounds for $\|K - K_n\|$. The derivation of bounds for the deviation of K_n from K in the case $k \in W_\infty^m([a, b] \times [a, b])$ can be found in [16]. This derivation essentially uses the fact that, because of the definiteness of K, this operator can be split into a product $K = LL$ using Mercer's theorem. Hence, for eigenvalues and for eigenfunctions convergence of the order $O(h^m)$ appears again. For a kernel $k \in C_\infty([a, b] \times [a, b])$ we therefore have an arbitrarily high speed of convergence.

To illustrate this appearance, just one among Prock's numerical examples in this class is given:

$$[a, b] := [0, 2\pi], \quad k(s, t) := \frac{3}{2\pi[5 - \cos(s - t)]};$$

$$\kappa_{2i} = (\frac{1}{2})^i, \quad \kappa_{2i+1} = (\frac{1}{2})^i, \quad (i = 0, 1, \ldots).$$

Here, all eigenvalues $\kappa_1, \kappa_2, \ldots$ of the non–negative integral operator K except κ_1 have multiplicity 2. The adjoint eigenfunctions are

$$\varphi_{2i+1}(s) = \gamma_1 \sin(\frac{i-1}{2}s) + \gamma_2 \cos(\frac{i-1}{2}s), \varphi_{2i}(s) = \gamma_1 \sin(\frac{i}{2}s) + \gamma_2 \cos(\frac{i}{2}s)$$

with γ_1, γ_2 constants.

n	$\kappa_1^{(n)}$	$\kappa_1 - \kappa_1^{(n)}$	NOC	n	$\|\varphi_1 - \varphi_1^{(n)}\|_2$	NOC
5	0.92510	$7.48985 \cdot 10^{-2}$		5	$1.56447 \cdot 10^{-1}$	
			3.33			2.36
10	0.99005	$9.94894 \cdot 10^{-3}$		10	$3.73755 \cdot 10^{-2}$	
			4.69			4.13
15	0.99829	$1.71432 \cdot 10^{-3}$		15	$7.94250 \cdot 10^{-3}$	
			6.36			5.94
20	0.99970	$3.03939 \cdot 10^{-4}$		20	$1.58051 \cdot 10^{-3}$	
			8.11			7.70
25	0.99995	$5.38271 \cdot 10^{-5}$		25	$3.05281 \cdot 10^{-4}$	
			9.85			9.44
30	0.99999	$9.52063 \cdot 10^{-6}$		30	$5.80620 \cdot 10^{-5}$	
			11.59			11.17
35	1.00000	$1.68324 \cdot 10^{-6}$		35	$1.09343 \cdot 10^{-5}$	
			13.32			12.89
40	1.00000	$2.97569 \cdot 10^{-7}$		40	$2.04434 \cdot 10^{-6}$	
			15.06			14.62
45	1.00000	$5.26029 \cdot 10^{-8}$		45	$3.80053 \cdot 10^{-7}$	
			16.80			16.35
50	1.00000	$9.29861 \cdot 10^{-9}$		50	$7.03292 \cdot 10^{-8}$	

Table 1

Green's kernels fall in the following class: Let

$$U_+ := \{(s,t) \in [a,b] \times [a,b] \mid s \le t\}, U_- := \{(s,t) \in [a,b] \times [a,b] \mid s > t\}.$$

Then $k : [a,b] \times [a,b] \to \mathbb{R}$, $k \in W_\infty^m([a,b] \times [a,b])$ and $k \in W_\infty^{m+1}(U_+)$, $k \in W_\infty^{m+1}(U_-)$ has the typical property of a Green's kernel. If k is again Hermitian, non–degenerate and non–negative, it has been shown in [20], that the convergence order $\|K - K_n\| = O(h^{m+1})$. We can conclude, that the appearance of a discontinuity in the derivatives of the kernel along the diagonal $s = t$ is completely balanced and does not influence the convergence quality. To achieve the proof, we have to treat those meshes of the grid over $[a,b] \times [a,b]$, which contain the diagonal, separately from the rest. Since the manifold of discontinuities is merely one–dimensional, it does not destroy the convergence order dictated by the smoothness of the kernel outside the diagonal.

Example: $[a,b] := [0,1]$, the kernel is the Green's function (5.4); the eigenvalues are $\kappa_i = \frac{1}{(i\pi)^2}$, with eigenfunctions $\varphi_i(s) = \gamma \sin(i\pi s)$, $(i = 1,2,\ldots)$. The kernel has the properties $k \in W_\infty^1([a,b] \times [a,b])$, $k \in W_\infty^2(U_+)$ and $k \in W_\infty^2(U_-)$. Hence, the expected convergence order is 2.

n	$\kappa_1^{(n)}$	$\kappa_1 - \kappa_1^{(n)}$	NOC	n	$\|\varphi_1 - \varphi_1^{(n)}\|$	NOC
5	0.09904	$2.28274 \cdot 10^{-3}$		5	$1.21914 \cdot 10^{-2}$	
			1.98			2.31
10	0.10064	$6.85887 \cdot 10^{-4}$		10	$3.00248 \cdot 10^{-3}$	
			1.99			1.90
15	0.10100	$3.24892 \cdot 10^{-4}$		15	$1.47605 \cdot 10^{-3}$	
			2.00			2.10
20	0.10113	$1.88753 \cdot 10^{-4}$		20	$8.33341 \cdot 10^{-4}$	
			2.00			1.94
25	0.10120	$1.23184 \cdot 10^{-4}$		25	$5.50782 \cdot 10^{-4}$	
			2.00			2.06
30	0.10123	$8.66707 \cdot 10^{-5}$		30	$3.83219 \cdot 10^{-4}$	
			2.00			1.96
35	0.10126	$6.42759 \cdot 10^{-5}$		35	$2.85985 \cdot 10^{-4}$	
			2.00			2.04
40	0.10127	$4.95591 \cdot 10^{-5}$		40	$2.19250 \cdot 10^{-4}$	
			2.00			1.97
45	0.10128	$3.93733 \cdot 10^{-5}$		45	$1.74798 \cdot 10^{-4}$	
			2.00			2.03
50	0.10129	$3.20329 \cdot 10^{-5}$		50	$1.41754 \cdot 10^{-4}$	

Table 2

As usual in calculating eigenvalue problems, the sequence of approximate eigenvalues attains the asymptotic convergence order 2 faster than the sequence of adjoint eigenfunctions.

As a final remark, we notice that Green's kernels behave in the same way when we treat inhomogeneous integral equations. The bound (5.2) gives us the information, that $\|K - K_n\|$ plays in this case the same decisive role for the order of convergence. Bateman's method shares this property with the spline–based procedures which we discussed in this paper (cf. (5.4) and the following comments).

References

[1]   K.E. Atkinson: A Survey of Numerical Methods for the Solution of Fredholm Integral Equations of the Second Kind. Soc. Ind. Appl. Math. Philadelphia 1976, VII + 230 p.

[2]   C.T.H. Baker: The Numerical Treatment of Integral Equations. Monographs on Numerical Analysis, Clarendon Press Oxford 1977, XII + 1034 p.

[3]   L. Bamberger and G. Hämmerlin: Spline–blended substitution kernels of optimal convergence. Treatment of Integral Equations by Numerical Methods, ed. by C.T.H. Baker and G.F. Miller, 47–57, Acad. Press London 1982.

[4]   H. Bateman: On the Numerical Solution of Linear Integral Equations. Proc. Roy. Soc. London, Ser. A 100, 441–449 (1922).

[5]   H. Brakhage: Über die numerische Behandlung von Integralgleichungen nach der Quadraturformelmethode. Numer. Math. 2, 183–196 (1960).

[6]   H. Brakhage: Zur Fehlerabschätzung für die numerische Eigenwertbestimmung bei Integralgleichungen, Numer. Math. 3, 174–179 (1961).

[7]   L.M. Delves and J.L. Mohamed: Computational methods for integral equations. Cambridge University Press 1985, XII + 376 p.

[8]   S. Fenyö–H.W. Stolle: Theorie und Praxis der linearen Integralgleichungen 4, Math. Reihe Bd. 77, Birkhäuser Verlag Basel 1984, 708 p.

[9]   W.J. Gordon: Spline–Blended Surface Interpolation through Curve Networks. J. of Math. and Mech. 18, 931–952 (1969).

[10]   G. Hämmerlin: Ein Ersatzkernverfahren zur numerischen Behandlung von Integralgleichungen 2. Art. Z. Angew. Math. Mech. 42, 439–463 (1962).

[11]   G. Hämmerlin: Zur numerischen Behandlung von homogenen Fredholmschen Integralgleichungen 2. Art mit Splines. Spline Functions Karlsruhe 1975, Lecture Notes in Math. vol. 501, ed. by K. Böhmer, G. Meinardus and W. Schempp, 92–98, Springer–Verlag 1976.

[12]   G. Hämmerlin and W. Lückemann: The Numerical Treatment of Integral Equations by a Substitution Kernel Method Using Blending–Splines. Memorie della Accademia Nazionale di Scienze, Lettere e Arti di Modena – Serie VI, Vol. XXI–1979, 1–15 (1981).

[13]   G. Hämmerlin and L.L. Schumaker: Error Bounds for the Approximation of Green's Kernels by Splines. Numer. Math. 33, 17–22 (1979).

[14]   G. Hämmerlin and L.L. Schumaker: Procedures for Kernel Approximation and Solution of Fredholm Integral Equations of the Second Kind. Handbook Series Approximations, Numer. Math. 34, 125–141 (1980).

[15]   A. Hammerstein: Nichtlineare Integralgleichungen nebst Anwendungen. Acta Mathematica 54, 117–176 (1930).

[16]   S. Joe and I.H. Sloan: On Bateman's Method for Second Kind Integral Equations. Numer. Math. 49, 499–510 (1986).

[17]   S. Kumar and I.H. Sloan: A New Collocation–Type Method for Hammerstein Integral Equations. Math. Comp. 48, 585–593 (1987).

[18]   H. Kaneko and Y. Xu: Degenerate Kernel Method for Hammerstein Equations. Math. Comp. 56, 141–148 (1991).

[19]   E.J. Nyström: Über die praktische Auflösung von linearen Integralgleichungen mit Anwendungen auf Randwertaufgaben der Potentialtheorie. Soc. Scient. Fenn. Comm. Phys.–Math. IV. 15, 1–52 (1928).

[20]   W. Prock: Zur Theorie des Bateman–Verfahrens für homogene Integralgleichungen zweiter Art. Master's thesis, Math. Inst. Univ. of Munich, 96 p. (1989).

[21]   E. Schäfer: Fehlerabschätzungen für Eigenwertnäherungen nach der Ersatzkernmethode bei Integralgleichungen. Numer. Math. 32, 281–290 (1979).

[22]   E. Schäfer: Spectral Approximation for Compact Integral Operators by Degenerate Kernel Methods, Numer. Funct. Anal. and Optimiz. 2, 43–63 (1980).

[23]   H. Weyl: Das asymptotische Verteilungsgesetz der Eigenwerte linearer partieller Differentialgleichungen (mit einer Anwendung auf die Hohlraumstrahlung). Math. Annalen 71, 441–479 (1912).

Numerical Integration of Singular and Hypersingular Integrals in Boundary Element Methods

Christoph Schwab [1]
Department of Mathematics & Statistics
University of Maryland Baltimore County
Baltimore, Maryland 21228
schwab@umbc1.umbc.edu

and

Wolfgang L. Wendland
Mathematisches Institut A
Universität Stuttgart
Pfaffenwaldring 57
D-7000 Stuttgart 80

ABSTRACT. For weakly-, Cauchy- and hypersingular surface integrals arising in three-dimensional boundary element methods, we present and analyze several numerical integration schemes. Asymptotic error estimates in terms of the size of the integration domain are given.

1. Introduction

In the past decade, the boundary element method (BEM) has become a reliable and efficient numerical method for the solution of boundary value problems, in particular of regular elliptic boundary value problems. The latter are first reduced to a system of integral equations on the boundary manifold (see e.g. [2]) which inherit ellipticity from the original boundary value problem. The boundary integral operators that arise in this way are in fact always pseudo-differential operators of integer orders, as was shown in [2].

---

[1]This author was partially supported under AFOSR grant No. 89-0252

T. O. Espelid and A. Genz (eds.), Numerical Integration, 203–218.
© 1992 *Kluwer Academic Publishers.*

Boundary element methods are discretizations of these boundary integral equations based on projections of the unknown boundary densities onto finite dimensional subspaces on the boundary manifold (see [16] for a survey). The stiffness matrix of the discretized integral equations involves, in its diagonal, integrals over a curved boundary patch F^h of a singular kernel K_n times a smooth density function ψ belonging to the finite dimensional space used in the discretization. In this paper we present numerical quadrature formulas for these integrals and give error estimates in terms of the size h of the boundary patch. They allow to select a priori the number of quadrature points needed to insure a prescribed consistency order of the approximations. Similar estimates were already obtained in [15]; here we review this work and prove generalizations for the case when the argument of the density function ψ is scaled by a factor h^{-1}.

The contents of this article are as follows: in section 2 we present the assumptions on the boundary manifold required for our results; we also collect results from [14] on the kernels in local coordinates. In section 3, we analyze three schemes for weakly singular integrals: a product scheme based on Gauss formulas with Duffy's triangular coordinates (see [4]), with polar coordinates and extrapolation schemes due to Lyness [10],[11]. Section 4 addresses Cauchy and hypersingular integrals which come up in BEM, e.g. in crack problems.

2. Surface Properties and Kernel Representations

Let Γ be an analytic surface in $\mathbb{R}^3$, with parametric representation $v = (v_1, v_2)^t$, i.e.

$$\Gamma = \chi(V)$$

where $\chi : V \to \Gamma$ is analytic and where $V \subset \mathbb{R}^2$ is a rectangular or, more generally, a polygonal domain. Consequently, the components of the vector-function χ are real analytic in (v_1, v_2) and we assume that they can be analytically continued to a complex neighborhood $\mathcal{U} \subset C^2$ of $\overline{V}$ so that for some $r^* > 0$,

$$(2.1) \qquad \left\{ v \in C^2 \mid |v_1 - u_1|^2 + |v_2 - u_2|^2 \le (r^*)^2 \right\} \subset \mathcal{U} \qquad \forall u \in \overline{V}.$$

We shall be interested in the numerical evaluation of surface integrals over so-called "boundary patches" $F_j^h \subset \Gamma$. They are supposed to be images under χ of a regular family of partitions of $\overline{V}$ (into quadrilaterals or triangles) with maximal meshwidth h, i.e.

$$\text{if} \quad \overline{V} = \overline{\cup_i \Omega_i}, \quad \text{diam } \Omega_i = O(h), \quad \text{then } \Gamma \supset F_i^h := \chi(\Omega_i).$$

We assume that each of the Ω_i is the image of a *reference element* Ω_0^h of size $O(h^2)$ under an affine map A_i which is *independent* of h. Then

$$(2.2) \qquad F_i^h = \chi \circ A_i(\Omega_0^h) =: \bar{\chi}_i(\Omega_0^h).$$

The normal vector field on Γ is given by

$$\nu(y) = \chi_{|1} \times \chi_{|2},$$

where $\chi_{|i}$ denotes the partial derivative with respect to v_i and the surface element is given by

$$ds_y = |\chi_{|1} \times \chi_{|2}|.$$

A consequence of the analyticity of χ is that both $\nu(\chi(v))$ and ds_y are analytic functions of the parameters v_i. The boundary integral operators occuring in three-dimensional BEM are usually of the following form:

$$(2.3) \qquad B_n \tilde{\psi}(x) = \text{p.f.} \int_\Gamma K_n(x, y - x) \tilde{\psi}(y) ds_y \quad \text{for } x \in \Gamma,$$

where

$$\tilde{\psi} ds_y \circ \chi = \psi \, dv$$

and $\psi(v_1, v_2)$ is a smooth density function of v_i. The selection of $\psi(v_1, v_2)$ depends on the finite dimensional subspace that is employed in the underlying BEM. We distinguish two basic cases:

a) the shape functions are global, smooth functions which are independent of h or

b) the shape functions are obtained by lifting piecewise polynomial functions in V to Γ via χ. They are usually normalized to have unit height and have support in F_j^h, where area $(F_j^h) = O(h^2)$.

The kernel $K_n(x, z)$ in (2.3) is homogeneous of degree $-2 - n$ in the second argument, i.e.

$$(2.4) \qquad K_n(x, tz) = t^{-2-n} K_n(x, z) \quad \forall x, z \in \mathbb{R}^3 \quad \text{if} \quad t \neq 0,$$

and n is the order of the operator B_n. In [2] it was shown that every boundary integral operator associated with elliptic boundary value problems is a pseudodifferential operator of integer order $n \in \mathbb{Z}$ that can be written in the form (2.3), (2.4). In current applications we have $n = -1, n = 0$ or $n = 1$, leading to weakly singular, Cauchy singular and hypersingular integrals in (2.3), respectively, if the source point x lies in the interior of F_j^h. This is the case we will focus on and we assume it in the sequel. Simple examples for $K_n(x, y - x)$ arising for the boundary value problems of the Laplacian, are

$$K_{-1}(x, y - x) = \frac{1}{\varrho} \quad \text{(single layer potential)},$$

$$K_{-1}(x, y - x) = \frac{\partial}{\partial \nu(y)} \left(\frac{1}{\varrho} \right) \quad \text{(double layer potential)},$$

$$K_1(x, y - x) = \frac{\partial}{\partial\nu(x)\partial\nu(y)}\left(\frac{1}{\varrho}\right) \quad \text{(normal derivative of the double layer potential).}$$

Since for $n \geq 0$, due to (2.4), the integrand $K_n\tilde{\psi}$ in (2.3) is not absolutely integrable, we define the "integral" in this case as the finite part of

$$\int_{|x-y|\geq\epsilon} K_n(x, y - x)\tilde{\psi}(y)\,ds_y,$$

in the asymptotic expansion with respect to $\epsilon > 0$ where $\varrho = |x - y|$ denotes the Euclidean distance in $\mathbb{R}^3$ and write "p.f. $\int$" then.

In boundary element calculations the domains of integration in (2.3) are the patches F_i^h. There we must evaluate

$$(2.5) \qquad\qquad I_i^h = \text{p.f.} \int_{F_i^h} K_n(x, y - x)\tilde{\psi}(y)\,ds_y$$

numerically up to a given order in the meshwidth h which arises from consistency analysis of the underlying boundary element scheme.

Since in applications often several hundred integrals of the form (2.5) are to be computed, we would like to apply the same quadrature rule to all these integrals. This is a nontrivial task for the following reasons:

1. Using the parameter representation $\bar{\chi}_i$ in (2.2), we would like to write I_i^h as an integral over Ω^h. If $n \leq -1$ this is a simple change of variables. However, if $n \geq 0$, we must change variables in a finite part integral, where additional point functionals may arise (see [14], Theorem 5).

2. Define

$$(2.6) \qquad\qquad \tilde{K}_n(u, v - u) := K_n\left(\bar{\chi}(u), \bar{\chi}(v) - \bar{\chi}(u)\right)$$

where we omit the index i of F_i^h. This kernel is not homogeneous any more in the local coordinates, but rather *pseudohomogeneous*, i.e. for every $L \in \mathbb{N}$ we have an asymptotic expansion

$$(2.7) \qquad\qquad \tilde{K}_n(u, v - u) = \sum_{j=0}^{L} k_{n-j}(u, v - u) + R_L(u, v - u)$$

where the remainder $R_L(u, v - u)$ satisfies [14]:

$$|R_L(u, v - u)| \leq C|v - u|^{-2-n+L+1}$$

$$R_L(u, \cdot) \in C^{-2-h+L}(\bar{V}) \quad \text{for} \quad L \geq n + 2.$$

The kernels k_{n-j} are given by

$$k_{n-j}(u, v - u) = \frac{f_{n-j}(u, \theta)}{r^{2+n-j}}$$

where f_{n-j}, the so-called characteristics are analytic functions of θ and

$$(2.8) \qquad v - u = r \begin{pmatrix} \cos \theta \\ \sin \theta \end{pmatrix}.$$

Clearly $\tilde{K}_n(u, z)$ is generally *not* of the form $(r^{-2-n} \cdot$ smooth function of z), hence e.g. for $n = -1$ simple weighted quadratures as in [3], cannot be applied, as was also observed in [1].

Therefore our approach to the numerical quadrature of (2.5) consists of
a) an analysis and classification of the kernels $K_n\left(\bar{\chi}_i(u), \bar{\chi}_i(v) - \bar{\chi}_i(u)\right)$ and
b) the construction of quadrature formulas which work well for the whole class.
The following result is the main tool for the analysis of $\tilde{K}_n(u, v - u)$ ([14, Theorem 4]).

Proposition 1 *Let the surface parametrization χ of Γ be analytic with domain $\mathcal{U}$ of analyticity, characterized by r^* as in (2.1). Then, for $x, y \in \Gamma$,*

$$(2.9) \qquad \varrho^2 = |x - y|^2 = \sum_{j=2}^{\infty} r^j l_j(\theta)$$

where the l_j are homogeneous polynomials of degree j in $\cos \theta$ and $\sin \theta$. Furthermore, for every integer a we have

$$(2.10) \qquad \varrho^{-2-a} = r^{-2-a} \sum_{\nu=0}^{\infty} r^\nu \left(l_2^{-\frac{a}{2}-\nu-1}(\theta) p_{3\nu}(\theta) \right)$$

and the series in (2.9) and (2.10) converge for $|r| < r^*$. Here $p_{3\nu}(\theta)$ denotes a homogeneous polynomial of degree 3ν in $\cos \theta$ and $\sin \theta$.

The relation (2.10) shows the significance of $l_2(\theta)$ and from (2.9) we find that

$$(2.11) \qquad l_2(\theta) = E \cos^2 \theta + 2F \cos \theta \sin \theta + G \sin^2 \theta$$

and E, F, G are the Gaussian entries of the Riemann tensor at u [6]. The complex zeros $\theta^* \in C$ of $l_2(\theta)$ are given by the solutions of

$$\cot \theta^* = \frac{-F \pm \sqrt{F^2 - EG}}{E}.$$

We are now ready to analyze the numerical quadrature in local coordinates for the integrals (2.5). In the following section we shall investigate three quadrature strategies for weakly singular (i.e. $n = -1$) operators: namely tensor product rules in connection with Duffy's triangular coordinates [4], with polar coordinates and, finally, also extrapolation based on the results of Lyness [10, 11, 12].

3. Weakly Singular Kernels

In the following two sections we study quadrature rules arising from special coordinates substituted into the kernel. Both times the substitution renders the transformed integrand *analytic* and we can estimate the domain of analyticity.

This in turn allows us to use the error estimates by Rabinowitz and Richter [13] to bound the quadrature error. We point out that we are interested in the dependence of this error on the size h of the integration domain F_j^h, rather than in the error as the number of quadrature points tends to infinity.

3.1. DUFFY'S TRIANGULAR COORDINATES

We assume that the reference element Ω_0^h in (2.2) is the triangle $(0,0)$, $(h,0)$, (h,h) and that the singularity u coincides with the local origin $(0,0)$. Then, after substituting variables, we must evaluate the integral

$$(3.1) \qquad I_i^h = \int_0^h \int_0^{v_1 - u_1} \tilde{K}_n(u, v - u)\psi_d(h^{-1}(v - u)) dv_1 dv_2$$

with $\tilde{K}_n$ as in (2.6) and containing the surface element ds_y, and $\psi_d \in \Pi_d$, the set of polynomials of degree d. Note that ψ_d has been scaled by h^{-1} in the argument, this corresponds to the case of a density with local support, i.e., case b) in the previous section. According to [14], the kernel $\tilde{K}_n$ takes the form

$$(3.2) \qquad \tilde{K}_n(u, v - u) = \frac{1}{\varrho^{\kappa+2}} \left\{ \sum_{|\alpha| \geq \kappa - n \geq 0} a_\alpha(u)(v - u)^\alpha \right\}$$

for some $\kappa \geq n$. Now Duffy's triangular coordinates [4] are defined by

$$(3.3) \qquad v_1 - u_1 = \xi, \quad v_2 - u_2 = \xi\eta,$$

hence

$$(3.4) \qquad dv_1 dv_2 = \xi \, d\xi d\eta.$$

We have the following result on the local behaviour of the transformed kernel near the source point.

Proposition 2 [15, Theorem 1]
Substituting (3.3) into (3.1) yields an integrand $\xi\tilde{K}_n(\xi, \xi\eta)\psi_d$ which, for every $\epsilon_0 > 0$ is complex analytic in ξ for $|\xi| \leq r^*/\epsilon_0$, provided $|\eta \pm i| \geq \epsilon_0$, $|\eta + \cot\theta^*| \geq \epsilon_0$. Moreover, the integrand may be continued analytically in $\eta \in C$ up to $\pm i$ and $\cot\theta^*$, respectively, if $\xi \in [0, h]$ and $|1 + \eta^2| \leq (r^*/h)$. On any compact subset C of the region of analyticity the extension of the integrand $\xi\tilde{K}_n(\xi, \xi\eta)\psi_d$ is uniformly bounded independent of h.

We point out that Proposition 2 holds for both cases, the scaled as well as the unscaled density functions ψ_d.

Now we can show the following bound on the quadrature error if N-point Gauss Legendre formulas G^N are used on the transformed integral

$$(3.5) \qquad I_i^h = \int_0^h \int_0^1 \xi \tilde{K}_n \left(u, \left(\begin{array}{c} \xi \\ \xi\eta \end{array} \right) \right) \psi_d \left(h^{-1} \left(\begin{array}{c} \xi \\ \xi\eta \end{array} \right) \right) d\xi d\eta.$$

Theorem 1 *Let* $t_i^{(N)}, w_i^{(N)}$ *denote knots and weights of* G^N *on* $(0,1)$. *Then there exists* $C_N > 0$, $\delta > 1$ *independent of* h *such that*

$$\left| \sum_{i=1}^N \sum_{k=1}^M w_i^{(N)} w_k^{(M)} \left[ht_i^{(N)} (\tilde{K}_n \psi_d) \right] \left(ht_i^{(N)}, ht_i^{(N)} t_k^{(M)} \right) - I_i^h \right|$$

$$(3.6) \qquad\qquad\qquad\qquad \leq C_N h \left\{ h^{2N-d} + e^{-2M \ln \delta} \right\}$$

If ψ_d *is not scaled by* h^{-1} *then* $d = 0$ *may be selected in* (3.6).

Proof: If ψ_d is not scaled, this is proved in [15, Theorem 2]. If $\psi_d = \psi_d(h^{-1}(v-u))$, we set

$$f(\xi, \eta) = \xi \tilde{K}_n \left(u, \left(\begin{array}{c} \xi \\ \xi\eta \end{array} \right) \right).$$

Note that f is independent of h and, by Proposition 2, analytic in ξ and η. We estimate

$$(3.7) \qquad \begin{aligned} E[f(\xi,\eta)\psi_d] &= \left| I_i^h - G_\xi^N G_\eta^M [f(\xi,\eta)\psi_d] \right| \\ &\leq \left| G_\xi^N (G_\eta^M - I_\eta)[f\psi_d] \right| + \left| I_\eta (G_\xi^N - I_\xi)[f\psi_d] \right| \\ &\leq h \sum_{i=1}^N w_i^{(N)} \left| (G_\eta^M - I_\eta)[f\psi_d] \right| + \max_{\eta \in [0,1]} \left| (G_\xi^N - I_\xi)[f\psi_d] \right| \end{aligned}$$

Here G_η^M denotes a Gauss Legendre formula of order M on $[0,1]$ and I_η denotes the integration $\int_0^1 d\eta$ with respect to η; G_ξ^N, I_ξ are defined analogously on $[0,h]$. The first term in (3.7) we estimate as in [15], which yields the bound

$$C h e^{-2M \ln \delta}, \qquad \delta = \delta(r^*) > 1$$

where C is independent of h and M.

For the second term in (3.7), we have the expansion (for $N \geq \frac{1}{2}(d-n)$)

$$f(\xi, \eta) = \sum_{j=-n-1}^{2N-1-d} \xi^j C_j(\eta) + R_N = f_N(\xi, \eta) + R_N$$

where
$$(3.8) \qquad\qquad\qquad |R_N(\xi, \eta)| \leq C_N |\xi|^{2N-d}.$$

If we split I_ξ accordingly, i.e.,

$$I_\xi[f\psi_d] = I^1 + I^2, \qquad I^2 = \int_0^h \int_0^1 R_N(\xi,\eta)\psi_d(h^{-1}\xi, h^{-1}\xi\eta)d\xi d\eta,$$

we estimate for fixed η

$$\left| \left(I_\xi - G_\xi^N \right) [f\psi_d] \right| \leq |I^1 - G_\xi^N [f_N\psi_d]| + |I^2 - G_\xi^N [R_N\psi_d]|.$$

Since

$$\psi_d(h^{-1}\xi, h^{-1}\xi\eta) = h^{-d}\xi^{-d} \sum_{|\alpha| \leq d} \psi_\alpha \eta^{\alpha_2},$$

the integrand

$$f_N\psi_d = \psi_d(h^{-1}\xi, h^{-1}\xi\eta) \sum_{j=-n-1}^{2N-1-d} \xi^j c_j(\eta)$$

is, for fixed η, a polynomial in ξ of degree $2N - 1$ which is integrated exactly by G_ξ^N, i.e. $I^1 - G_\xi^N[f_N\psi_d] = 0$. It remains to estimate for fixed η

$$|I^2| = \left| \int_0^h R_N(\xi, \eta)\psi_d(h^{-1}\xi, h^{-1}\xi\eta)d\xi \right|.$$

Since the coefficients ψ_α are independent of h, the term $|\psi_d(h^{-1}\xi, h^{-1}\xi\eta)|$ is bounded for $0 \leq \xi \leq h$, and, from (3.8), we have

$$|G_\xi^N[R_N\psi_d]| \quad \text{and} \quad |I^2| \leq C_N h^{2N+1-d}.$$

This proves (3.6). $\qquad\qquad\qquad\qquad\qquad\qquad\qquad\qquad\qquad\qquad\qquad\qquad\square$

Remark: The bound $C_N h$ in (3.6) can be replaced by $C_N h^{-n}$ if appropriate *weighted* Gaussian formulas G_ξ^N are used for $n < -1$.

3.2. LOCAL POLAR COORDINATES

Here we rewrite (3.5) using (2.8) as

$$(3.9) \qquad I_i^h = \int_0^{hR(\theta)} \int_{\theta_1}^{\theta_2} r\tilde{K}_n\left(u, r\begin{pmatrix} \cos\theta \\ \sin\theta \end{pmatrix}\right) \psi_d\left(h^{-1}r\begin{pmatrix} \cos\theta \\ \sin\theta \end{pmatrix}\right) dr d\theta.$$

The function $R(\theta)$ is independent of h and analytic in θ. For an approximate quadrature based on Gauss Legendre formulas of degree N (properly scaled) in r and of degree M in θ we have the following result.

Theorem 2 [15] *There exists C_N independent of h and M such that*

$$(3.10) \qquad \left| I_i^h - G_\theta^M G_r^N[r\tilde{K}_n\psi_d] \right| \leq C_N h \left\{ h^{2N-d} + e^{-2M\ln\delta} \right\}.$$

If ψ_d is independent of h, we can choose $d = 0$ here.

Hence, in comparison, the use of triangular coordinates yields the same degree of accuracy as polar coordinates. The latter approach, however, can be extended to Cauchy- and hypersingular integrals, as we shall show in Section 4.

3.3. EXTRAPOLATION BASED BOUNDARY ELEMENT QUADRATURE

In this section we demonstrate that *all* integrands arising from weakly singular boundary integral operators on arbitrary smooth surfaces Γ with kernels (2.7) are admissible integrands for the Lyness extrapolation–based integration scheme. This extrapolation scheme was introduced in [10]. For a survey and references see [12]. There, however, it is left open which class of weakly singular boundary integral operators is admissible for the extrapolation technique and what terms appear in general in asymptotic expansion of the quadrature error. We present here a result which states that for any weakly singular integral arising in BEM

$$(3.11) \qquad I_i^h = \int_{\Omega_0^h} \tilde{K}_n(u, v - u)\psi_d(h^{-1}(v - u))dv_1 dv_2$$

is an admissible integral for extrapolation based quadrature and identify the terms in the error expansion.

Lyness in [10, 11] introduces the following class of integrands.

Definition 1 [10, Definition 5.2] *Let*

$$\Omega_0^h = [0, h] \times [0, h] \ or \ \Omega_0^h = \{v \mid 0 \le v_1 - u_1 \le h, 0 \le v_2 - u_2 \le h - (v_1 - u_1)\}.$$

Then $f(v - u) \in H_{-1}^{(2)}(\Omega_0^h)$ *if f is of the form*

$$(3.12) \qquad f(v - u) = r^{-1}\phi(\theta)h(r)g(v - u)$$

where $\phi(\theta)$ *is analytic in* θ *for* $0 \le \theta \le \pi/2$, $h(r)$ *is analytic in r and* $g(\cdot)$ *is analytic in each variable in* $\overline{\Omega_0^h}$.

The significance of $H_{-1}^{(2)}$ is that the quadrature error of I_i^h in (3.11) for $Q^{(m)}$, an m-copy of the mid-point rule on Ω_0^h (see [10, 11] or [12] for a definition), admits an asymptotic expansion upon which extrapolation can be based. We show now that the integral (3.11) is, indeed, in $H_{-1}^{(2)}(\Omega_0^h)$ and present the corresponding asymptotic quadrature error expansion.

Theorem 3 *For all weakly singular integrands* (3.11) *we have*

$$\tilde{K}_n(u, v - u)\psi_d(h^{-1}(v - u)) \in H_{-1}^{(2)},$$

$$(3.13) \qquad I_i^h - Q^{(m)}[\tilde{K}_n\psi_d] = \sum_{s=1}^{l-1} \frac{\Lambda_s + B_s}{m^s} + \frac{C_s}{m^s}\ln m + O(m^{-l}\ln m)$$

Proof: Using the kernel expansion (2.7) and [10], Lemma 5.3], the first assertion readily follows. The expansion (3.13) is then a consequence of [10, Theorem 5.14], if $\Omega_0^h = [0, h]^2$ and in case Ω_0^h is a triangle, of [11, Theorem 5.4]. $\qquad\qquad\qquad\qquad\square$

The result (3.13) gives precise information or the exponents occurring in the quadrature error expansion for *any* weakly singular kernel. Hence, we have proved that using extrapolation based on (3.13) as suggested in [12] will provide an efficient quadrature scheme for all weakly singular kernels arising in BEM.

4. Strongly Singular Integrals

Let us now address the numerical integration of (2.3), (2.5) for $n \geq 0$, i.e. the case when the kernel K_n is not absolutely integrable anymore.

We distinguish two basic approaches to these integrals:
a) reduction to a weakly singular integral which can be treated by the above methods and
b) direct integration of (2.3), (2.5) via suitable quadrature schemes.

We address a) in section 4.1 and b) in section 4.2 below.

It is well known that in general one can not formally substitute variables in integrals as (2.5). In fact, for such integrals a smooth change of variables yields a finite part integral once more plus point functionals of the density ψ_d at the source point.

$$(4.1) \quad I_i^h = \text{p.f.} \int_{F_i^h} K_n(x, y - x)\tilde{\psi}_d(y)ds_y$$

$$= \text{p.f.} \int_{\Omega_0^h} \tilde{K}_n(u, v - u)\psi_d(h^{-1}(v - u))dv + \sum_{0 \leq \alpha \leq n} c_\alpha \frac{1}{\alpha!}\left(D^\alpha \psi_d\right)(0)$$

where the constants c_α depend on $\bar{\chi}$ and its derivatives at 0, see [14, Theorem 3]. There also explicit formulas for the calculation of c_α can be found. If, however, $x \in \text{int}(F_i^h)$ in (4.1), then R. Kieser in [7] has shown that in fact the c_α in (4.1) are zero for all kernels arising in BEM. This is the basis for numerical quadrature of (4.1) without any additional explicit analytical calculations.

4.1. REGULARIZATION AND REDUCTION TO A WEAKLY SINGULAR INTEGRAL

The basic idea behind the regularization used here is the subtraction of a Taylor polynomial of order n at the source point off the density function and to treat the finite-part integral of the Taylor polynomial analytically. We have [14, Section 3]

$$(4.2) \quad I_i^h = \sum_{0 \leq |\alpha| \leq n} (c_\alpha + d_\alpha)\frac{1}{\alpha!}D^\alpha \psi_d(0)$$

$$+ \int_{\Omega_0^h} \tilde{K}_n(u, v - u)R_n[\psi_d]\left(h^{-1}(v - u)\right) dv,$$

where $\tilde{K}_n$ is defined in (2.6), and

$$R_n[\psi_d] = \psi_d\left(h^{-1}(v-u)\right) - \sum_{|\beta|\le n} \frac{h^{-|\beta|}}{\beta!}(v-u)^\beta \left(D^\beta\psi_d\right)(0)$$

is the remainder of the Taylor expansion of ψ_d about 0, and

$$d_\alpha(h) = \sum_{0\le j\le h} \int_0^\omega f_{n-j}(u,\theta)\left(\begin{array}{c}\cos\theta\\\sin\theta\end{array}\right)^\alpha \cdot$$
$$\left[\left\{\begin{array}{ll}\ln\left(hR(\theta)\right) & \text{if } |\alpha| = n-j\\(|\alpha|-n+j)^{-1}\left(hR(\theta)\right)^{|\alpha|-n+j} & \text{else}\end{array}\right\} + \frac{l_2(\theta)}{2}\right]d\theta.$$

Again we recall that $(\theta, hR(\theta))$ for $0 \le \theta \le \omega \le 2\pi$ is an analytic parameter representation of $\partial\Omega_0^h$. For piecewise analytic $R(\theta)$ and analytic Γ we break up the domain of integration into a sum with analytic $R(\theta)$ on each piece, respectively, and, there, confine ourselves to the analytic case. Now we obtain with the general kernel $\tilde{K}_n$ in (3.2):

Theorem 4 *For $\psi_d \in \Pi_d$, the integrand*

$$\tilde{K}_n(u, v-u)R_n[\psi_d]\left(h^{-1}(v-u)\right) = \frac{1}{\varrho^{\kappa+2}}\left\{\sum_{|\alpha|\ge\kappa-n} a_\alpha(u)(v-u)^\alpha\right\} R_n[\psi_d]\left(h^{-1}(v-u)\right)$$

of (4.2) is weakly singular and of the form (3.2). Consequently, all methods discussed in Section 2 can be applied to the numerical evaluation of (4.2) (however, some of the error estimates must then be modified).

We point out that for the evaluation of the coefficients c_α in (4.2) one needs the characteristics $f_{n-j}(u,\theta)$ in (2.7) explicitly. They were obtained recently for hypersingular operators which arise in 3-d crack problems by Guiggiani et al. in [5]. Since $f_{n-j}(u,\theta)$ is analytic in θ, the calculation of d_α can be done efficiently by means of Gaussian formulas.

Let us now describe the construction of integration formulas for I_i^h in (4.2) which do *not* require the analytical evaluation of the f_{n-j}.

4.2. TENSOR PRODUCT FORMULAS IN POLAR COORDINATES

In this section we consider the numerical integration of the integral term in (4.1), i.e., of

(4.3) $$I^h = \text{p.f.} \int_{\Omega_0^h} \tilde{K}_n(u, v-u)\psi_d\left(h^{-1}(v-u)\right)dv.$$

We assume that $\partial\Omega_0^h$ is parameterized in polar coordinates (2.8) as $(\theta, hR(\theta))$, $0 \le \theta \le \omega$, where $R(\theta)$ is independent of h and piecewise analytic with domains of analyticity

$\mathcal{B}_i \subset C$. Based on the kernel expansion (2.7) we write the contribution from each piece of analyticity of $R(\theta)$ in the form

$$I^h = \int_0^\omega \text{p.f} \int_0^{hR(\theta)} r^{-2-n} \left\{ \sum_{j=0}^L r^j f_{n-j}(u,\theta) \right\} \psi_d \left(h^{-1} r \left(\begin{array}{c} \cos\theta \\ \sin\theta \end{array} \right) \right) r\, dr d\theta$$

$$(4.4) \qquad + \int_{\Omega_0^h} R_L(u, v-u) \psi_d \left(h^{-1}(v-u) \right) dv$$

and we assume that $R(\theta)$ admits an analytic extension to $[0,\omega] \subset \mathcal{B} \subset C$. Again I^h may be viewed as the contribution from a single piece of analyticity of $R(\theta)$. Let us describe how we construct quadrature formulas for (4.4). Denote

$$(4.5) \qquad g(\theta) := \text{p.f.} \int_0^{hR(\theta)} r^{-2-n} k(u, v-u) \psi_d \left(h^{-1} r \left(\begin{array}{c} \cos\theta \\ \sin\theta \end{array} \right) \right) r\, dr d\theta$$

where

$$k(u, v-u) = \sum_{j=0}^L r^j f_{n-j}(u,\theta).$$

To approximate the outer integral in (4.4), we use an M-point Gauss-Legendre or Gauss Lobatto formula on $[0,\omega]$:

$$I^k \sim Q_\theta^M[g] = \sum_{i=1}^M w_i^{(M)} g(\theta_i^{(M)}).$$

This leaves us with the approximate evaluation of $g(\theta_i^{(M)})$ defined in (4.5). We use here an interpolatory finite part formula as proposed by Kutt in [8, 9]. Due to the variation in length of the integration interval $[0, hR(\theta_i^{(M)})]$ we must determine a distinct finite part formula for each $\theta_i^{(M)}$. To this end, for $\theta \in [0,\omega]$, we select

$$0 < r_1(\theta) < \cdots\cdots < r_N(\theta) \le R(\theta)$$

and find $w_i^{(N)}(\theta)$ so that

$$Q_r^N(\theta)[p] - \text{p.f.} \int_0^{R(\theta)} r^{-1-n} p(r) dr = 0 \quad \text{for all } p \in \Pi_{N-1}$$

where

$$Q_r^N(\theta) = \sum_{i=1}^N w_i^{(N)}(\theta) p(r_i).$$

A scaling argument shows that also

$$(4.6) \qquad \tilde{Q}_r^N(\theta)[p] := \sum_{i=1}^N h^{-n} w_i^{(N)} p(h r_i) = \text{p.f.} \int_0^{hR(\theta)} r^{-1-n} p(r) dr,$$

if we ignore point functionals arising through scaling. As we remarked above, if the source point lies in the *interior* of F_j^h such point functionals do not occur and we are justified in scaling the formula. We have

Theorem 5 *With $\tilde{Q}_r^N(\theta)$ as in (4.6) and I^h as in (4.3), for N sufficiently large, we have*

$$\left| I^h - Q_\theta^M \tilde{Q}_r^N(\theta) \left[r^{2+n} \tilde{K}_n(u, \bullet)\psi_d(h^{-1}\bullet) \right] - \sum_{0 \le |\alpha| \le n} c_\alpha \frac{1}{\alpha!}(D^\alpha \psi_d)(0) \right|$$
$$\le C_N h^{-n} \left\{ |\ln h| e^{-2M \ln \delta} + h^{N-d} \right\}.$$

where $\delta > 1$ depends only on $\bar{\chi}$ in Ω_0^h.

Here the c_α are the coefficients of the point functionals from (4.1) which are given explicitly in [14]. If the source point $x = \bar{\chi}(u)$ lies in the interior of F^h then $c_\alpha = 0$ [7]. *Proof:* We estimate the second term in (4.4). With (2.7) we have

(4.7)
$$\left| \int_{\Omega_0^h} R_L(u, v - u)\psi_d(h^{-1}(v - u))dv \right| \le C_L h^{-n+L+1},$$

and $L > 0$ is at our disposal. This shows that the second term in (4.4) can be neglected by choosing L large enough. Hence we need only estimate the difference between the first term and the quadrature formula applied to it. Using (4.5), we set $\mathcal{I}^h := \int_0^\omega g(\theta)d\theta$ and estimate

(4.8)
$$\left| \mathcal{I}^h - Q_\theta^M \tilde{Q}_r^N(\theta)[k(u, \cdot)\psi_d(h^{-1}\cdot)] \right|$$
$$\le \left| \mathcal{I}^h - Q_\theta^M[g] \right| + \left| Q_\theta^M \left[g - \tilde{Q}_r^N(\theta)[k(u, \cdot)\psi_d(h^{-1}\cdot)] \right] \right|.$$

Consider the first term in (4.8). Observe that

$$g(\theta) = \sum_{j=0}^L g_{n-j}(\theta)$$

where

$$
\begin{aligned}
g_{n-j}(\theta) &= \text{p.f.} \int_0^{hR(\theta)} r^{-1-n} f_{n-j}(u, \theta)\psi_d\left(h^{-1}r \begin{pmatrix} \cos\theta \\ \sin\theta \end{pmatrix} \right) dr \\
&= f_{n-j}(u, \theta) \sum_{k=0}^d \psi_k(\theta) \cdot \text{p.f.} \int_0^{hR(\theta)} r^{-1-n+j+k} dr \cdot h^{-k} \\
&= f_{n-j}(u, \theta) \sum_{k=0}^d \psi_k(\theta) h^{-k} \left\{ \begin{array}{ll} \ln(hR(\theta)) & , j+k = n \\ \frac{1}{j+k-n}(hR(\theta))^{j+k-n} & , j+k \ne n \end{array} \right\}.
\end{aligned}
$$

Based on this representation we infer that $g_{n-j}(\theta)$ is analytic in θ with domain of analyticity $\mathcal{B} \subset C$ being the intersection of that of $R(\theta)$ and of $f_{n-j}(u, \theta)$ provided that $h > 0$ is sufficiently small. Moreover, for all $\theta \in \mathcal{C} \subset\subset \mathcal{B}$, $\mathcal{C}$ independent of h,

$$|g_{n-j}(\theta)| \le C(\mathcal{C})h^{j-n}|\ln h|,$$

hence also $g(\theta)$ is analytic in $\mathcal{B}$ and

$$|g(\theta)| \le C(\mathcal{C})h^{-n}|\ln h| \quad \text{for all } \theta \in \mathcal{C} \subset\subset \mathcal{B}.$$

Consequently we may use the derivative–free error estimates in [13] for the first term in (4.8) and, as in (4.8), get for some $\delta(\Gamma) > 1$ that

$$(4.9) \qquad \left|\mathcal{I}^h - Q_\theta^M[g]\right| \le Ch^{-n}|\ln h|e^{-2M\ln\delta}.$$

The second term in (4.8) is estimated as follows.

$$\left|Q_\theta^M\left[g - \tilde{Q}_r^N(\theta)[k(u,\cdot)\psi_d(h^{-1}\cdot)]\right]\right|$$
$$\le \sum_{i=1}^{M} w_i^{(M)} \max_\theta \left|g(\theta) - \tilde{Q}_r^N(\theta)[k(u,\cdot)\psi_d(h^{-1}\cdot)]\right|.$$

Since from (4.5) we have

$$k\left(u, r\begin{pmatrix}\cos\theta\\\sin\theta\end{pmatrix}\right)\psi_d\left(h^{-1}r\begin{pmatrix}\cos\theta\\\sin\theta\end{pmatrix}\right) = \sum_{j=0}^{L} r^j f_{n-j}(u,\theta)\sum_{k=0}^{d} h^{-k}r^k \psi_k(\theta)$$

and since $\tilde{Q}_r^N(\theta)$ defined in (4.6) is exact for polynomials in r of degree $N-1$, we have for

$$k_1 = \sum_{j=0}^{N-d-1} r^j f_{n-j}(u,\theta), \quad k_2 := k - k_1$$

that

$$\epsilon := \left|g(\theta) - \tilde{Q}_r^N(\theta)\left[k(u,\cdot)\psi_d(h^{-1}\cdot)\right]\right|$$
$$\le \left|\text{p.f.} \int_0^{hR(\theta)} r^{-1-n}k_2(u,\cdot)\psi_d(h^{-1}\cdot)dr\right| + \left|\tilde{Q}_r^N(\theta)\left[k_2(u,\cdot)\psi_d(h^{-1}\cdot)\right]\right|.$$

On $[0, hR]$, $\psi_d(h^{-1}\cdot)$ is bounded and $\left|k_2\left(u, r\begin{pmatrix}\cos\theta\\\sin\theta\end{pmatrix}\right)\right| \le C_N r^{N-d}$. Hence, for N sufficiently large ($N \ge d+n$) we have the bound

$$\epsilon \le C_N\left\{(hR(\theta))^{N-d-n} + h^{-n+N-d}\right\} \le C_N h^{N-d-n}$$

Combining this with (4.9) and selecting L in (4.7) large enough completes the proof.□

Remarks:

1. Exactly the same proof is also valid for $n = -1$; if $\tilde{Q}_r^N(\theta)$ is replaced by a properly scaled, N-point Gauss Legendre formula, we obtain estimate (3.10).

2. If the density function ψ_d is not scaled by a factor h^{-1} in the argument, the term h^{N-d} in Theorem 5 can be replaced by h^N, as was shown in [15].

3. In order for the integration error E in Theorem 5 to satisfy a consistency estimate of the form

$$|E| \leq C h^a$$

with given $a > 0$, one should choose

$$N \geq a + n + d \text{ and } M \geq \frac{|\ln h|}{2 \ln \delta}(N - d).$$

Analogously, Theorems 1 and 2 imply corresponding minimal numbers of integration points.

References

[1] Aliabadi, M.H. and Hall, W.S.:
Weigthed Gaussian methods for three dimensional boundary element kernel integrations.
Comm. Appl. Num. Meth. 30, pp. 89-96 (1987)

[2] Costabel, M. and Wendland, W.L.:
Strong Ellipticity of Boundary Integral Operators.
Journ. Reine Angew. Math. 372, 34-63 (1986).

[3] Cristescu, M. and Laubignac, G.:
Gaussian Quadrature Formulas for Functions with Singularities in $1/R$ over Triangles and Quadrangles.
in: Recent Advances in Boundary Element Methods, C.A. Brebbia (Ed.), pp. 375-390. London, Pentech Press 1978.

[4] Duffy, M.G.:
Quadrature over a pyramid or cube of integrals with a singularity at a vertex.
SIAM J. Numer. Anal. 19, pp. 1260-1262 (1982)

[5] Guiggiani, M., Krishnasamy, G., Rudolphi, T.J. and Rizzo, F.J.:
A general algorithm for numerical solution of hypersingular boundary integral equations.
to appear in ASME J. Applied Mechanics.

[6] Haack, W.:
Elementare Differentialgeometrie
Birkhäuser - Verlag, Basel and Stuttgart (1955)

[7] Kieser, R.:
Über einseitige Sprungrelationen und hypersinguläre Operatoren in der Methode der Randelemente.
Doctoral Thesis, University Stuttgart (1991)

218

[8] Kutt, H.R.:
On the numerical evaluation of finite part integrals involving an algebraic singularity.
WISK-report No. 179, Univ. Stellenbosch, Pretoria, South Africa (1975)

[9] Kutt, H.R.:
The numerical evaluation of principal value integrals by finite part integration.
Numer. Math. 24, pp. 205-210 (1975)

[10] Lyness, J.N.:
An error functional expansion for N-dimensional quadrature with an integrand function singular at a point.
Math. Comp. 30, pp. 1-23, (1976)

[11] Lyness, J.N.:
Quadrature error functional expansions for the simplex when the integrand function has singularities at vertices.
Math. Comp. 34, pp. 213-225, (1980)

[12] Lyness, J.N.:
On Handling Singularities In Finite Elements.
These proceedings.

[13] Rabinowitz, P. and Richter, N.:
New error coefficients for estimating quadrature errors for analytic functions.
Math. Comp. 24, pp. 561-570 (1970)

[14] Schwab, C. and Wendland, W.L.:
Kernel properties and representations of boundary integral operators.
Preprint 91-2, Math. Inst. A, Univ. Stuttgart, Germany (1991)

[15] Schwab, C. and Wendland, W.L.:
On Numerical Cubatures of Singular Surface Integrals in Boundary Element Methods.
Preprint 91-3, Math. Inst. A, Univ. Stuttgart, Germany (1991)

[16] Wendland, W.L.:
Strongly elliptic boundary integral equations.
in: The State of the Art in Numerical Analysis. A. Iserles, M. Powell (Eds.), pp. 511-561, Oxford, Clarendon Press 1987.

ON HANDLING SINGULARITIES IN FINITE ELEMENTS*

J. N. LYNESS
Mathematics and Computer Science Division
Argonne National Laboratory
9700 South Cass Avenue
Argonne, Illinois 60439 U.S.A.

ABSTRACT. In the practice of the Boundary Element Method, a basic task involves the quadrature over a quadrilateral or triangle of an integrand function which has a singularity of known form at a vertex. A not uncommon situation is that this quadrature has already been studied in depth for the standard triangle or the square, and all that is now necessary is to apply the known results in the context of a different triangle or parallelogram, one that has been obtained from the standard region by an affine transformation.

It can be surprising to someone who has not done it himself, how difficult this task can be.

This article provides an account of how easy it is to be misled in this area. Besides describing an apparently cost effective approach which turns out to be a disaster, I discuss some of the advantages and disadvantages of using rules based on extrapolation either as an alternative to, or in conjunction with Gaussian rules. This article is anecdotal in character.

1. Introduction

In books and research articles, the seeker after knowledge will find plenty of information about how to integrate regular or "well behaved" integrand functions over various standard regions. These will include in great detail, a standard square, such as $H:[0,1)^2$ and possibly in less detail, a standard triangle, usually an isosceles right-angled one

$$T_2: x \geq 0; \ y \geq 0; \ x+y < 1$$

and, of course, higher-dimensional analogues of these.

If our knowledge seeker is interested in implementing a finite element program, he will want more than this. He has to deal with large number of elements. Many will be similar to one another. The majority will involve integrating regular integrand functions over nonstandard triangles and quadrilaterals. This is not particularly difficult. But there will be a small but significant proportion which are more difficult, having one or another of the features mentioned below. He may have regions which only approximate to triangles or squares. He may have, for example, a plane curvilinear triangle, such as a quadrant of a circle, or a more general curvilinear triangle such as the surface of an octant of a sphere. These boundaries and surfaces may be specified either in a convenient form, or in some inconvenient possibly highly implicit form. His integrand function may have singularities. These are usually quite simple singularities and the user is usually well aware of their nature and location. But, while the singularity may be simple in structure, integrating over it may be tedious. In any single problem, it is most unlikely that a single element will have all these inconvenient features, but one might.

* This work was supported by the Applied Mathematical Sciences subprogram of the Office of Energy Research, U.S. Department of Energy, under Contract W-31-109-Eng-38.

T. O. Espelid and A. Genz (eds.), Numerical Integration, 219–233.

This article is about some of the problems encountered by such a user. My feeling is that a user spends nearly all of the time which he devotes to numerical quadrature to attempting to adapt the results given in textbooks for standard regions and regular integrands to his problem. He finds, to his dismay, that this topic is not discussed in textbooks, and he turns for help to his local quadrature expert. All too often he finds the quadrature expert, while sympathetic and ready to help, to be of little use. This is because the quadrature expert has not previously encountered this sort of problem in detail. His attempts to help may be hampered by misconceptions. Sometimes he imagines that all these problems can be handled by scaling, and tries to prove this incorrect hypothesis. Other times he thinks that nothing can be scaled which may also be wrong. Either misconception can lead to seemingly endless discussion and unnecessarily inefficient programs.

My hope is that this article, which is written for the quadrature expert and which is anecdotal in nature, may be helpful in directing attention to some of the pitfalls in this area.

It is worth stating at the outset that in a one-dimensional context these problems are very rare, and when one is encountered, there is usually a quick remedy that only works in one dimension. This problem is essentially multidimensional.

In this article, we shall treat principally two quadrature methods. These are Gaussian Quadrature, and Linear Extrapolation Quadrature of which Romberg Integration is a special case. We shall look at the effect of Affine Transformations of the coordinate system on these integration procedures. We shall discuss briefly the Duffy transformation (of a triangle into a square).

2. Extrapolation Quadrature

It is well known that polynomials are basic to Gaussian Quadrature. A corresponding role in the theory underlying Extrapolation Quadrature is played by Homogeneous functions. As a preliminary, we remind the reader of the definition and simple properties of these functions.

A function $f(x,y)$ is said to be homogeneous (about the origin) of degree α if

$$f(\lambda x, \lambda y) = \lambda^\alpha f(x,y) \quad \text{for all } \lambda > 0 .$$

We shall often denote such a function by $f_\alpha(x,y)$.

A monomial $x^p y^q$ is homogeneous of degree $p+q$, and many properties relating to the polynomial degree of functions of monomials have direct analogues in the context of homogeneous degree. Thus, $(f_\alpha)^\beta$ and $f_\alpha f_\beta$ are of degree $\alpha\beta$ and $\alpha+\beta$, respectively, and $f_\alpha(M\mathbf{x})$ when $|M| \neq 0$ is also of degree α; $\partial^s f_\alpha / \partial x^s$ is of degree $\alpha-s$.

In more than one dimension, many more interesting functions are homogeneous. For example, $r = (x^2+y^2)^{1/2}$ is homogeneous of degree 1, and $\theta = \arctan y/x$ is homogeneous of degree zero, as is any function $\Phi(\theta)$.

Extrapolation Quadrature, abbreviated here to EQ, is a natural development of Richardson's deferred approach to the limit. It is a technique designed for integration over hypercubes or simplices. Romberg integration is a special and important one-dimensional example of Extrapolation Quadrature. In this section, we shall restrict the discussion to integration over the square $[0,1)^2$. However, all results below are valid for the triangle T_2 also. See Lyness (1991) for a brief elaboration of this remark.

In this paper we adopt the following convention for the polynomial degree of a quadrature rule. A rule of degree d is one which integrates all polynomials of degree d exactly. A rule of *strict* degree d is one of degree d but not of degree $d+1$.

In this paper we treat only degree zero quadrature rules. These are rules which integrate the constant function correctly. We denote by Q any degree zero quadrature rule for $[0,1)^2$; that is,

$$Qf = \sum w_j f(x_j, y_j) \quad \text{with } \sum w_j = 1 , \tag{2.1}$$

and we define its m-copy $Q^{(m)}f$ in the standard way as the approximation to the exact integral, If, obtained by subdividing the square into m^2 squares each of side $1/m$ and applying a properly scaled version of Q to each. In our context it is usually advantageous to use either the mid-rectangle rule or the vertex trapezoidal rule for Q. The theory allows any rule which integrates the constant function correctly.

In general, one would expect the rule $Q^{(m)}f$ with a large value of m to be a better approximation to If than the rule with a small value of m. A measure of this effect would be provided by an asymptotic expansion of the error functional $E^{(m)}f = Q^{(m)}f - If$ in inverse powers of m or other suitable expansion functions. We can justify the use of Extrapolation Quadrature if such an expansion exists and its form is known. Whether or not there is one available depends on the nature of $f(x)$.

Briefly, these error functional expansions are built up from two basic asymptotic expansions given in the following two theorems.

THEOREM 2.2. *When $f(x)$ together with its partial derivatives of order p or less are integrable over H, and Q is a degree zero quadrature rule for H, then*

$$Q^{(m)}f - If = \frac{B_1}{m} + \frac{B_2}{m^2} + \cdots + \frac{B_{p-1}}{m^{p-1}} + O(m^{-p}) , \tag{2.2}$$

where $B_s = B_s(H;Q;f)$ are independent of m.

This is a straightforward generalization of the classical Euler Maclaurin formula.

THEOREM 2.3. *When $f(x)$ is homogeneous of degree α and has no singularity in H except possibly at the origin, and Q is a degree zero quadrature rule for H, then*

$$Q^{(m)}f - If = \frac{A_{2+\alpha}}{m^{2+\alpha}} + \frac{C_{2+\alpha}\ln m}{m^{2+\alpha}} \tag{2.3}$$

$$+ \frac{B_1}{m} + \frac{B_2}{m^2} + \cdots + \frac{B_{p-1}}{m^{p-1}} + O(m^{-p}),$$

where the coefficients A, B, and C are independent of m, and $C_{2+\alpha} = 0$ unless α is an integer.

A detailed proof of this is given in Lyness (1976i).

We may construct an expansion for any function which is a linear sum of any number of component functions, so long as each component satisfies the hypotheses of one or the other of these theorems. This is easy to do when

$$f(x,y) = f_\alpha(x,y)g(x,y) , \tag{2.4}$$

where g is regular. Here we may expand $g(x,y)$ in a Taylor expansion about the origin retaining only monomial terms of degree $p-1$ or less and deferring the rest to the remainder term. This gives rise to an expression for $f(x,y)$ of the form

$$f_\alpha(x,y)g(x,y) = g(0,0)f_\alpha(x,y) + f_{\alpha+1}(x,y) + ... + f_{\alpha+p-1}(x,y) + g_{\alpha+p}(x,y) . \tag{2.5}$$

The final term g is not a homogeneous function but satisfies the hypothesis of Theorem 2.2 above. All the other terms in this expansion are homogeneous of the indicated degree and so satisfy the hypotheses of Theorem 2.3 above. Thus, (2.5) may be used to establish error functional expansions valid for classes of familiar functions. For example, in Lyness (1976i) the approach outlined above is used to show the following theorem.

THEOREM 2.7. *Let $F(x,y)$ be of the form*

$$F(x,y) = r^\alpha \Phi(\theta) h(r) g(x,y),$$ (2.6)

where (r,θ) are the polar coordinates of (x,y) and Φ, h, and g are analytic functions; and let Q be a degree zero quadrature rule for H. Then

$$Q^{(m)}F - IF = \sum_{t=0} \frac{A_{2+\alpha+t}}{m^{2+\alpha+t}} + \sum_{s=1} \frac{B_s}{m^s} \qquad \alpha \neq \text{integer}$$ (2.7)

$$\approx \sum_{s=1} \frac{A_s + B_s + C_s \ln m}{m^s} \qquad \alpha = \text{integer}.$$

Some logarithmic singularities can also be treated. The corresponding expansions are obtained by differentiating already-available expansions with respect to some incidental parameter, such as α. We may exploit the identity

$$\frac{\partial}{\partial \alpha}(r^\alpha g(x,y)) = r^\alpha \ln r \ g(x,y)$$ (2.8)

to obtain an expansion like (2.7) but having additional terms $\log m / m^s$ and, when these are already present, terms $(\ln m)^2 / m^s$. For more detailed information about these and other expansions, the reader may refer to Lyness (1976ii). However, this is a continuing research area; other expansions have been discovered since then and more may remain to be discovered. See Sidi (1983).

The user of Linear Extrapolation Quadrature need not concern himself about the derivation of the expansion. Once he has satisfied himself that it exists and knows its form, he can proceed to apply extrapolation. This is done by constructing linear sums of values of $Q^{(m)}f$ in such a way as to eliminate the early terms of the relevant expansion. The Neville algorithm can be used when the expansion is a simple one, like the one in Theorem 2.2 above. But, in general, all that one has to do is to solve a set of linear equations.

We close this section by stating the corresponding results for Gaussian Quadrature. It is convenient to present the following definition as a theorem.

THEOREM 2.9. *When $f(x,y)$ is a polynomial of degree d and Q is a degree d quadrature rule, with respect to a specified region R and a specified weight function $w(x,y)$, then*

$$Qf - I(R)f = 0,$$ (2.9)

where $I(R)f = \int_R w(x,y)f(x,y)dxdy$.

Note that, in Gaussian Quadrature, the singularity enters through a weight function, and in general, no m-copy rule is treated. The user of Gaussian Quadrature improves his accuracy by using a sequence of different rules of successively higher degree. On the other hand, the user of Extrapolation Quadrature achieves the same end using the same basic rule with a sequence of successively higher mesh ratios m.

Again, the user need not concern himself about where the weights and abscissas came from. If they are available, he simply has to use them in a straightforward rule evaluation program.

3. Affine Transformation

As mentioned in the introduction, Quadrature rules and theory is conventionally discussed in the literature in the context of standard regions. In this article, we are particularly interested in nonstandard regions of the same general character. To this end we employ the Affine Transformation (represented by a nonsingular $N \times N$ constant matrix A). The mapping

$$\mathbf{x} = A\mathbf{x}' \tag{3.1}$$

takes any parallelogram (or triangle) R into another parallelogram (or triangle) R'. To be specific, when R is defined by inequalities involving the components of $\mathbf{x}$, the region R' is defined by the same set of inequalities, but with each component of $\mathbf{x}$ replaced by the corresponding component of $A\mathbf{x}$. It is readily established that, given any triangle R' having one vertex at the origin, there exists an Affine transformation A which takes the standard triangle R into R'. This remark is valid when R' is any parallelogram and R the standard square. But one cannot obtain a general quadrilateral in this way. Besides transforming regions, an Affine transformation transforms associated Quadrature rules.

DEFINITION 3.2. *The affine transform of the rule*

$$Qf = \sum w_j f(\mathbf{x}_j) \tag{3.2}$$

with respect to A is

$$Q'f = \sum W_j f(\mathbf{X}_j), \tag{3.3}$$

where $W_j = w_j / |\det A|$ and $\mathbf{X}_j = A^{-1}\mathbf{x}_j$.

The abscissas of the new rule are in precisely the same positions relative to the region R' as the abscissas of Q are relative to R. The weights have been scaled uniformly to account for a possibly different area. It is readily verified that the Affine transform of a degree zero quadrature rule is also a degree zero quadrature rule. (See Theorem 3.8 below.) We now relate the error functionals of these two rules.

Lemma 3.4. *Let A be an affine transformation that takes R into R'; let $Q'f$ be the affine transform of Qf, and let the respective error functionals be*

$$Ef = Qf - \int_R w(\mathbf{x})f(\mathbf{x})d^N x, \tag{3.4}$$

$$E'\phi = Q'\phi - \int_{R'} W(\mathbf{x})\phi(\mathbf{x})d^N x, \tag{3.5}$$

where $W(\mathbf{x}) = w(A\mathbf{x})$. Then, when

$$\phi(\mathbf{x}) = f(A\mathbf{x}), \tag{3.6}$$

it follows that

$$Ef = |\det A| E'\phi. \tag{3.7}$$

Proof. This is a matter of elementary algebraic substitution.

Note that R may be a general region but we only use it for triangles or squares. The important part of this result is that it states that two numbers are equal. It is not a result about form. That comes later.

This lemma can be used to establish results about Extrapolation Quadrature. But, as a preliminary, we use it to confirm a well known result.

THEOREM 3.8. *When Q is a rule of polynomial degree d for a region R with weight function $w(\mathbf{x})$, then its Affine transform Q' is a rule of the same polynomial degree for the region R'*

with weight function w (A x).

Proof. Let $\phi\epsilon\Pi_d$. It follows immediately from (3.6) that $f\,\epsilon\Pi_d$. Since by hypothesis Q is of polynomial degree d, it follows that Ef given by (3.4) is zero; and then, from (3.7) that $E'\phi = 0$. Consequently $E'\phi = 0$ for all $\phi\epsilon\Pi_d$ and this establishes the theorem.

We may now use the same approach to derive the less trivial analogues about extrapolation.

THEOREM 3.9. *Theorem 2.2 above is valid precisely as written when H is replaced by R, a general parallelogram.*

We restate Theorem 3.9 in a notation which allows us to derive it from Theorem 2.2.

THEOREM 3.9′. *Let R′ be a parallelogram and Q′ be a degree zero quadrature rule for R′. Then when $\phi(x)$, together with its partial derivatives of order p or less, are integrable over R′,*

$$Q'^{(m)}\phi-I(R')\phi = \frac{B_1}{m} + \frac{B_2}{m^2} +...+ \frac{B_{p-1}}{m^{p-1}} + O(m^{-p}),\qquad(3.9)$$

where $B_s = B_s(R';Q';\phi)$ are independent of m.

Proof. Let $\phi(x)$ satisfy the hypothesis in the theorem. It follows immediately that $f(x)$ given by (3.6) satisfies the same hypothesis with respect to R. Thus Theorem 2.2 may be applied to $f(x)$ establishing that Ef has expansion (2.2). However, from (3.7) it follows that the same expansion applies to $E'\phi|\det A|$. This is precisely the statement in (3.9) above, with

$$B_s(R';Q';\phi) = B_s(r;Q;f)/|\det A|,\qquad(3.10)$$

which establishes the theorem.

The key to the proofs of the last two theorems is that $f(x)$ and $\phi(x)$ share some property. In the first theorem, this property is that they are both polynomials of the same degree. In the previous theorem, both have continuous partial derivatives of order p. However, in the present context the really important shared property is the following.

LEMMA 3.11. *Let $\phi(x) = f(Ax)$ with $\det A \neq 0$. Then if one of f or ϕ is homogeneous of degree α, so is the other, and if one has no singularity except at the origin, the same is true about the other.*

The proof is trivial. This is displayed as a lemma simply because of its importance.

THEOREM 3.12. *Theorem 2.3 is valid precisely as written when H is replaced by R, a general parallelogram.*

The proof is logically similar to that of the previous theorem, the property shared by f and ϕ being the one described in the lemma.

However, this approach to handling the parallelograms and triangles denoted by $R′$ is not general. One may be interested in a singularity of the sort encountered in Theorem 2.7, whose principal component is r^a. When one applies Lemma 3.4 directly, one finds in just the same way that the error functional asymptotic expansion (2.7) applies when $F(x,y)$ has a singularity whose principal component is of the form $(Ax^2 + 2Hxy + By^2)^{\alpha/2}$, this being the particular homogeneous function into which r^a is transformed by the transformation which takes R into $R′$. The geometrically inclined reader can visualize a plot of R containing the circular contours of r^{-1}. It is this whole picture which is transformed into a plot of R'; the circular contours become elliptical. Nevertheless, we have the following theorem.

THEOREM 3.13. *Theorem 2.7 is valid, precisely as written when H is replaced by R , a general parallelogram.*

Proof. We recall that Theorem 2.7 was derived from Theorems 2.2 and 2.3 by means of expansion (2.5), which developed $F(x,y)$ as a series of homogeneous functions, to which Theorem 2.3 was applied, together with a remainder term, to which Theorem 2.2 was applied. We need only employ the same expansion, but apply Theorems 3.11 and 3.12 to the respective terms instead. This gives the result in the theorem.

Theorem 3.9 is technically new and I believe Theorem 3.13 is new. The author has attempted to establish these results before by transforming each term in the expansion separately. In the simpler case, it is possible but very tedious to do this. In the singular case, the integral representations of the coefficients are too formidable. The proofs given above avoid this by not providing direct formulas for the coefficients. This is no real hardship as, in the practice of Linear Extrapolation Quadrature, one needs only the form of the error functional expansion and details about the coefficients are not required.

Theorems 3.9, 3.12, and 3.13 illustrate a major convenience of using Extrapolation Quadrature. This is that the effect of the singularity is taken care of by using the proper expansion. For the same singularity, this is the same for all triangles and for all parallelograms. Once this expansion is known, one may go ahead and carry out extrapolation using a linear equation solver. On the other hand, Theorem 3.8 confirms that in Gaussian Quadrature the situation is quite different. The affine transform rule Q' applies to a different weight function $w(Ax)$ and not to $w(x)$. Unless these two weight functions happen to be closely related, one will need a completely new set of Gaussian rules for each new triangle. In the familiar case in which $w(x) = 1$, clearly $w(Ax) = w(x)$ and one can use the affine transformed rule. There are other special cases described in the next section. But, in general, one cannot expect a relationship, so, when there is a singularity, separate sets of rules are needed for separate triangles.

4. Gaussian Quadrature with Singularities

It is conventional wisdom that, whether singularities are present or not, the proper use of Gaussian Quadrature is generally more cost effective than the proper use of Extrapolation Quadrature by a factor of about two or even more in the number of function values needed to attain a particular accuracy. So one's natural inclination is to prefer to use Gaussian Quadrature. In the regular case, in which no nontrivial weight function is involved, it is straightforward to obtain weights and abscissas from standard texts such as Stroud (1971) or Davis and Rabinowitz (1984). And, as mentioned above, an affine transformed rule can be used when the region is an affine transform of a standard region.

The term "proper use" in the first sentence above is vital. This implies that in the singular case, the appropriate weight function is identified, and Gaussian Quadrature is carried out using the weights and abscissas corresponding to that weight function; or in the EQ case extrapolation is based on the correct error functional expansion. Improper use of these techniques usually has the effect of utterly compromising the accuracy or reducing the rate of convergence to a snails pace. As usual, one would find that the accuracy increases with the number of function values used, however unwisely the abscissas and weights are chosen. The symptom of misuse here is not lack of convergence, but extremely slow convergence.

The identification of the appropriate weight function in the context of finite element methods is usually no problem whatever. However, finding lists of weights and abscissas is a different matter. The common experience seems to be that one cannot locate lists of weights and abscissas for Gaussian Quadrature rules having genuine two dimensional singular weight functions.

To proceed, there are several possibilities. The two most attractive are the following.

(a)   Look for some analytic transformation that may reduce the present numerical problem into another less intractable or more familiar numerical problem. The Duffy transformation described in Section 5 is an example.

(b)   If an error functional expansion is available, use Extrapolation Quadrature.

A looming disaster overhangs the following approach.

(c)   Try to get by using a different Gaussian rule, perhaps one pertaining to a nearby weight function for which weights and abscissas are available. The lurking dangers here are discussed in detail in Section 6.

We pursue a somewhat obvious approach in the rest of this section, which we could also characterize as being both self serving and altruistic. If these rules are not listed and we need them, surely others will need them from time to time. What we ought do is to calculate them ourselves, use them in our problem, and also publish them for others to use later. Naturally, this is a not insignificant task. To do this calculation one needs accurate numerical values of the moments, one needs to design a quadrature rule structure, and, if one wants an optimal rule, one will almost certainly have to solve systems of nonlinear equations. If we take short cuts to simplify the calculation, the resulting rules will be less cost effective, possibly leaving EQ a more attractive choice. Let us suppose that all these problems have been successfully tackled, and we now have a Gaussian type quadrature rule for a standard triangle R and a weight function $w(x)$.

The trouble is that one can use this rule only for integration over this triangle R. If one wants to integrate over a different triangle R', we have already noted that the affine transformation that changes R into R' also changes the weight function to $w(Ax)$. In general, if one wants the same weight function $w(x)$, but for a different region R', one has to calculate a new set of quadrature rules. The immediate answer to the suggestion that we publish this list of rules is that this list is not general enough. A triangle with a fixed vertex at the origin requires four parameters to specify it. Our list treats only a single choice of these parameters. To be useful, the list should cover a relatively general problem and not one extremely special case.

This is the general situation, but there are exceptions. In fact, so many familiar weight functions are exceptional in some way, that the user has to be forgiven for imagining that all are. The critical point is to note whether $w(Ax)$ and $w(x)$ are closely related. This depends on $w(x)$ and A. Specifically, if for some k we have

$$w(x) = kw(Ax),$$

then one can obtain one set of rules directly from the other set. One case in which this happens is when $w(x)$, reexpressed as a function of r and θ, turns out to be independent of θ, and in addition the transformation A is a rotation about the origin. A second case occurs when $w(x)$ is homogeneous (about the origin of degree α) and A is a uniform magnification. In either case, to obtain the second set one takes the affine transform of each rule of the first set (which incorporates the factor det A into each weight) and then multiplies each weight by $k^{2+\alpha}$.

Examples of the first case include $1/r$ and $\ln r/r$. Examples of the second case include $1/r$ and $1/(\lambda x + \mu y)$. Thus, if one has available a rule of polynomial degree d for $w(x) = r^\alpha$ for the standard triangle T_2, one may rotate the triangle and rotate each abscissa by the same angle, keeping the weights constant; then one may magnify the triangle by a linear factor k, moving the abscissas accordingly, but also multiplying the weights by $k^{2+\alpha}$. In this very favorable case one has reduced to two the number of parameters needed for the list of rules. However, this particular example can be handled more elegantly using the Duffy transformation.

5. The Duffy Transformation

If one has a "product" singularity which fits conveniently into the integration region, for example

$$x^\alpha y^\beta g(x,y) \tag{5.1}$$

for the unit square $[0,1]^2$, then one can handle the problem using cartesian product formulas involving one-dimensional weight functions, such as the Gauss Jacobi Quadrature Rules. This suggests that it may be useful to look for a transformation that takes an apparently intransigent "singularity-region" pair into an easier one, like the one above.

A somewhat sophisticated example of this is the Duffy transformation technique for the triangle. In 1982, apropos of nothing, Duffy published a paper about integrating over triangles and tetrahedra. He suggested the following transformation.

$$\int_0^1 dx \int_0^x dy \; f(x,y) = \int_0^1 dx \int_0^1 dt \; xf(x,tx) . \tag{5.2}$$

For example, instead of integrating

$$f(x,y) = (x^2+y^2)^{\alpha/2} g(x,y) \tag{5.3}$$

over the triangle

$$\tilde{T}: \; x < 1; \; y > 0; \; x-y > 0 , \tag{5.4}$$

one might prefer to integrate

$$xf(x,tx) = x^{1+\alpha}(1+t^2)^{\alpha/2} g(x,tx) \tag{5.5}$$

over the rectangle $[0,1]^2$.

The vertex singularity has been smeared out to make a more conventional line singularity. When $\alpha = -1$, what was originally a weak singularity has disappeared, leaving an analytic function, which can be handled using a product Gauss Legendre quadrature rule. For general noninteger α, a product of a Gauss Jacobi rule with a Gauss Legendre rule is appropriate.

In fact, this transformation is much more powerful and more useful than was at first realized, particularly as a tool for use in the Boundary Element Method.

Given a rectangle and a function that is singular at the origin, one can always divide it into two triangles and apply the Duffy transformation to each.   In some cases one may be lucky.

For example, suppose for $\tilde{T}$

$$f(x,y) = x^\lambda y^\mu f_\alpha(x,y) g(x,y) , \tag{5.6}$$

where $f_\alpha(x,y)$ is homogeneous of degree α and $g(x,y)$ is regular. Applying the transformation leads to the problem of integrating over a rectangle the function

$$xf(x,tx) = x^{1+\lambda+\mu+\alpha} t^\mu f_\alpha(1,t) g(x,tx) . \tag{5.7}$$

If the only singularity of $f_\alpha(x,y)$ is at the origin, then $f_\alpha(1,t)$ is regular and one can use here the cartesian product of a pair of Gauss Jacobi rules.

Duffy's method for integrating over a square is to subdivide it into two triangles and then use the transformation above to transform each into a square. In some cases it is useful to iterate the whole procedure.

The same technique is available in any number of dimensions. In three dimensions one first splits the cube into three square-based pyramids and separately transform each pyramid back into a cube using the three-dimensional Duffy transformation.

6. A Misuse of Gaussian Quadrature

In view of the difficulty in finding lists of Gaussian rules for singular weight functions, it is very tempting to use weights and abscissas for a nearby weight function $W(x)$ instead, particularly when these are available and the ones for $w(x)$ are not. We suppose that Gaussian rules are available for a triangle R with weight function $w(x)$, singular at a vertex $x=0$ of R; but, unfortunately, we are evaluating an integral whose integrand has the singular behavior described by $w(x)$ over R'. The regions R and R', although close to each other, are related by an affine transformation. There is a mismatch here. Since

$$\int_R w(x)g(x)dx = |\det A| \int_{R'} w(Ax)g(Ax)dx , \tag{6.1}$$

we have Gaussian rules available for R with $w(x)$ or for R' with $w(Ax)$ but not for R' with $w(x)$. The quantity we want to evaluate may be reexpressed as

$$\int_{R'} w(x)f(x)dx = \int_{R'} w(Ax)\left[\frac{w(x)}{w(Ax)} f(Ax)\right]dx . \tag{6.2}$$

An initially promising but in fact deceptive approach is to use the available Gaussian rule on the right-hand member of (6.2). $f(Ax)$ like $f(x)$ is regular, so whether it is worth using this rule depends on the smoothness of $w(x)/w(Ax)$. We pursue this example by taking $w(x) = r^\alpha$. Then $W(Ax)$ has the form $(Ax^2+2Hxy+By^2)^{\alpha/2}$. The function

$$w(x)/w(Ax) = \left|\frac{x^2+y^2}{Ax^2+2Hxy+By^2}\right|^{\alpha/2} \tag{6.3}$$

is somewhat deceptive. It can be expressed as a function of y/x and so is constant along any radius vector. But these constants are different for different radius vectors. For example, on the x-axis, $w/W = A^{-\alpha/2}$, while on the y-axis, $w/W = B^{-\alpha/2}$. Technically, this means that the function is not Hölder continuous at the origin. But one needs only minimal intuition to see that such a function is simply not readily approximated by a polynomial or entire function and so Gaussian Quadrature with this component in the integrand seems pointless.

If a user remains to be convinced that this misuse of Gaussian Quadrature is expensive, he should be invited to carry out minor numerical experiments. To do this he does not need to attempt his problem for which presumably results are not available. At issue is how well a Gaussian Rule integrates a function with this sort of behavior at the origin. A numerical example is given in Section 8.

There are two further points to be made en passant. It might have happened that, in the example above, the various limits of w/W as x approached the origin were identical. In the particular example, this would happen if A were orthogonal. Such cases are precisely those discussed toward the end of Section 4 in which one may, in any case, construct a Gaussian Rule for R' from one for R.

The other point is that w/W is, in fact, a homogeneous function of degree zero. Unless it is constant, such a function causes trouble in Gaussian Quadrature. But Extrapolation Quadrature handles these with no trouble at all. (See Subsection 7.2 below.)

7. Further Remarks about Extrapolation Quadrature

Up to this point in this article, I have taken the view that most users, given the choice, would prefer Gaussian Quadrature to Extrapolation Quadrature. The first reason for believing this is that most users have never heard of EQ. However, the ones that know anything about it know that Gaussian Quadrature is generally more cost effective. My belief is that this attitude will change as more and more users find how easy it is to handle EQ. In this section we look at some problems in which even the most intrepid Gaussian Integrator would concede that a role exists for EQ.

7.1. MIXED SINGULARITIES

One can envisage all sorts of really complicated singular behavior of an integrand function at a point. In this subsection we consider a singularity only one stage removed in difficulty from a standard singularity. Let us suppose that the integrand has a double singularity form, such as

$$f(x,y) = r^{-1}g_1(x,y) + r^{-1/2}g_2(x,y)$$

or

$$f(x,y) = r^{-1}g_1(x,y) + r^{-1}\ln r\, g_2(x,y)$$

or

$$f(x,y) = r^{-1}g_1(x,y) + g_2(x,y),$$

where g_1 and g_2 are both regular. It is necessary to distinguish between the case in which function values of g_1 and of g_2 are both available and the case in which function values of f are available, f is known to be of this form, but function values of g_1 and g_2 are not available separately. In the former case, the problem is straightforward. One can simply evaluate the two parts of the integral separately. Two applications of Gaussian Quadrature instead of one may increase the cost sufficiently to make Extrapolation Quadrature competitive. In the latter case one requires a double type Gaussian rule, that is, one that preserves the integrity of two components of a singularity. There is presently no developed theory for constructing this type of quadrature rule, though some individual one-dimensional rules of this type are known. However, in either case, EQ may be used simply based on the concatenation of the two error functional expansions. In the second and third example, this concatenation coincides with one of the component expansions.

7.2. HOMOGENEOUS SINGULARITIES

EQ handles some functions of low Hölder continuity quite well. Typical of these is $\Phi(\theta)$, a nontrivial homogeneous function of degree zero. For example, let A, B, C, and D be positive.

The function

$$\Phi(\theta) = (Ax+By)/(Cx+Dy)$$

is regular in $[0,1]^2$ except at the origin, where one finds a singularity which occurs because the limit as (x,y) approaches the origin depends on the angle of approach. In spite of the fact that it is bounded wherever defined, it cannot be ignored for Gaussian Quadrature. However, since it is homogeneous (of degree zero), EQ handles this integrand function with no difficulty. A function of this type was mentioned in an example in Section 6.

Much more sophisticated singularities in which this sort of singularity is a component can be handled in a simple manner using EQ, but are virtually out of reach of Gaussian Quadrature. Take, for example, the result given in Theorem 2.7 about the integrand function being of the form

$$F(x,y) = r^{\alpha}\Phi(\theta)h(r)g(x,y) . \tag{7.1}$$

A Gaussian rule having weight r^{α} will handle $r^{\alpha}p_1(x,y)$. A double-type Gaussian rule will handle $r^{\alpha}p_1(x,y) + r^{\alpha+1}p_2(x,y)$ which covers these functions in (7.1) when $\Phi(\theta)$ is constant. However, when $\Phi(\theta)$ is not specified individually, but is known to be non trivial, one is led to the conclusion that Gaussian Quadrature cannot be applied and EQ is the obvious choice.

7.3. ROLE IN CONSTRUCTING GAUSSIAN RULES

Towards the end of Section 4, we considered very briefly the possibility of constructing ones own set of Gaussian Rules. We mentioned that to do this one requires accurate numerical values of the moments. If the weight function is one for which no convenient analytic form for the moments is available, one may have to calculate these numerically for oneself. In one dimension, one thinks in terms of a sledgehammer approach to get the moments used to construct the elegant rule.

In two or more dimensions, the sledgehammer approach can be unexpectedly expensive. If EQ is available, it can be effectively used in this subsidiary role.

8. A Numerical Example

The following example deals with approximations to

$$If = \int_0^1\int_0^1 \cos\theta \, dxdy = \frac{1}{2}\left[\log(\sqrt{2}+1) + \sqrt{2} - 1\right],$$

where $\cos\theta = x/r$ and (r,θ) are standard polar coordinates. In this contrived and stylized example of a Hölder discontinuous function, we use principally the product midpoint trapezoidal rule $Qf = f(1/2, 1/2)$ and its m-copy version.

The first task when using Extrapolation Quadrature is to be absolutely assured that one is using the correct expansion. Since $\cos\theta$ is a homogeneous function of degree zero, expansion (2.3) with $\alpha = 0$ is appropriate. Only even powers are needed since the rule is symmetric. Nevertheless, Table 1 illustrates some expensive numerical tests one might make if one wanted to convince oneself — or someone else. The convention used here about expansion exponents is that they represent negative exponents of m, they are in order of magnitude, and if two (or more) are equal, corresponding log (or log power) terms occur in the expansion. Since the exact result is available, we have presented in these tables the error, $Qf - If$. The reader will agree that at a cost of about 21,000 function values, Table 1 presents prima facie numerical evidence that the second sequence treated may be the most appropriate.

TABLE 1

EXPENSIVE NUMERICAL EXPERIMENT TO VERIFY THE CORRECT EXPANSION

THE BASIC PRODUCT TRAPEZOIDAL RULE RESULTS

```
mesh =   1 Rf =    0.7071   If =    0.6478   Rf -If = 0.5931D-01
mesh =   2 Rf =    0.6698   If =    0.6478   Rf -If = 0.2199D-01
mesh =   4 Rf =    0.6551   If =    0.6478   Rf -If = 0.7296D-02
mesh =   8 Rf =    0.6501   If =    0.6478   Rf -If = 0.2275D-02
mesh =  16 Rf =    0.6485   If =    0.6478   Rf -If = 0.6815D-03
mesh =  32 Rf =    0.6480   If =    0.6478   Rf -If = 0.1986D-03
mesh =  64 Rf =    0.6479   If =    0.6478   Rf -If = 0.5669D-04
mesh =128 Rf =    0.6478   If =    0.6478   Rf -If = 0.1594D-04
```

THE EXTRAPOLATION TABLE; expansion exponents are; 2 4 6 8 10

k	m	(k,1,1)	(k,2,1)	(k,3,1)	(k,4,1)	(k,5,1)	(k,6,1)
1	1	0.059313	0.9546D-02	0.1922D-02	0.4584D-03	0.1133D-03	0.2823D-04
2	2	0.021988	0.2399D-02	0.4813D-03	0.1146D-03	0.2831D-04	0.7058D-05
3	4	0.007296	0.6011D-03	0.1203D-03	0.2865D-04	0.7079D-05	0.1764D-05
4	8	0.002275	0.1504D-03	0.3008D-04	0.7163D-05	0.1770D-05	0.0000D+00
5	16	0.000681	0.3760D-04	0.7521D-05	0.1791D-05	0.0000D+00	0.0000D+00
6	32	0.000199	0.9401D-05	0.1880D-05	0.0000D+00	0.0000D+00	0.0000D+00
7	64	0.000057	0.2350D-05	0.0000D+00	0.0000D+00	0.0000D+00	0.0000D+00
8	128	0.000016	0.0000D+00	0.0000D+00	0.0000D+00	0.0000D+00	0.0000D+00

THE EXTRAPOLATION TABLE; expansion exponents are; 2 2 4 6 8

k	m	(k,1,1)	(k,2,1)	(k,3,1)	(k,4,1)	(k,5,1)	(k,6,1)
1	1	0.059313	0.5931D-01	0.1621D-04	0.9690D-06	0.1579D-07	-0.5428D-10
2	2	0.021988	-0.7396D-02	0.1922D-05	0.3069D-07	0.7634D-11	0.5071D-13
3	4	0.007296	-0.7379D-03	0.1489D-06	0.4870D-09	0.8034D-13	0.8663D-16
4	8	0.002275	-0.1152D-03	0.9761D-08	0.7689D-11	0.3996D-15	0.0000D+00
5	16	0.000681	-0.2093D-04	0.6173D-09	0.1205D-12	0.0000D+00	0.0000D+00
6	32	0.000199	-0.4111D-05	0.3869D-10	0.0000D+00	0.0000D+00	0.0000D+00
7	64	0.000057	-0.8463D-06	0.0000D+00	0.0000D+00	0.0000D+00	0.0000D+00
8	128	0.000016	0.0000D+00	0.0000D+00	0.0000D+00	0.0000D+00	0.0000D+00

THE EXTRAPOLATION TABLE; expansion exponents are; 2 2 4 4 6

k	m	(k,1,1)	(k,2,1)	(k,3,1)	(k,4,1)	(k,5,1)	(k,6,1)
1	1	0.059313	0.5931D-01	0.1621D-04	0.2354D-05	-0.3187D-07	-0.1045D-08
2	2	0.021988	-0.7396D-02	0.1922D-05	-0.4420D-06	-0.1526D-08	-0.4233D-12
3	4	0.007296	-0.7379D-03	0.1489D-06	-0.7322D-08	-0.2427D-10	-0.4930D-14
4	8	0.002275	-0.1152D-03	0.9761D-08	-0.2792D-09	-0.3840D-12	0.0000D+00
5	16	0.000681	-0.2093D-04	0.6173D-09	-0.1247D-10	0.0000D+00	0.0000D+00
6	32	0.000199	-0.4111D-05	0.3869D-10	0.0000D+00	0.0000D+00	0.0000D+00
7	64	0.000057	-0.8463D-06	0.0000D+00	0.0000D+00	0.0000D+00	0.0000D+00
8	128	0.000016	0.0000D+00	0.0000D+00	0.0000D+00	0.0000D+00	0.0000D+00

In Table 2, three approaches which are realistic are illustrated. Extrapolation Quadrature based on both an appropriate and an inappropriate error functional expansion are given. These cost 204 function values. For the comparison, some product Gauss-Legendre approximations using up to 400 function values are given. These results speak for themselves as to the effect of the proper use of Gaussian and Extrapolation Quadrature.

TABLE 2

THE BASIC PRODUCT TRAPEZOIDAL RULE RESULTS

```
mesh =   1 Rf =    0.7071   If =   0.6478   Rf -If = 0.5931D-01
mesh =   2 Rf =    0.6698   If =   0.6478   Rf -If = 0.2199D-01
mesh =   3 Rf =    0.6594   If =   0.6478   Rf -If = 0.1164D-01
mesh =   4 Rf =    0.6551   If =   0.6478   Rf -If = 0.7296D-02
mesh =   5 Rf =    0.6528   If =   0.6478   Rf -If = 0.5041D-02
mesh =   6 Rf =    0.6515   If =   0.6478   Rf -If = 0.3711D-02
mesh =   7 Rf =    0.6507   If =   0.6478   Rf -If = 0.2858D-02
mesh =   8 Rf =    0.6501   If =   0.6478   Rf -If = 0.2275D-02
```

THE EXTRAPOLATION TABLE; expansion exponents are; 2 2 4 6 8

k	m	(k,1,1)	(k,2,1)	(k,3,1)	(k,4,1)	(k,5,1)	(k,6,1)
1	1	0.059313	0.5931D-01	0.2242D-04	0.3222D-05	0.4052D-06	-0.2092D-07
2	2	0.021988	-0.1302D-01	0.6228D-05	0.6401D-06	0.6766D-09	0.1642D-09
3	3	0.011642	-0.3333D-02	0.2285D-05	0.1211D-06	0.2305D-09	0.7314D-10
4	4	0.007296	-0.1480D-02	0.9798D-06	0.3381D-07	0.1052D-09	0.0000D+00
5	5	0.005041	-0.8186D-03	0.4824D-06	0.1196D-07	0.0000D+00	0.0000D+00
6	6	0.003711	-0.5126D-03	0.2633D-06	0.0000D+00	0.0000D+00	0.0000D+00
7	7	0.002858	-0.3478D-03	0.0000D+00	0.0000D+00	0.0000D+00	0.0000D+00
8	8	0.002275	0.0000D+00	0.0000D+00	0.0000D+00	0.0000D+00	0.0000D+00

THE EXTRAPOLATION TABLE; expansion exponents are; 2 4 6 8 10

k	m	(k,1,1)	(k,2,1)	(k,3,1)	(k,4,1)	(k,5,1)	(k,6,1)
1	1	0.059313	0.9546D-02	0.2592D-02	0.1060D-02	0.5335D-03	0.3052D-03
2	2	0.021988	0.3365D-02	0.1156D-02	0.5546D-03	0.3115D-03	0.1930D-03
3	3	0.011642	0.1708D-02	0.6508D-03	0.3385D-03	0.2027D-03	0.1320D-03
4	4	0.007296	0.1032D-02	0.4166D-03	0.2276D-03	0.1420D-03	0.0000D+00
5	5	0.005041	0.6899D-03	0.2893D-03	0.1634D-03	0.0000D+00	0.0000D+00
6	6	0.003711	0.4937D-03	0.2126D-03	0.0000D+00	0.0000D+00	0.0000D+00
7	7	0.002858	0.3707D-03	0.0000D+00	0.0000D+00	0.0000D+00	0.0000D+00
8	8	0.002275	0.0000D+00	0.0000D+00	0.0000D+00	0.0000D+00	0.0000D+00

PRODUCT GAUSS-LEGENDRE RESULTS

```
npg =  4   Gf =    0.6489   If =   0.6478 Gf - If =   0.1130D-02
npg =  8   Gf =    0.6479   If =   0.6478 Gf - If =   0.9276D-04
npg = 12   Gf =    0.6478   If =   0.6478 Gf - If =   0.2008D-04
npg = 16   Gf =    0.6478   If =   0.6478 Gf - If =   0.6646D-05
npg = 20   Gf =    0.6478   If =   0.6478 Gf - If =   0.2796D-05
```

In fact, example (8.1) is one which Duffy's transformation renders trivial. Dividing the square into two equal triangles, and then using Duffy's transformation on both, leads to

$$If = \int_0^1 \int_0^1 \frac{x(1+t)}{\sqrt{(1+t^2)}} \, dx \, dt \; ,$$

which is too straightforward to pursue numerically.

Acknowledgments

It is a pleasure to acknowledge continuing encouragement and help in my work in this area from Professor Wolfgang Wendland, and also to acknowledge several conversations with Dr. Ch. Schwab in which he was able to clarify my ideas on some of these topics.

References

Davis, P. J. and Rabinowitz, P. (1984). Methods of Numerical Integration, 2nd Edition, Academic Press, New York.

Duffy, M. G. (1982). 'Quadrature over a pyramid or cube of integrands with a singularity at a vertex,' J. Numer. Anal. 19, 1260-1262.

Lyness, J. N. (1976i). 'An error functional expansion for N-dimensional quadrature with an integrand function singular at a point,' Math. Comp. 30, 1-23.

Lyness, J. N. (1976ii). 'Applications of extrapolation techniques to multidimensional quadrature of some integrand functions with a singularity,' J. Comp. Phys. 20, 346-364.

Lyness, J. N. (1991). 'Extrapolation-based boundary element quadrature,' to appear in Numerical Methods, Rend. Mat. Univ. Pol. Torino, Fascicolo Speciale.

Sidi, A. (1983). 'Euler-Maclaurin expansions for integrals over triangles of functions having algebraic/logarithmic singularities along an edge,' J. Approx. Theory 39, 39-53.

Stroud, A. H. (1971). Approximate Calculation of Multiple Integrals, Prentice-Hall, Englewood Cliffs, New Jersey.

A ROBUST NUMERICAL INTEGRATION METHOD FOR 3-D BOUNDARY ELEMENT ANALYSIS AND ITS ERROR ANALYSIS USING COMPLEX FUNCTION THEORY

KEN HAYAMI
C&C Information Technology Research Laboratories
NEC Corporation
4-1-1, Miyazaki, Miyamae, Kawasaki, 213 Japan
Email: hayami@ibl.cl.nec.co.jp

ABSTRACT. A robust numerical integration method is proposed for the accurate calculation of nearly singular integrals over general curved surfaces occurring in the three dimensional boundary element method. The method which is referred to as the Projection and Angular & Radial Transformation (PART) method, approximately projects the curved surface S to a polygon $\bar{S}$ in the plane tangent to S at the point $\bar{\mathbf{x}}_s$ nearest to the source point $\mathbf{x}_s$ which is causing the near singularity. Then, polar coordinates (ρ, θ) centred at $\bar{\mathbf{x}}_s$ are introduced in $\bar{S}$. Next, a radial variable transformation $R(\rho)$ is introduced in order to weaken the near singularity of the integral kernel. The $\log\text{-}L_2$: $R(\rho) = \log\sqrt{\rho^2 + d^2}$ and the $\log\text{-}L_1$: $R(\rho) = \log(\rho + d)$ radial variable transformations are introduced and the latter is shown to be robust and efficient by numerical experiments. Theoretical error estimates are also derived using complex function theory and are shown to correspond well with the numerical results.

1 Introduction

The Boundary Element Method (BEM) has long been established as a powerful method for numerical simulation in engineering and science. Numerical integration plays a key role in BEM, since the accurate and efficient calculation of the boundary integrals governs the accuracy and computation time of the whole analysis, and analytical integration is not possible for general integral kernels over curved boundary elements.

The numerical integration becomes particularly challenging for nearly singular integrals which arise when the source point is very near the boundary over which the integration is performed. This situation frequently occurs in engineering when analysing thin structures or gaps, or when using boundary elements with high aspect ratio and when calculating the potential or flux at a point very near the boundary. The author proposed the Projection and Angular & Radial Transformation (PART) method for the accurate and efficient calculation of these nearly singular integrals [1,2]. In this paper, we will give theoretical error estimates of the proposed method, using complex function theory.

235

T. O. Espelid and A. Genz (eds.), Numerical Integration, 235–248.
© *1992 Kluwer Academic Publishers.*

2   Boundary Element Formulation of 3-D Potential Problems

For simplicity, let us take the three dimensional potential problem as an example. (The method can also be applied to general problems.) The boundary integral equation is given by

$$c(\mathbf{x}_s)\, u(\mathbf{x}_s) = \int_S (q\ u^* - u\ q^*) dS \tag{1}$$

where $\mathbf{x}_s$ is the source point, $u(\mathbf{x})$ is the potential, $q(\mathbf{x}) \equiv \partial u/\partial n$ is the derivative of u along the unit outward normal $\mathbf{n}$ at $\mathbf{x}$ on the boundary S. S is the boundary of the region V of interest and boundary conditions concerning u and q are given on S. $c(\mathbf{x}_s) = 1$ when $\mathbf{x}_s \in V$ and $c(\mathbf{x}_s) = 1/2$ when $\mathbf{x}_s$ is on a smooth boundary. The fundamental solution u^* and q^* are defined by

$$u^*(\mathbf{x},\mathbf{x}_s) = \frac{1}{4\pi r}, \quad q^*(\mathbf{x},\mathbf{x}_s) = -\frac{(\mathbf{r},\mathbf{n})}{4\pi r^3} \tag{2}$$

where $\mathbf{r} = \mathbf{x} - \mathbf{x}_s$ and $r = |\mathbf{r}|$.

The flux at a point $\mathbf{x}_s \in V$ is given from the potential gradient

$$\frac{\partial u}{\partial \mathbf{x}_s} = \int_S \left(q\frac{\partial u^*}{\partial \mathbf{x}_s} - u\frac{\partial q^*}{\partial \mathbf{x}_s} \right) dS \tag{3}$$

where

$$\frac{\partial u^*}{\partial \mathbf{x}_s} = \frac{\mathbf{r}}{4\pi r^3}, \quad \frac{\partial q^*}{\partial \mathbf{x}_s} = \frac{1}{4\pi}\left\{ \frac{\mathbf{n}}{r^3} - \frac{3\mathbf{r}(\mathbf{r},\mathbf{n})}{r^5} \right\} \tag{4}$$

Equations (1) and (3) are discretized on the boundary S by boundary elements defined by interpolation functions. The integral kernels of equations (1) and (3) become nearly singular when the distance d between $\mathbf{x}_s$ and S is small compared to the size of the boundary elements which discretize the boundary S.

3   Nature of Nearly Singular Integral Kernels in 3-D Potential Problems

We will first analyze the nature of nearly singular integrals occurring in the boundary element formulation of 3-D potential problems. Since near singularity becomes significant in the neighbourhood of the source point $\mathbf{x}_s$, one can approximate the curved boundary element S by the planar element $\overline{S}$ tangent to S at the point $\overline{\mathbf{x}}_s$ which is nearest to $\mathbf{x}_s$ on S. Then, Cartesian coordinates (x, y, z) and polar coordinates (ρ, θ) centered at $\overline{\mathbf{x}}_s$ are introduced with $\overline{S}$ in the xy-plane.

Since

$$\mathbf{x}_s = (0,0,d), \quad \mathbf{x} = (x,y,0) = (\rho\cos\theta, \rho\sin\theta, 0), \tag{5}$$

$$\mathbf{r} = (\rho\cos\theta, \rho\sin\theta, -d), \quad \mathbf{n} = (0,0,-1), \quad (\mathbf{r},\mathbf{n}) = d \ , \tag{6}$$

equations (2) and (4) can be expressed as

$$u^* = \frac{1}{4\pi r}, \quad q^* = -\frac{d}{4\pi r^3} \tag{7}$$

$$\frac{\partial u^*}{\partial \mathbf{x}_s} = \frac{1}{4\pi}\left(\frac{\rho\cos\theta}{r^3}, \frac{\rho\sin\theta}{r^3}, -\frac{d}{r^3}\right) \tag{8}$$

$$\frac{\partial q^*}{\partial \mathbf{x}_s} = \frac{1}{4\pi}\left(-3d\frac{\rho\cos\theta}{r^5}, -3d\frac{\rho\sin\theta}{r^5}, -\frac{1}{r^3}+\frac{3d^2}{r^5}\right) \tag{9}$$

Considering that the near singularity is essentially related to r and the radial variable ρ , the nature of the integral kernels in equations (1) and (3) can be summarized as in Table 1 .

Table 1: Nature of near singularity ($0 < d \ll 1$) for 3-D potential problems

	order of near singularity		
u^*	$1/r$		
q^*	$1/r^3$		
$\partial u^*/\partial \mathbf{x}_s$	$1/r^3$	ρ/r^3	
$\partial q^*/\partial \mathbf{x}_s$	$1/r^3$	$1/r^5$	ρ/r^5

4   The PART Method

The PART method is briefly described as follows:
For the nearly singular integral

$$I = \int_S \frac{f}{r^\alpha}dS \ , \quad (\alpha \in \mathbf{N}) \tag{10}$$

1. <u>Source projection</u>

Find the nearest point $\bar{\mathbf{x}}_s = \mathbf{x}(\bar{\eta}_1, \bar{\eta}_2)$ on S to $\mathbf{x}_s$ (e.g. using the Newton-Raphson method).

2. <u>Approximate projection of S</u>

Approximately project the curved element S onto the polygon $\bar{S}$ in the plane tangent to S at $\bar{\mathbf{x}}_s$. (cf. Fig.1)

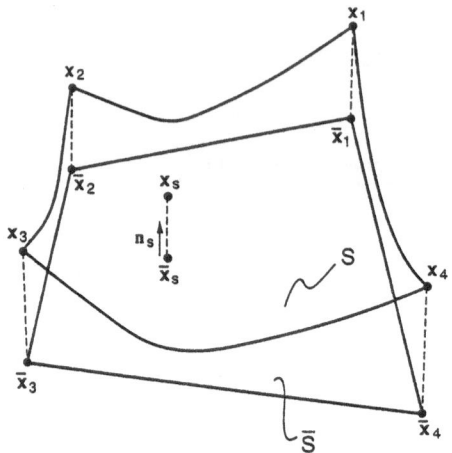

Figure 1: Approximate projection $\bar{S}$ of the curved element S.

3. Introduce <u>polar coordinates</u> (ρ, θ) in $\bar{S}$, centred at $\bar{\mathbf{x}}_s$ to get

$$I = \int_0^{2\pi} d\,\theta \int_0^{\rho_{max}(\theta)} \frac{f}{r^\alpha} J\rho d\,\rho \tag{11}$$

where J is the Jacobian of the mapping from Cartesian coordinates on $\bar{S}$ to curvilinear coordinates on S .

4. Apply <u>Radial variable transformation</u> : $R(\rho)$ in order to weaken the near singularity due to $1/r^\alpha$, which is essentially related to the radial variable ρ only .

5. Apply the <u>Angular variable transformation</u>:

$$t(\theta) = \frac{h_j}{2} \log \left\{ \frac{1 + \sin(\theta - \alpha_j)}{1 - \sin(\theta - \alpha_j)} \right\} \tag{12}$$

in order to weaken the angular near singularity which arises when $\bar{\mathbf{x}}_s$ is near the edge of the polygon $\bar{S}$. This uses the fact that

$$\frac{d\,\theta}{d\,t} = \frac{1}{\rho_{max}(\theta)} = \frac{\cos(\theta - \alpha_j)}{h_j} \tag{13}$$

6. Use Gauss-Legendre's formula to perform <u>numerical integration</u> in the transformed variables R and t in

$$I = \int_{t(0)}^{t(2\pi)} \frac{dt}{\rho_{max}(\theta)} \int_{R(0)}^{R\{\rho_{max}(\theta)\}} \frac{fJ\rho}{r^\alpha} \frac{d\rho}{dR} dR \tag{14}$$

5 Radial Variable Transformation

The choice of the radial variable transformation governs the efficiency of the PART method.

For constant planar elements,

$$\rho \, d\rho = r'^\alpha \, dR \quad \text{or} \quad R(\rho) = \int \frac{\rho}{r'^\alpha} \, d\rho \tag{15}$$

where $r' = \sqrt{\rho^2 + d^2}$, is equivalent to analytical integration in the radial variable in equation (11) (only one integration point is required for the exact integration in the radial variable), since $r = r'$.

However, for general curved elements, the <u>$\log$-L_2 transformation</u>:

$$\rho \, d\rho = r'^2 \, dR \quad \text{or} \quad R(\rho) = \log \sqrt{\rho^2 + d^2} \tag{16}$$

turns out to be robust and efficient , except for the calculation of the flux, as shown in Fig.2.

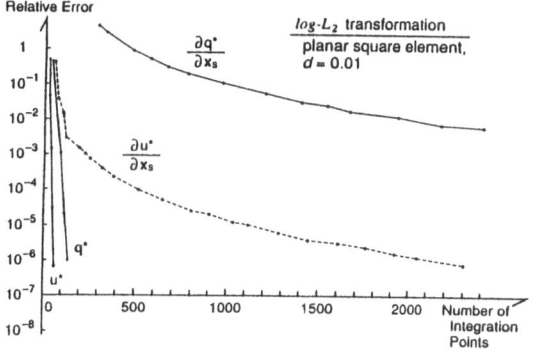

Figure 2: Convergence graph for log-L_2 transformation

The near optimum radial variable transformation which works efficiently all round, seems to be the <u>$\log$-L_1 transformation</u>:

$$R(\rho) = \log(\rho + d) \tag{17}$$

as shown in Fig.3.

240

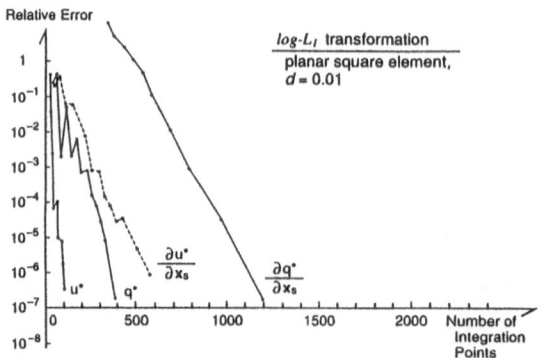

Figure 3: Convergence graph for log-L_1 transformation

6 Error Analysis using Complex Function Theory

6.1 RADIAL COMPONENT INTEGRAL

The essential nature of the integration in the radial variable which appear in the potential integrals $\int_S u^* dS$, $\int_S q^* dS$ and flux integrals $\int_S \partial u^*/\partial x_s dS$, $\int_S \partial q^*/\partial x_s dS$ in equations (1) and (3) can be modelled by

$$I = \int_0^{\rho_j} \rho^\delta / r'^\alpha d\rho \tag{18}$$

where $\rho_j \equiv \rho_{max}(\theta)$ in equation (11), and $r = r' \equiv \sqrt{\rho^2 + d^2}$ for planar elements.

Potential calculation of equation (1) gives rise to

- $\alpha = \delta = 1$ for the integration of u^*
- $\alpha = 3, \delta = 1$ for the integration of q^*

Flux calculation of equation (3) gives rise to

- $\alpha = 3; \ \delta = 1, 2$ for the integration of $\partial u^*/\partial x_s$
- $\alpha = 3, \delta = 1$ and $\alpha = 5; \ \delta = 1, 2$ for the integration of $\partial q^*/\partial x_s$

Equation (18) is transformed by the radial transformation $R(\rho)$ as

$$I = \int_{R(0)}^{R(\rho_j)} \frac{\rho^\delta}{r^\alpha} \frac{d\rho}{dR} dR \tag{19}$$

which, in turn, can be transformed as

$$I = \int_{-1}^{1} \frac{\rho^\delta}{r^\alpha} \frac{d\rho}{dR} \frac{dR}{dx} dx \equiv \int_{-1}^{1} f(x) dx \tag{20}$$

where

$$R = \frac{\{R(\rho_j) - R(0)\}x + R(\rho_j) + R(0)}{2} \tag{21}$$

6.2 BASIC THEOREM AND ASYMPTOTIC EXPRESSION

The following theorem gives the error $E_n = I - I_n$ of the numerical integration $I_n = \sum_{j=1}^{n} A_j f(a_j)$ of the integral $I = \int_{-1}^{1} f(x)dx$ over the interval $J = (-1, 1)$.
Theorem 1 [3,5,4]

If $f(z)$ is regular on $K = [-1, 1]$,

$$E_n(f) = \frac{1}{2\pi i} \oint_C \Phi_n(z) f(z) dz \tag{22}$$

where

$$\Phi_n = \int_{-1}^{1} \frac{dx}{z - x} - \sum_{j=1}^{n} \frac{A_j}{z - a_j} \tag{23}$$

and the contour C is taken so that it encircles the integration points $a_1, a_2, ..., a_n$ in the positive direction, and $f(z)$ is regular inside C.

The following asymptotic expressions are known for the error characteristic function $\Phi_n(z)$ of equation (23) for the Gauss-Legendre rule.

1. For $\mid z \mid \gg 1$ [7]

$$\Phi_n(z) = \frac{c_n}{z^{2n+1}} \{1 + O(z^{-2})\} \tag{24}$$

where

$$c_n = \frac{2^{2n+1}(n!)^4}{(2n)!(2n+1)!} \tag{25}$$

and $c_n \sim \pi 2^{-2n}$ for $n \gg 1$.

2. For $n \gg 1$ [3,5]

- For all $z \in \mathbf{C}$ except for an arbitrary neighbourhood of $K = [-1, 1]$:

$$\Phi_n(z) \sim 2\pi(z + \sqrt{z^2 - 1})^{-2n-1} \tag{26}$$

- For all $z \in \mathbf{C}$ except for an arbitrary neighbourhood of $z = 1$:

$$\Phi_n(z) \sim 2e^{-i\pi} \frac{K_0(2k\zeta)}{I_0(2k\zeta)} \tag{27}$$

where $z = e^{i\pi} \cosh(2\zeta)$, $k = n + \frac{1}{2}$ and $I_0(z), K_0(z)$ are the modified Bessel functions of the first and second kind, respectively.

6.3 ERROR ANALYSIS FOR THE log-L_2 TRANSFORMATION

For the log-L_2 radial variable transformation $R(\rho) = \log \sqrt{\rho^2 + d^2}$ of equation(16),

$$R(0) = \ln d \,, \quad R(\rho_j) = \ln r_j \,, \quad r_j = \sqrt{\rho_j^2 + d^2} \tag{28}$$

and

$$\rho(R) = (e^{2R} - d^2)^{\frac{1}{2}} \tag{29}$$

so that we obtain

$$f(z) = b \left(e^{z \ln a} - \frac{1}{a} \right)^{\frac{\delta-1}{2}} e^{\left(\frac{2-\alpha}{2} \ln a \right) z} \tag{30}$$

where

$$b \equiv \frac{\ln a}{2}(r_j d)^{\frac{\delta-\alpha+1}{2}} > 0 \,, \quad a \equiv \frac{r_j}{d} \tag{31}$$

6.3.1 *Case : $\delta = odd.$* Since $\frac{\delta-1}{2}$ is a non-negative integer, $f(z)$ is regular except for $z = \infty$. Hence, taking $C = \{z \mid \mid z \mid = R, R \to \infty\}$ as the contour in *Theorem 1* and using the asymptotic expression of equation (24) for $\mid z \mid \gg 1$, we obtain

$$E_n(f) = \frac{c_n}{2\pi i} \oint_C f(z) z^{-2n-1} = c_n a_{2n} \tag{32}$$

where

$$f(z) = \sum_{k=1}^{\infty} a_k z^k \tag{33}$$

In conclusion, we obtain

$$E_n(f) \sim D^{\frac{\delta+1-\alpha}{2}} \left(\frac{\ln D}{n} \right)^{2n} \sim n^{-2n} \tag{34}$$

where $D = d/\rho_j$ is the relative source distance. This corresponds well with numerical results for the integration of potential kernels using the log-L_2 transformation in Fig.2.

6.3.2 *Case : $\delta = even.$* When δ is even, as in the case of flux kernels, $f(z)$ of equation(30) has a branching point singularity at

$$z_m = -1 + i\frac{2\pi m}{\ln(r_j/d)} \,, \quad (m : integer) \tag{35}$$

as shown in Fig.4. In this case, we can modify *Theorem 1* by taking the contour as

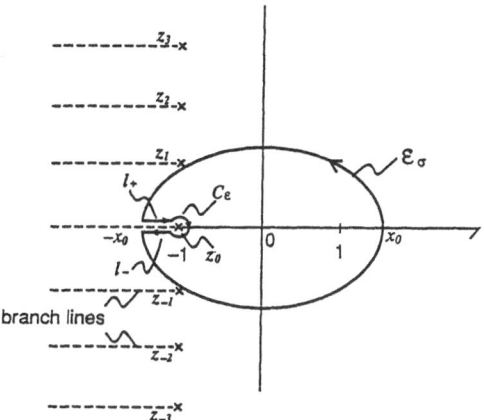

Figure 4: Contour for $\log$-L_2 transformation (δ =even)

$$C = \varepsilon_\sigma + \ell_+ + C_\epsilon + \ell_- \tag{36}$$

as shown in Fig.4.

It turns out that the most significant contribution to $E_n(f)$ of equation (22) comes from the branch lines ℓ_+ and ℓ_-, which is

$$E_{\ell_+, \ell_-} = \frac{1}{2\pi i} \int_{-x_0}^{-1-\epsilon} \{f(z_+)\Phi_n(z_+) - f(z_-)\Phi_n(z_-)\}dz \tag{37}$$

Let

$$z = z_+ \equiv e^{i\pi} \cosh 2\zeta \quad on \ \ell_+ \tag{38}$$

and

$$z = z_- \equiv e^{-i\pi} \cosh 2\zeta \quad on \ \ell_- \tag{39}$$

where

$$0 \leq \zeta \leq \zeta_0 \equiv \frac{1}{2} \cosh^{-1} x_0 \ll 1 \tag{40}$$

Then, we obtain

$$f(z_\pm) \sim B \, e^{\frac{\alpha-2}{2} \ln a} \, e^{\pm i \frac{\pi(\delta-1)}{2}} \zeta^{\delta-1} \tag{41}$$

where

$$B \equiv 2^{\frac{\delta-3}{2}} (\ln a)^{\frac{\delta+1}{2}} r_j^{\frac{2-\alpha}{2}} d^{\frac{2\delta-\alpha}{2}} \tag{42}$$

244

Hence,

$$f(z_+) - f(z_-) \sim B\, e^{\frac{\alpha-2}{2}\ln a}\, 2i \sin\left\{\frac{\pi(\delta-1)}{2}\right\} \zeta^{\delta-1} \tag{43}$$

for $|\zeta| \ll 1$.

Since $z = -\cosh 2\zeta \sim -1 - 2\zeta^2$ for $|\zeta| \ll 1$, the asymptotic expression of equation (27) gives

$$E_{\ell_+,\ell_-} \sim \frac{8B}{\pi} e^{\frac{\alpha-2}{2}\ln a} \sin\left\{\frac{\pi(\delta-1)}{2}\right\} \int_{\sqrt{\frac{\epsilon}{2}}}^{\zeta_0} \zeta^\delta \frac{K_0(2k\zeta)}{I_0(2k\zeta)} d\zeta \tag{44}$$

for $n \gg 1$.

If we let $2k\zeta \geq 2k\zeta_0 \gg 1$,

$$\frac{K_0(2k\zeta)}{I_0(2k\zeta)} \sim \pi e^{-4k\zeta} \tag{45}$$

so that

$$\int_{\zeta_0}^{\infty} \zeta^\delta \frac{K_0(2k\zeta)}{I_0(2k\zeta)} d\zeta \sim \frac{\pi \zeta_0^\delta e^{-4k\zeta_0}}{4k} \tag{46}$$

From equations (44) and (46), if we choose x_0 of the ellipse ε_σ such that

$$\frac{1}{2n+1} = \frac{1}{2k} \ll \zeta_0 \ll 1, \tag{47}$$

and take the limit as $\epsilon \to 0$ so that

$$\int_0^{\sqrt{\frac{\epsilon}{2}}} \zeta^\delta \frac{K_0(2k\zeta)}{I_0(2k\zeta)} d\zeta \sim O(\epsilon^{\frac{\delta+1}{2}} \log \epsilon) \to 0, \tag{48}$$

we obtain

$$E_{\ell_+,\ell_-} \sim \frac{\left\{\sin\frac{\pi(\delta-1)}{2}\right\} \int_0^\infty t^\delta \frac{K_0(t)}{I_0(t)} dt \, (\ln\frac{r_j}{d})^{\frac{\delta+1}{2}} d^{\delta-\alpha+1}}{\pi 2^{\frac{\delta-1}{2}} (n+\frac{1}{2})^{\delta+1}} \tag{49}$$

The contribution from C_ϵ as $\epsilon \to 0$ is zero. The contribution from the ellipse ε_σ is of order σ^{-2n} where $\sigma = 2.7 \sim 5.9$ for $D \equiv d/\rho_j = 10^{-3} \sim 10^{-1}$. Hence, for $\delta = $ even, we have for $n \gg 1$, $D \ll 1$,

$$E_n(f) \sim E_{\ell_+,\ell_-} \sim O\left\{(-\ln D)^{\frac{\delta+1}{2}} D^{\delta+1-\alpha} n^{-\delta-1}\right\} \sim O(n^{-\delta-1}) \tag{50}$$

This matches well with numerical results for the integration of the flux kernels, which give $E_n(f) \sim O(n^{-3})$, where $\delta = 2$.

6.4 ERROR ANALYSIS FOR THE log-L_1 TRANSFORMATION

For the $\log$-L_1 transformation $R(\rho) = \log(\rho + d)$, we have

$$R(0) = \ln d , \quad R(\rho_j) = \ln(\rho_j + d) \tag{51}$$

and

$$\rho(R) = e^R - d , \quad \frac{d\rho}{dR} = e^R \tag{52}$$

so that $f(x)$ of equation (20) is given by

$$f(z) = \frac{b'(w - 1)^\delta w}{\{w - (1 - i)\}^{\frac{\alpha}{2}}\{w - (1 + i)\}^{\frac{\alpha}{2}}} \tag{53}$$

where

$$w \equiv e^{\frac{z+1}{2}\ln a'} , \quad b' \equiv \frac{\ln a'}{2}d^{\delta - \alpha + 1} , \quad a' \equiv 1 + \frac{\rho_j}{d} > 1 \tag{54}$$

$f(z)$ has singularities (branching when $\alpha = odd$) at

$$z = z_m^{\pm} \equiv -1 + \frac{\ln 2}{\ln a'} + i\frac{(4m \pm \frac{1}{2})\pi}{\ln a'} , \quad (m : integer) \tag{55}$$

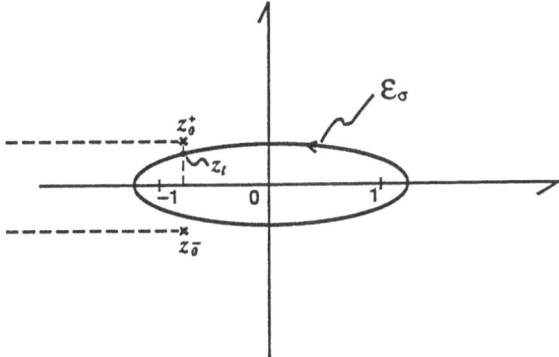

Figure 5: Contour for the $\log$-L_1 transformation

As the contour C in *Theorem 1*, we will take the ellipse ε_σ :

$$\mid z + \sqrt{z^2 - 1} \mid = \sigma , \quad (\sigma > 1) \tag{56}$$

which has its foci at $z = \pm 1$ and passes through the point

$$z_t \equiv -1 + \frac{\ln 2}{\ln a'} + i\frac{\pi t}{2\ln a'}, \quad (0 < t < 1) \tag{57}$$

as shown in Fig.5. Note that there are no singularities of $f(z)$ inside ε_σ.

Using the asymptotic expression of equation (26) for $n \gg 1$ in equation (22), we obtain

$$|E_n(f)| \le \frac{\ell(\varepsilon_\sigma)}{\sigma^{2n+1}} \max_{z \in \varepsilon_\sigma} |f(z)| < 2\pi\sigma^{-2n} \max_{z \in \varepsilon_\sigma} |f(z)| \tag{58}$$

where $\ell(\varepsilon_\sigma)$ is the length of the ellipse ε_σ [6].

For the ellipse ε_σ passing through z^t, we have

$$\sigma = \frac{c}{2}p + \sqrt{\frac{c^2}{4}p^2 - p\ln 2 + 1 + \sqrt{\frac{c^2}{2}p^2 - p\ln 2 + \sqrt{\frac{c^2}{4}p^2 - p\ln 2 + 1}}} \tag{59}$$

where

$$p \equiv \frac{1}{\ln\left(1 + \frac{1}{D}\right)}, \quad D \equiv \frac{d}{\rho_j} \tag{60}$$

and

$$c \equiv \sqrt{(\ln 2)^2 + \left(\frac{\pi t}{2}\right)^2}, \quad (0 < t < 1) \tag{61}$$

$\sigma(D)$ is a strictly increasing function of D.

Next, we will estimate $\max_{z \in \varepsilon_\sigma} |f(z)|$. Since $|f(\bar{z})| = |f(z)|$, it suffices to consider the region $Im(z) \ge 0$. Also, since $|f(z_1)| = +\infty$ where $z_1 \equiv z_0^+$ of equation (54), if $|1 - t| \ll 1$, we may assume that

$$|f(z_t)| \sim \max_{z \in \varepsilon_\sigma} |f(z)| \tag{62}$$

Let,

$$\Delta z \equiv z_t - z_1 = -\frac{(1 - t)\pi}{\ln a'}i \tag{63}$$

Then, for $|\Delta z| \ll 1$,

$$w(z_t) \sim (1 + i)\left\{1 + \frac{\ln a'}{2}\Delta z + O(\Delta z^2)\right\} \tag{64}$$

so that

$$\max_{z \in \varepsilon_\sigma} |f(z)| \sim 2^{\frac{\alpha-2}{4}}\pi^{-\frac{\alpha}{2}}d^{\delta-\alpha+1}\left\{\ln\left(1 + \frac{\rho_j}{d}\right)\right\}(1 - t)^{-\frac{\alpha}{2}} \tag{65}$$

Since we are interested in cases $\alpha = 1, 3, 5$, $(1 - t)^{-\frac{\alpha}{2}} \leq 10$ implies $t \leq 0.6$. Hence, equation (58) gives $\sigma = 1.31, 1.40, 1.63$ for the nearly singular cases $D = 10^{-3}, 10^{-2}, 10^{-1}$, respectively.

To sum up, for the log-L_1 transformation: $R(\rho) = \log(\rho + d)$, the numerical integration error is given by

$$E_n(f) \sim (-\ln D) D^{\delta + 1 - \alpha} \sigma^{-2n} \qquad (66)$$

where $\sigma = 1.31 \sim 1.63$ for $D = 10^{-3} \sim 10^{-1}$.

This estimate corresponds well with numerical experiment results e.g. in Fig.3.

Similar analysis shows that for the identity radial variable transformation: $R(\rho) = \rho$, the error is given by

$$E_n(f) \sim D^{\delta - \frac{3}{2}\alpha} \sigma^{-2n} \qquad (67)$$

where $\sigma = 1.04, 1.12, 1.42$ for $D = 10^{-3}, 10^{-2}, 10^{-1}$.

This shows, together with numerical experiment results that the log-L_1 radial variable transformation is a robust and efficient candidate for the calculation of nearly singular integrals arising in three dimensional boundary element analysis. The error analysis using complex analysis also renders a clear insight when searching for even better variable transformations.

Acknowledgements
The author would like to thank Dr. Masaaki Sugihara for valuable advice, and the anonymous referee who suggested some improvements on the original draft.

References

[1] Hayami, K. and Brebbia, C.A. (1988) 'Quadrature methods for singular and nearly singular integrals in 3-D boundary element method' (Invited Paper), in C.A. Brebbia (ed.), Boundary Elements X, Vol.1, pp.237-264, Computational Mechanics Publications with Springer-Verlag.

[2] Hayami, K. (1990) 'A robust numerical integration method for three dimensional boundary element analysis', in M. Tanaka, C.A. Brebbia and T. Honma (eds.), Boundary Elements XII, Vol.1, pp.33-51, Computational Mechanics Publications with Springer-Verlag.

[3] Barret, W. (1960) 'Convergence properties of Gaussian quadrature formulae', Comput. J., Vol.3, pp.272-277.

[4] Takahasi,H. and Mori,M. (1970) 'Error estimation in the numerical integration of analytic functions', Report of the Computer Centre University of Tokyo, Vol.3, pp.41-108.

[5] Donaldson, J.D. and Elliot, D. (1972) 'A unified approach to quadrature rules with asymptotic estimates of their remainders', SIAM J. Numer. Anal. Vol.9, No.4, 573-602.

[6] Davis, P.J. and Rabinowitz P. (1984) Methods of Numerical Integration, Academic Press.

[7] McNamee, J. (1964) 'Error-bounds for the evaluation of integrals by the Euler-Maclaurin formula and by Gauss-type formulae', Math.Comp., Vol.18, pp.368-381.

ON THE NUMERICAL CALCULATION OF MULTIDIMENSIONAL INTEGRALS APPEARING IN THE THEORY OF UNDERWATER ACOUSTICS

Jarle Berntsen
Institute of Marine Research
P.O.Box 1870
N-5024 Bergen-Nordnes
Norway

ABSTRACT. Multidimensional integrals appear frequently in applied sciences. In this paper we will present some integrals from the theory of underwater acoustics. Automatic adaptive integration routines may be useful tools for computing approximations to many of these integrals, and we will focus on the properties that should be implemented in such routines in order to produce software that will meet the requirements of applied scientists. We will describe how some of these properties are implemented in recently developed routines for hyperrectangular regions, triangles and tetrahedrons. The routines to be described are now being used by a number of scientists in acoustics, and the feedback acquired has given us many ideas on how to further improve the software. Based on the needs of these scientists we will discuss which features that should be implemented in the next generation of automatic adaptive multidimensional integration routines.

1. INTRODUCTION

Many of the multidimensional integrals appearing in the theory of acoustics present hard challenges to automatic integration routines. The first class of integrals was presented to this author in 1981, see [22, 31, 32]. The difficulties met when trying to approximate these integrals, initiated a still ongoing activity within numerical integration at the University of Bergen. Terje Espelid has been a major contributor to the activity all the way and early on Tor Sørevik also became involved in our work.

The development of numerical software may easily develop into a research activity separated from the applied sciences that gave birth to the numerical research. The last 10 years I have been involved in the numerical part of the work performed by the scientists connected to the acoustics group led by Jacqueline Naze Tjøtta and Sigve Tjøtta at the Department of Mathematics, University of Bergen. A number of automatic integration routines have been used by these scientists. On one hand the routines have helped the scientists in acoustics to compute approximations to the integrals appearing in their area of research, see [22, 23, 28, 29, 30, 31, 32, 38, 39]. On the other hand the feedback has helped us to gradually improve the quality of our software, see [2, 5, 12, 16, 20, 21].

Two integrals from acoustics will be presented in section 2. In section 3 a brief description of adaptive integration routines will be given. In section 4 we will focus on the

T. O. Espelid and A. Genz (eds.), Numerical Integration, 249–265.

reliability of such routines when applied to some integrals from acoustics. Some algorithmic considerations will be discussed in section 5. Three new routines that already are being used by a number scientists in acoustics will be described in section 6.

2. INTEGRALS FROM ACOUSTICS

In this section two classes of integrals will be presented. The first one is also the first class of integrals presented to me by Jacqueline Naze Tjøtta in 1981. The integrals

$$q_+(z,x) = -i\frac{k_1 k_2 k_+}{2\pi z \rho_0 c_0^2}\beta \int_0^\infty \int_0^\infty \int_0^{2\pi} \left[E_1 \left(i\frac{k_1 k_2}{2k_+ z}\left(x_1^2 + x_2^2 - 2x_1 x_2 cos(\phi)\right)\right)\right]^* \cdot$$

$$J_0\left(\frac{xF^{\frac{1}{2}}}{z}\right) exp\left(\frac{i}{2k_+ z}\left[(k_+ x)^2 + F\right]\right) \cdot q_1(0,x_1)q_2(0,x_2)d\phi x_1 x_2 dx_1 dx_2 \qquad (1)$$

describe the sum frequency pressure from a circular monochromatic sound source with frequencies f_1 and f_2.

q_+ is the sum frequency sound pressure,
z is the distance from the source along the axis,
x is the distance transverse to the axis,
$k_j = 2\pi f_j/c_0$, $j = 1,2$,
$k_+ = 2\pi(f_1 + f_2)/c_0$,
c_0 is the under water speed of sound,
ρ_0 is the density,
β is a nonlinearity parameter,
E_1 is the exponential integral,
J_0 is the Bessel function of order zero,
$q_j(0,x_j)$, $j = 1,2$, define the on source pressure,
$*$ denotes the complex conjugate and
$F = (k_1 x_1)^2 + (k_2 x_2)^2 + 2k_1 x_1 k_2 x_2 cos(\phi)$.

We notice that the integral is complex. For uniform piston sources, that is $q_j(0,x_j) = 1$ if $x_j \leq$ the radii of the piston and 0 otherwise, $j = 1,2$, the region of integration will be a hyperrectangle. For piston sources there are logarithmic singularities along the diagonal of one of the faces of the hyperrectangle. We note that the integral is highly oscillatory and that the number of oscillations depend on the frequencies and size of the source. If we introduce dissipation into the theory, E_1 becomes a one dimensional integral with an infinite oscillating tail. If we in addition apply non-axis symmetric sources, the region of integration becomes 4 dimensional. For problems where the frequencies are close to 1 megahertz, problems which include dissipation and for non-axis symmetric sources, we have had severe problems approximating the corresponding integrals. For further details see [22, 31, 32].

The last class of integrals from acoustics brought to my attention appears in the PhD thesis of Kjell-Eivind Frøysa [29]. The integrals

$$p_2(\tau, z, x) = \frac{\beta \rho_0 c_0^2}{2\pi} \left(\frac{\omega_0 a}{c_0}\right)^2 \int_0^\infty \int_{\omega/2}^\infty \int_0^z \text{Re} \{ \cdot$$

$$\omega^2 s(\omega - s) \widetilde{F}(s) \widetilde{F}(\omega - s) e^{-i\omega\tau - L_i(\omega^2 z - 2s(\omega - s)z')}.$$

$$\exp\left[\frac{-i\omega x^2(s(\omega - s)(2 + i\omega d_i(1 - zd_i)) + \omega z'(i - \omega d_i))}{s(\omega - s)(i\omega(1 - zd_i)(1 - z'd_i) - 2(z - z')) - z'\omega(\omega(1 - zd_i) + zi)}\right] \cdot$$

$$\frac{1}{s(\omega - s)(i\omega(1 - zd_i)(1 - z'd_i) - 2(z - z')) - z'\omega(\omega(1 - zd_i) + zi)} \} dz'dsd\omega \qquad (2)$$

describe the quasilinear sound field from a pulsed Gaussian source which radiates into a dissipative fluid. The parameters not described in connection to integral (1) are:

p_2 is the quasilinear part of the sound pressure,
τ is non-dimensional time,
ω is the characteristic frequency,
a is the radius of the source,
$\widetilde{F}$ is the Fourier transform of the pulsed signal,
L_i is the Rayleigh distance in absorption distances and
d_i is the Rayleigh distance in focal distances.

Again we have a rather complicated and highly oscillatory integrand. The region of integration for this integral deserves special attention. The integration domain is infinite in ω and s, and the projection of the domain on to the $\omega - s$ plane is an infinite triangle. Even if we may truncate the region of integration at some large value of ω with a negligible error, no automatic routine known to this author can handle these integrals in a straightforward manner. Frøysa transformed the region for $\omega > 1$ on to a hyperrectangle. Transforming the whole domain this way will introduce a strong singularity. The region for $\omega < 1$ is split into a number of hyperrectangles over which the integrand is defined to be 0 outside the actual region of integration. For all subregions, including the hyperrectangle with the transformed integrand, the routine DCUHRE [21] is applied to approximate the integral. Frøysa describes in detail the algorithm for computing a reasonable splitting of the domain and also the algorithm for computing reasonable error requests when DCUHRE is applied to one subregion at the time.

When Frøysa was about to finish his degree, the work with the routine DCUTET [12] which may integrate over a collection of tetrahedrons in one call to the routine, was completed. He then divided his region into a number of tetrahedrons, taking care that all difficult spots were placed at the vertices of tetrahedrons, and applied DCUTET to the collection. With DCUTET available from the start of his work, he could have saved computer time, produced a safer error estimate and last but not least saved a lot of human time.

In general all integrals from acoustics that I have met upon have this general form

$$q(\underline{x}) = \int_{R_n} f(\underline{x}, \underline{y}) d\underline{y} \,. \tag{3}$$

q is some physical property (q is very often complex),
$\underline{x}$ is the observation point,
$\underline{y}$ is the n-dimensional integration variables,
f is typically a complex function and
R_n is a n-dimensional region of integration.

The applied scientists typically want to approximate q for a large number of observation points.

We should also note that most of the published routines attack the problem

$$I[f] = \int_{R_n} f(\underline{x}) d\underline{x} \,. \tag{4}$$

We will see later on that it may make a great difference to an applied scientist whether the routine is written for problem (4) or written with problem (3) in mind.

3. ADAPTIVE INTEGRATION ROUTINES

Let m be the number of subregions in our collection of regions and M the number of subregions in the data structure during the subdivision process. Let the error tolerance be ϵ. Let $\hat{Q}$ be a global estimate of the integral over the collection of regions and $\hat{E}$ the corresponding error estimate. Let $\hat{Q}_k$ be the local estimate of the integral over subregion k and $\hat{E}_k$ the corresponding local error estimate. A globally adaptive cubature algorithm may then be written:

$\quad$ *Initialize*: $\quad$ Initialize the collection of subregions: $M = m$;
$\qquad\qquad\qquad$ Produce $\hat{Q}_k$ and $\hat{E}_k$ for $k = 1, 2, \ldots, M$;
$\qquad\qquad\qquad$ Put $\hat{Q} = \sum_{k=1}^{M} \hat{Q}_k$ and $\hat{E} = \sum_{k=1}^{M} \hat{E}_1$;
$\quad$ *Control*: $\quad$ **while** $\hat{E} > \epsilon$ **do**
$\qquad\qquad\qquad$ **begin**
$\qquad\qquad\qquad$ Pick region k from the collection;
$\qquad\qquad\qquad$ Divide region k into P subregions;
$\quad$ *Process*: $\quad$ Compute $\hat{Q}_k^{(i)}$ and $\hat{E}_k^{(i)}$, $i = 1, \ldots, P$;
$\quad$ *Update*: $\quad$ $\hat{Q} = \hat{Q} + (\sum_{i=1}^{P} \hat{Q}_k^{(i)} - \hat{Q}_k)$;
$\qquad\qquad\qquad$ $\hat{E} = \hat{E} + (\sum_{i=1}^{P} \hat{E}_k^{(i)} - \hat{E}_k)$;
$\qquad\qquad\qquad$ Let the new subregions replace
$\qquad\qquad\qquad$ the old region and
$\qquad\qquad\qquad$ put $M = M + (P - 1)$;
$\qquad\qquad\qquad$ **end**

The main ingredients of such an algorithm are:

1. A basic rule for estimating the integral over each subregion.
2. A procedure for estimating the error over each subregion.
3. The algorithm for dividing the region of integration into subregions.

The construction of basic rules has been studied by a number of authors, see [24, 40]. There has been less attention to the last two ingredients, and I will therefore focus on these. The decisions in automatic integration are based upon the error estimates and therefore I will focus especially on the error estimating procedure.

4. RELIABILITY

Let Q_i denote a cubature rule of degree i. In many automatic cubature routines the following error estimating procedure:

$$E[f] = |Q_{2m+1}[f] - Q_{2m-1}[f]| \tag{5}$$

is used. The hope is of course that the actual error is bounded by this estimate. That is

$$|I[f] - Q_{2m+1}[f]| \leq E[f]. \tag{6}$$

The assumption behind this procedure is that $Q_{2m+1}[f]$ is a much better approximation to $I[f]$ than $Q_{2m-1}[f]$. We will then have

$$|I[f] - Q_{2m-1}[f]| \sim |Q_{2m+1}[f] - Q_{2m-1}[f]|. \tag{7}$$

It is therefore expected that $E[f]$ will be a good approximation to the error in $Q_{2m-1}[f]$ and that the error in $Q_{2m+1}[f]$ will be much smaller than $E[f]$ most of the time.

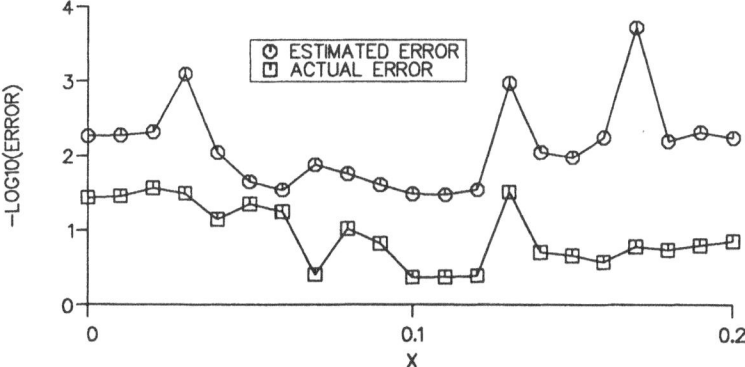

Figure 1. Absolute errors for $E[f] = |Q_9[f] - Q_7[f]|$.

However, when applying routines using error estimating procedure (5) to the integrals from acoustics, we may often be misled by the error estimate. In order to illustrate this we approximate integral 1 for the parameters $q_i = 1$ if $x_i \leq 0.9$ and 0 otherwise, $i = 1, 2$, $f_1 = 16000.$, $f_2 = 11000.$, $c_0 = 1500.$, $z = 30.$ and $x = 0(0.01)0.2$ with a globally adaptive routine requesting an absolute error less or equal to 0.1. The parameters above are typical parameters for a large piston used in numerous experiments in a lake near Austin, Texas. The order of size of the the integral is 1. Figure 1 shows the actual absolute errors and the estimated absolute errors when approximating the real part of integral 1 with the routine DCUHRE[21]. In this experiment the error estimating procedure $E[f] = |Q_9[f] - Q_7[f]|$ is implemented into DCUHRE. The exact values of the integrals that we need to compute the absolute actual errors, are approximated by letting DCUHRE apply 10^6 integrand evaluations for each x.

We note that the actual absolute error may or may not be smaller than our error request. We also note that we always are misled by the error estimate. The actual error is greater than the estimated error in all cases.

In order to improve the error estimating procedure applied above, we have found it very useful to apply null rules, see Lyness [35]. A null rule approximation of degree d may be given by:

$$N_d[f] = Q_{d\prime}[f] - Q_d[f] \,, \ d\prime > d \,. \tag{8}$$

This rule integrates to zero all monomials of degree less or equal to d. Our error estimating procedure may then be written:

$$E[f] = |N_7[f]| \,. \tag{9}$$

In Figure 2 we have plotted $N_7[f]$ as a function of x. We note that $N_7[f]$ is an oscillatory function of x for our integral. For values of x giving $N_7[f] = 0$, the error will always be estimated to zero regardless of the size of the actual error, and no simple scaling of the error estimate will help.

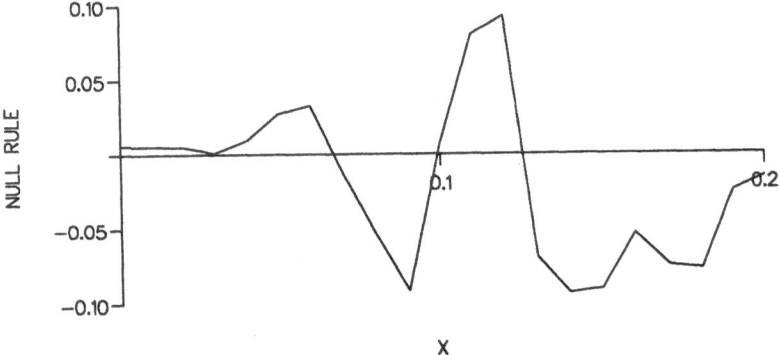

Figure 2. $N_7[f]$.

In order to improve our error estimating procedure (9), we may try to apply more than one null rule. In Figure 3 we have plotted two null rule approximations, $N_{7,1}[f]$ and $N_{7,2}[f]$, as functions of x, and we note that these functions may or may not have the same zeroes. Both null rules have the same 1-norm. That is, $\| N_d \|_1 \equiv \sum_{j=1}^{p} |w_j| = 2^3$ for both rules. p

is the number of evaluation points used by the null rule and $w_j, j = 1, ..., p$, are the weights of the null rule.

Let $N_7^*[f]$ be the the greatest error that can be produced by a null rule in the space spanned by $N_{7,1}$ and $N_{7,2}$ and with the same 1-norm as $N_{7,1}$ and $N_{7,2}$. By using $N_7^*[f]$ as our error estimate, we may hope to improve the reliability of our routine.

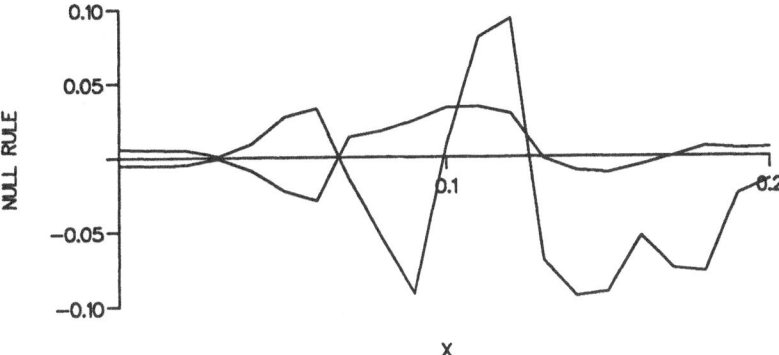

Figure 3. $N_{7,1}[f]$ and $N_{7,2}[f]$.

Figure 4 shows the actual absolute errors and the estimated absolute errors when approximating the real part of integral 1 with $N_7^*[f]$ as our error estimating procedure in DCUHRE. The absolute error request is still 0.1. From Figure 4 we see that now the actual error is less than 0.1 most of the time. The actual errors and the estimated errors are of more comparable size, but we are still being clearly misled by the error estimate 5 out of 21 times.

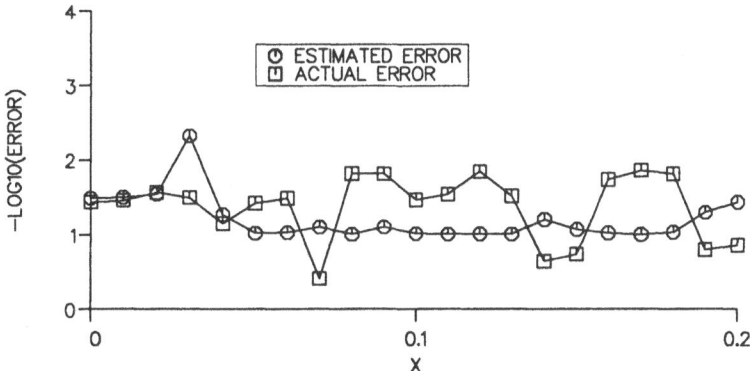

Figure 4. Absolute errors for $E[f] = N_7^*[f]$.

We may produce a sequence of null rules, $N_{7,1}, N_{7,2}, N_5$ and N_3, all having the same 1-norm. Let $N_5^*[f]$ be the greatest error that can be produced by a null rule in the space spanned by $N_{7,2}$ and N_5 and let $N_3^*[f]$ be the greatest error that can be produced by a null rule in the space spanned by N_5 and N_3. The following error estimating procedure is then implemented in order to further improve the reliability.

256

If $N_7^*[f] < N_5^*[f] < N_3^*[f]$, then
  the asymptotic assumption behind (6) is believed to be satisfied
  and we accept
  $E[f] = N_7^*[f]$ as our error estimate.
Otherwise
  $E[f] = max(N_7^*[f], N_5^*[f], N_3^*[f])$
  which is the greatest error that can be produced by
  a null rule in the space spanned by all our null rules.

Figure 5 shows the actual absolute errors and the estimated absolute errors when we approximate integral 1 with the algorithm above as the error estimating procedure in DCUHRE.

By looking at the sequence of figures presented in this section we see that we may trust the error estimates to a far greater extent when our new error estimating procedure is implemented. This is earlier documented for several test families of integrands in [4, 8, 9, 18].

The use of sequences of null rule approximations in automatic cubature routines was first implemented into the routine CADCUB[5]. CADCUB could only integrate over 3-dimensional hyperrectangles and was used by the scientists in acoustics at the University of Bergen in the period before DCUHRE became available.

In this paper we show that our new error estimating procedure also may help to considerably improve the reliability when our cubature routines are asked to approximate the more complicated integrals that appear in applied sciences. We must admit that we may have to pay for the increased reliability in terms of an increase in the number of integrand evaluations we need in order to satisfy our error requests. However, the cost of CPU cycles has been considerably reduced the last few years and most applied scientists are willing to spend the extra computer time to get a result they can trust. The cost of introducing the new error estimating procedures is also greatest in those cases where it is most needed. We will emphasize that the same reliability could <u>not</u> have been achieved to the same cost by scaling the 'old' error estimate.

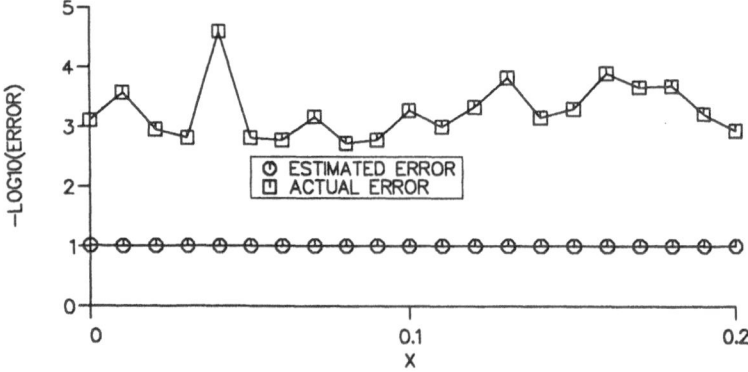

Figure 5. Absolute errors when applying DCUHRE.

In the more recent routines DCUTRI[16] and DCUTET we try to extrapolate the error estimates to the order of the actual errors in the cubature rule approximations when we are asymptotic. This way our routines may be even more economic than routines based on the error estimating procedure (5) for some problems. See the testing reports [10, 11, 14, 18] for further details on the performance.

5. THE ALGORITHM

In this section we will focus on some features that should be implemented into automatic integration routines. First of all a routine must be able to integrate a vector integrand in one call to the routine. One of the reasons for this is that we may save a lot of computer time by utilizing such a feature. If we for instance look at the integrals (1), we see that the argument of E_1 is independent of x. This means that when we are producing beam patterns (z constant, x variable), and apply a routine with this feature, we need to evaluate E_1 only once for all integrand evaluation points when we vary x. For this particular integral E_1 is also the most time consuming part of the integrand. There are assembler codes for the other parts of the integrand on most machines, but in order to approximate E_1 we must typically apply a FORTRAN code.

On the other hand , this feature must be used with caution. For instance, if we utilize this feature when approximating integral (3) for a number of observation points, the algorithm chooses the same subdivision for all values of $\underline{x}$. This means that we may easily spend far more integrand evaluations than strictly necessary to satisfy our error request if the difficulty of the integrand varies significantly with $\underline{x}$.

Secondly, Lyness and Kaganove [36] have shown that even if integrals of type (3) are continuous functions of $\underline{x}$, the approximations produced by automatic integrators may not be continuous. In order to illustrate this also for problems from acoustics we show in Figure 6 the actual error produced by DCUHRE when approximating integral (1) for the parameters used in section 4, except we now approximate the integral for $x = 0.(0.001)0.2$. We call DCUHRE separately for each value of x requesting an absolute error of 0.1.

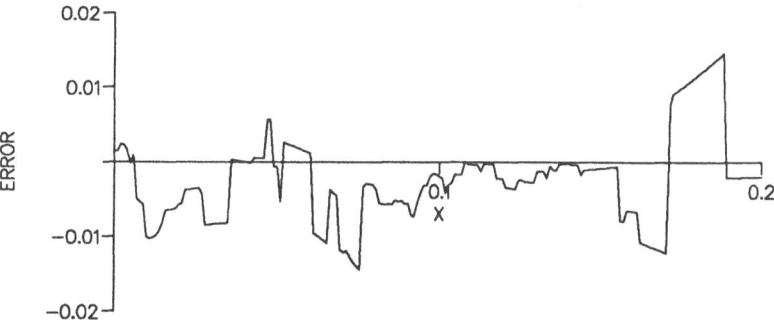

Figure 6. The actual error with 201 separate calls to DCUHRE.

We know that $q_+(z, x)$ is a continuous function of x. However, because of the discontinuous behaviour of the actual error, the approximations computed this way will be discontinuous as functions of the observation points. Very often the approximations to $q_+(z, x)$ are used as input into further computations. For instance when we follow the sound beam defined by (1) from one medium to another, $q_+(z,x)$ will appear as a source function in the integrand of a new multidimensional integral. If the approximations to $q_+(z,x)$ then are discontinuous, it may be very hard to approximate the second integral. We may therefore try to approximate $q_+(z,x)$ for the 201 values of x by calling DCUHRE once for the whole collection of integrals. Figure 7 shows the actual absolute errors when approximating the integrals this way.

We note that as expected the actual error appears to be a continuous function of x. Measured with a fine enough measure it will of course be discontinuous as long as the numbers come from a digital computer. However, compared to the accuracy we are requesting in the next step of the computation, the actual error will appear to be continuous.

I hope that the discussion above has illustrated that if we produce software for problem (4), we are taking the integrals out of the context where they appear and we may end up with software that does not meet the requirements of applied scientists.

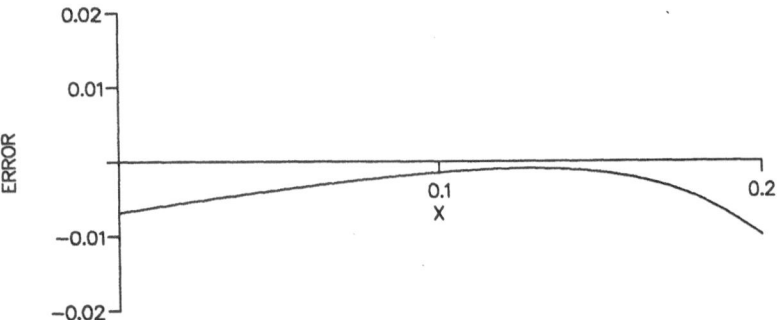

Figure 7. The actual error with one call to DCUHRE.

Secondly, automatic integration routines should be able to integrate over a collection of subregions in one call to the routine. For the second class of integrals discussed in section 2 we have already motivated the need for routines that can integrate over a collection of hyperrectangles and/or a collection of tetrahedrons.

For the integrals (1) it will also be an advantage to have a routine with this feature. For piston sources the integrals have singularities along the diagonal of one of the faces of the hyperrectangle. An automatic routine for hyperrectangles therefore chooses a subdivision with numerous small hyperrectangles along this diagonal. We may divide our hyperrectangle into a number of subhyperrectangles and subtetrahedrons such that the singularity is placed at the edges of subtetrahedrons. We believe that a routine that may integrate over such a collection of regions, would be able to integrate the integrals (1) more efficiently than the present codes. Also for the integrals (2) it would be very interesting to try a routine having this feature. Therefore it would be a great advantage in many applications to have a rouutine that may integrate over a mixture of n-dimensional simplices and n-dimensional hyperrectangles.

6. THREE NEW SUBROUTINES

In this section we will present three automatic adaptive cubature routines and in particular we will focus on the error estimating procedures of these routines.

The work with the first routine, DCUHRE [20, 21], started in 1986. DCUHRE can compute approximations of integrals over one n-dimensional hyperrectangular region at the time. The number of dimensions n may vary from 2 to 15, and the upper limit can be increased by changing one internal parameter in the code. DCUHRE can integrate a vector integrand in one call to the routine. In each subdivision step of the algorithm DCUHRE divides the subregion with largest estimated error along the direction with largest fourth divided difference. The users of DCUHRE are offered 4 different cubature rules:

1. Degree 7 Genz and Malik rules [33] in all dimensions.
2. Degree 9 Genz and Malik rules in all dimensions.
3. A 65 point degree 13 rule due to Eriksen [26] in 2 dimensions.
4. A 127 point degree 11 rule due to Berntsen and Espelid [13] in 3 dimensions.

DCUHRE represent our first attempt to introduce sequences of null rules into cubature routines published in international journals. DCUHRE apply quadrature rules of degree $2m + 1$ for $m = 3, 4, 5$ and 6. For each rule we have constructed a sequence of null rules of degree $N_{2m-1,1}, N_{2m-1,2}, N_{2m-3}$ and N_{2m-5}. The 1-norm for all null rules is equal to 2^n. Two and two null rule approximations are combined in the same way as described for the degree 9 rule in section 4 to produce $N^*_{2m-1}[f], N^*_{2m-3}[f]$ and $N^*_{2m-5}[f]$. Furthermore we compute:

$r_i = N^*_{2m-i}[f]/N^*_{2m-2-i}[f], i = 1, 3.$
$r = max(r_1, r_3)$

The basic idea of the error estimating procedure can then be described by the following algorithm:

If $r > c < 1$ (non-asymptotic case) then
 $E[f] = max(N^*_{2m-1}[f], N^*_{2m-3}[f], N^*_{2m-5}[f]).$
elseif $r \leq c$ (asymptotic case) then
 $E[f] = N^*_{2m-1}[f]$

For further details see [20]. DCUHRE has been presented to the acoustic society at an international conference in 1990, see [19], and is now being used by a number of scientists in acoustics.

The second routine, DCUTRI[16], can integrate over a collection of triangles. This feature has previously been implemented by Kahaner and Rechard into TWODQD[34]. DCUTRI can integrate a vector integrand and at each subdivision step of the algorithm DCUTRI divides the subregion with largest estimated error into four congruent sub-triangles.

The cubature rule used in DCUTRI is of degree 13 and uses 37 evaluation points. The weights are positive. The rule was constructed by Berntsen and Espelid [15] in 1990. However, a rule in the same class of rules was constructed by Dunavant[25] in 1985.

Before we started the work on DCUTRI, Espelid[27] studied the use of null rules. He showed that if we compute orthogonal null rules scaled according to the sizes of their 2-

norms, the computation of the gratest null rule approximation that can be produced by a null rule in the linear space spanned by two orthogonal null rules is considerably simplified. Before starting on DCUTRI we also studied error estimating in one dimensional quadrature routines [17], and the error estimating procedure in DCUTRI is to a great extent inspired by our one dimensional studies.

Based on the same evaluation points as used by the cubature rule, we can produce a sequence of orthogonal null rules, $N_i, i = 1, 2, ..., 8$. The 2-norm, $\| N_i \|_2 = \sum_{j=1}^{p} w_j^2$, is equal to the corresponding 2-norm of the cubature rule for all null rules. N_1 and N_2 are of degree 7, N_3 and N_4 are of degree 5, N_5 is of degree 4, N_6 is of degree 3, N_7 is of degree 2 and N_8 are of degree 1. For a given triangle and function we compute:

$$e_i = N_i[f], i = 1, 2, ..., 8.$$
$$E_j = (e_{2j-1}^2 + e_{2j}^2)^{\frac{1}{2}}, j = 1, 2, 3, 4,$$

will then by the greatest errors that can be produced by a rule in the space spanned by N_{2j-1} and N_{2j} and with the same norm as N_{2j-1} and N_{2j}. Next we define the reduction factors:

$$r_i = E_i/E_{i+1}, i = 1, 2, 3.$$
$$r = max(r_1, r_2, r_3)$$

Our error estimating procedure can then be written:

If $r > 1$ (non-asymptotic case) then
$\qquad E = 10 * max(E_1, E_2, E_3, E_4).$
elseif $1/2 < r \leq 1$ (weak-asymptotic case) then
$\qquad E = 10 * r * E_1$
elseif $r \leq 1/2$ (asymptotic case) then
$\qquad E = 40 * r^3 * E_1$

The results of the test reports [10] and [14] indicate that DCUTRI is as economic in terms of the number of integrand evaluations needed to satisfy a given error request as previously published routines, and at the same time more reliable. One major reason for this is that in the asymptotic case we extrapolate to the order of the actual error in the cubature rule approximation.

More general regions can be more easily approximately represented by a union of triangles than by a union of rectangles. Therefore, for many 2-dimensional integration problems, including those arising in acoustics, DCUTRI will be a more useful integration routine than DCUHRE.

The third routine, DCUTET[12], can integrate over a collection of tetrahedrons. DCUTET can integrate a vector integrand, and at each subdivision step DCUTET divides the subregion with largest estimated error into 8 congruent sub-tetrahedrons. The cubature rule of degree 8 is due to Beckers and Haegemans[1] and uses 43 evaluation points. The error estimating procedure is similar to the procedure used in DCUTRI, but we are now only able to produce a sequence of 6 null rules based on the same evaluation points as the cubature rule. N_1 is of degree 5, N_2 and N_3 are of degree 4, N_4 is of degree 3, N_5 is of degree 2 and N_6 is of degree 1. For a given tetrahedron and function we compute:

$$e_i = N_i[f], i = 1, 2, ..., 6.$$

$$E_j = (e_{2j-1}^2 + e_{2j}^2)^{\frac{1}{2}}, j = 1, 2, 3.$$

Again we define the reduction factors:

$$r_i = E_i/E_{i+1}, i = 1, 2.$$
$$r = max(r_1, r_2)$$

Our error estimating procedure may then be written:

If $r \geq 1/2$ (non-asymptotic case) then
$$E = 5 * r * E_1$$
elseif $r < 1/2$ (asymptotic case) then
$$E = 10 * r^2 * E_1$$

DCUTET is to our knowledge the first globally adaptive cubature routine that can integrate over 3-D simplices. Therefore, as demonstrated in section 2, DCUTET can be applied to approximate integrals that can not be handled by any other presently available routine in a straightforward manner.

In literature there has been some focus on the construction of pairs of cubature rules of degree $2m+1/2m-1$, see [13, 33]. The construction of such pairs is a more difficult task than to construct rules of degree $2m+1$ without further requirements, and we typically need more evaluation points when constructing the pairs. The reason for wanting a rule of reference with degree $2m-1$ is connected to the use of the absolute value of the difference between two cubature rule approximations as an error estimating procedure. If the difference in degree becomes too large, the error estimate will very often be many orders of magnitude larger than the actual error and the cost of using the error estimate described above will be prohibitive.

In DCUTRI and DCUTET we are filling the gap in degree between the cubature rule approximation and the rule of reference of highest degree by using extrapolation. Therefore, with the new error estimating procedures the need for pairs of rules of specific degrees is considerably diminished. For each cubature rule we produce a sequence of null rules based on the same set of points as used by the cubature rule.

All three routines are structured to allow efficient implementation on shared memory parallel computers and good speedups are reported on such computers, see [11, 14, 18].

7. FINAL REMARKS

Finally I will try to point at some areas of research that need further attention if we want to develop software that may meet the requirements of applied scientists.

First of all there is a great need for general purpose automatic integration routines for a variety of regions. Applied scientists tend to come up with integrals over all kinds of regions. Therefore we need routines for the general n-simplex, for the n-sphere, for the surface of the n-sphere etc.

It has been documented that routines based on the simple error estimating procedure (5) may be very unreliable. We have shown that one way of improving the reliability is to apply sequences of null rule approximations in the error estimating procedure. Therefore, we believe that to introduce error estimating procedures similar to those used in DCUTRI and DCUTET will help improving the reliability also for automatic routines over other regions.

There are two algorithmic features that as we have seen, should be implemented in new pieces of software and that are fairly easy to include. The routines must be able to integrate vector integrands in <u>one</u> call to the routine. An automatic routine should be able to integrate over a collection of subregions in <u>one</u> call to the routine.

When developing software for automatic integration, the state of the art way of providing numerical evidence that one piece of software works better than another in some respect is to run Lyness and Kaganove [37] type performance profile tests. To run such tests requires a lot of both computer time and human time. However, by applying this technique we have been able to improve considerably the understanding of how many well known automatic integration routines perform (see [3, 6, 7, 10, 11, 14, 18]). The insight gained has helped us reveal the least developed ingredient of automatic adaptive algorithms, the error estimating procedure, and when trying to improve this ingredient or any other ingredient, testing routines based on this technique are neccessary working tools.

In this paper I have focussed on the numerical calculation of multidimensional integrals appearing in acoustics by applying general purpose globally adaptive cubature routines. By having high quality general purpose routines available applied scientists at least have some pieces of software that they may try to apply to approximate their integrals. However, for specific classes of integrals it will of course be much more efficient to apply integration routines that take advantage of the special nature of that class. If we for instance go back to integral (1), we have not been able to approximate these integrals for small values of z, large values of x and for frequencies in the megahertz range. By taking into account the error expansion for the cubature rule approximation and by using extrapolation techniques, we would probably be able to approximate integral (1) for a larger class of parameters than we do with present techniques.

ACKNOWLEDGEMENTS

There are a number of persons that have contributed very much to the work I have been doing the last 10 years. I will thank Terje Espelid for the support during the work with my master thesis and PhD thesis. I have also appreciated very much the close and inspiring collaboration we have had. I am greatful to Jacqueline Naze Tjøtta and Sigve Tjøtta for their continued interest in our work and for their support during these years. Alan Genz was the first non-Norwegian scientist I got in contact with and I will thank him for the help and advice he so freely has offered. I will thank all the scientists that have visited our group in Bergen during these years for contributing with their experience and knowledge.

References

[1] M. Beckers and A. Haegemans. The construction of cubature formulae for the tetrahedron. Report TW 128, Katholieke Universiteit Leuven, Belgium, 1990.

[2] J. Berntsen. User documentation, Program HALF, A subroutine for numerical evaluation of three-dimensional complex integrals. Dept. of Math., Univ. of Bergen, 1983.

[3] J. Berntsen. On the subdivision strategy in numerical adaptive integration over the cube. Reports in Informatics 11, Dept. of Inf., Univ. of Bergen, Norway, 1984.

[4] J. Berntsen. Cautious adaptive numerical integration over the 3-cube. Reports in Informatics 17, Dept. of Inf., Univ. of Bergen, Norway, 1985.

[5] J. Berntsen. Program CADCUB, A program for numerical evaluation of three-dimensional integrals. Dept. of Inf., Univ. of Bergen, 1985.

[6] J. Berntsen. A test of the NAG-software for automatic integration over the 3-cube. Reports in Informatics 15, Dept. of Inf., Univ. of Bergen, Norway, 1985.

[7] J. Berntsen. A test of some wellknown quadrature routines. Reports in Informatics 20, Dept. of Inf., Univ. of Bergen, Norway, 1986.

[8] J. Berntsen. Practical error estimation in adaptive multidimensional quadrature routines. Reports in Informatics 30, Dept. of Inf., Univ. of Bergen, Norway, 1988.

[9] J. Berntsen. Practical error estimation in adaptive multidimensional quadrature routines. J.Comp.Appl.Math., 25:327–340, 1989.

[10] J. Berntsen. TRITST: A Subroutine for Evaluating the Performance of Subroutines for Automatic Integration over Triangles. Reports in Informatics 34, Dept. of Inf., Univ. of Bergen, 1989.

[11] J. Berntsen, R. Cools, and T.O. Espelid. A Test of DCUTET. Reports in Informatics 46, Dept. of Inf., Univ. of Bergen, 1990.

[12] J. Berntsen, R. Cools, and T.O. Espelid. An algorithm for automatic integration over a collection of 3-dimensional simplices. To appear, 1991.

[13] J. Berntsen and T.O. Espelid. On the construction of higher degree three-dimensional embedded integration rules. SIAM J. Numer. Anal., 25:222–234, 1988.

[14] J. Berntsen and T.O. Espelid. A Test of DCUTRI and TWODQD. Reports in Informatics 39, Dept. of Inf., Univ. of Bergen, 1989.

[15] J. Berntsen and T.O. Espelid. Degree 13 symmetric quadrature rules for the triangle. Reports in Informatics 44, Dept. of Inf., Univ. of Bergen, 1990.

[16] J. Berntsen and T.O. Espelid. DCUTRI:An Algorithm for Adaptive Cubature over a Collection of Triangles. To appear in ACM Trans. Math. Softw., 1991.

[17] J. Berntsen and T.O. Espelid. Error estimation in automatic quadrature routines. To appear in ACM Trans. Math. Softw., 1991.

[18] J. Berntsen, T.O. Espelid, and A. Genz. A Test of ADMINT. Reports in Informatics 31, Dept. of Inf., Univ. of Bergen, 1988.

[19] J. Berntsen, T.O. Espelid, and A. Genz. An automatic integration routine applicable in linear and nonlinear acoustics. In M.F.Hamilton and D.T.Blackstock, editors, *Frontiers of Nonlinear Acoustics:Proceedings of the 12th ISNA*. Elsevier Science Publishers Ltd., 1990.

[20] J. Berntsen, T.O. Espelid, and A.C. Genz. An Adaptive Algorithm for the Approximate Calculation of Multiple Integrals. To appear in ACM Trans. Math. Softw., 1991.

[21] J. Berntsen, T.O. Espelid, and A.C. Genz. An Adaptive Multidimensional Integration Routine for a Vector of Integrals. To appear in ACM Trans. Math. Softw., 1991.

[22] J. Berntsen, J. Naze Tjøtta, and S. Tjøtta. Nearfield of a large acoustic transducer. Part IV: Second harmonic and sum frequency radiation. *J.Acoust.Soc.Am.*, 75:1383–1391, 1984.

[23] J. Berntsen, J. Naze Tjøtta, and S. Tjøtta. Interaction of sound waves. Part IV: Scattering of sound by sound. *J.Acoust.Soc.Am.*, 86:1968–1983, 1989.

[24] P.J. Davis and P. Rabinowitz. *Methods of Numerical Integration*. Academic Press, 1984.

[25] D.A. Dunavant. High degree efficient symmetrical gaussian quadrature rules for the triangle. *Int.J.Numer.Methods Eng.*, 21:1129–1148, 1985.

[26] S.S. Eriksen. On the development of embedded fully symmetric quadrature-rules for the square. Master's thesis, Dept. of Inf., Univ. of Bergen, 1986.

[27] T.O. Espelid. Integration rules, null rules and error estimation. Reports in Informatics 33, Dept. of Inf., Univ. of Bergen, 1988.

[28] K.G. Foote, J. Naze Tjøtta, and S. Tjøtta. Performance of the parametric receiving array. Effects of misalignment. *J.Acoust.Soc.Am.*, 82:1753–1757, 1987.

[29] K.E. Frøysa. *Linear and weakly nonlinear propagation of a pulsed sound beam*. PhD thesis, Dept. of Applied Math., Univ. of Bergen, Norway, 1991.

[30] G.S Garrett, J. Naze Tjøtta, R.L. Rolleigh, and S. Tjøtta. Reflection of parametric radiation from a finite planar target. *J.Acoust.Soc.Am.*, 75:1462–1472, 1984.

[31] G.S Garrett, J. Naze Tjøtta, and S. Tjøtta. Nearfield of a large acoustic transducer. Part II: Parametric radiation. *J.Acoust.Soc.Am.*, 74:1013–1020, 1983.

[32] G.S Garrett, J. Naze Tjøtta, and S. Tjøtta. Nearfield of a large acoustic transducer. Part III: General results. *J.Acoust.Soc.Am.*, 75:769–779, 1984.

[33] A.C. Genz and A.A. Malik. An imbedded family of fully symmetric numerical integration routines. *SIAM J. Numer. Anal.*, 20:580–588, 1983.

[34] D.K. Kahaner and O.W. Rechard. TWODQD an adaptive routine for two-dimensional integration. *J.Comp.Appl.Math.*, 17:215–234, 1987.

[35] J.N. Lyness. Symmetric integration rules for hypercubes III. Construction of integration rules using null rules. *Math.Comp.*, 19:625–637, 1965.

[36] J.N. Lyness and J.J. Kaganove. Comments on the nature of automatic quadrature routines. *TOMS*, vol 2, no. 1:65–81, 1976.

[37] J.N. Lyness and J.J. Kaganove. A technique for comparing automatic quadrature routines. *Computer J.*, vol 20:170–177, 1977.

[38] J. Naze Tjøtta, J.A. TenCate, and S. Tjøtta. Effects of boundary conditions on the nonlinear interaction of sound beams. To appear in J.Acoust.Soc.Am., 1990.

[39] J. Naze Tjøtta and S. Tjøtta. Sound field of a parametric focusing source. *J.Acoust.Soc.Am.*, 75:1392–1394, 1984.

[40] A.H. Stroud. *Approximate calculation of multiple integrals.* Prentice-Hall, 1971.

STATISTICS APPLICATIONS OF SUBREGION ADAPTIVE MULTIPLE NUMERICAL INTEGRATION

ALAN GENZ
School of Electrical Engineering and Computer Science
Washington State University
Pullman, WA 99164-2752 USA
Email: acg@yoda.eecs.wsu.edu

ABSTRACT. Subregion adaptive integration algorithms can be used for the accurate and efficient solution of numerical multiple integration problems in statistics. The key to good solutions for these problems is the choice of an appropriate transformation from the infinite integration region for the original problem to a suitable finite region for the adaptive algorithm. After a discussion of some different types of transformations, several examples are presented to illustrate the effectiveness of the combination of a good transformation choice with a subregion adaptive integration algorithm for solving statistics integration problems.

1 Introduction

Statistical analysis in a wide variety of applications requires evaluation of multidimensional integrals in the form

$$I(f) = \int_R f(\boldsymbol{\theta})p(\boldsymbol{\theta})d\boldsymbol{\theta}$$

The integration region R is usually infinite, and the number of dimensions m can be large. The probability density function $p(\boldsymbol{\theta})$ is assumed nonnegative. In many cases it has a global maximum at a unique $\boldsymbol{\theta}$ in R. When p is unnormalized, I(1) is needed. The f function is usually uncomplicated, and the most common f of interest is the vector $\mathbf{f} = \boldsymbol{\theta}$, which can be used to obtain expected values for the $\boldsymbol{\theta}$ variables, $\bar{\boldsymbol{\theta}} = I(\boldsymbol{\theta})/I(1)$. Another common f is the matrix $F = \boldsymbol{\theta}\boldsymbol{\theta}^t$, which is used to obtain the covariance matrix for $\boldsymbol{\theta}$ (here, and throughout this paper, vectors are column vectors, so that $\boldsymbol{\theta} = (\theta_1, \theta_2, ..., \theta_m)^t$). Some good review references for statistics integration problems are the papers by Naylor and Smith (1982 and 1988) and Shaw (1988).

For statisticians, Monte-Carlo methods have been the most popular numerical integration methods for these integrals. However, many of these problems are difficult, so crude Monte-Carlo methods often cannot provide timely and accurate results. Therefore a considerable amount of research work has gone into the development of better Monte-Carlo methods. Most of this research has studied different types of importance sampling techniques (see Davis and Rabinowitz, 1984, for a good overview of these). A lot of recent work has focussed on Gibb's sampling and related methods (see Metropolis et. al., 1953, Geweke, 1989, Ogata, 1989, Gelfand and Smith, 1990 and Wolpert, 1991).

T. O. Espelid and A. Genz (eds.), Numerical Integration, 267–280.

Running somewhat in parallel with the development of Monte-Carlo methods, research work by numerical analysts has focussed more on the development of better integration rules and the incorporation of these rules into efficient algorithms. The most successful of these algorithms are the *subregion adaptive* algorithms (see Genz, 1991a, for an overview), where the initial integration region is dynamically subdivided to adapt to the local irregularities of the integrand. This approach is to be contrasted with the adaptive algorithms often preferred by statisticians, which might be called the *distribution adaptive* algorithms. In these algorithms, the sample points for the integrand evaluations are dynamically selected to approximate the statistical distribution defined using the integrand (or part of the integrand) as a probability density function.

A prevailing assumption among both numerical analysts and statisticians has been that the subregion adaptive algorithms could be effective only for problems with low dimensionality (e.g. $m = 1 - 5$), and that the distribution adaptive algorithms were the algorithms of choice for higher dimensional problems. While this assumption is to some extent still valid, the purpose of this paper is to show how subregion adaptive algorithms can be used more effectively for the solution of statistics problems. A key part of the solution of a statistics integration problem using a subregion adaptive algorithm is the careful selection of a transformation from the infinite integration region R for the original problem, to an appropriate finite region for the adaptive algorithm. This paper begins with a brief overview of subregion adaptive integration. This is followed with a discussion of transformation methods. Finally, three examples are used to show how the combination of an appropriate transformation and a subregion adaptive algorithm can provide efficient and accurate solutions to some multivariate integration problems in statistics.

2   Subregion Adaptive Integration

Subregion adaptive integration methods are based on the fundamental assumption that the integrand in the problem of interest can be accurately approximated locally by a low degree multivariate polynomial. The basic strategy for a subregion adaptive algorithm is to dynamically subdivide the initial integration region R into smaller and smaller subregions that are concentrated in the parts of R where the integrand is more irregular. The hope is that at some stage in this process the region R is sufficiently well-partitioned that the combined integrated polynomial approximations for all of the subregions provide an accurate approximation to the initial integral. Typical input for this type of algorithm consists of (i) a description of the initial integration region R, (ii) the integrand, (iii) an error tolerance ϵ and (iv) a limit k_{max} on the total number of subregions allowed (this can either be used to limit the maximum time allowed or to limit the maximum workspace allowed).

The adaptive algorithm itself also requires a basic integration rule (or formula) B and an associated error estimation rule E. Let B_i be the approximation to the integral in a subregion R_i obtained using the basic rule, and let E_i be the estimate for the absolute error in B_i. If at some stage in the algorithm R has been subdivided into k subregions, the relevant pieces of information are kept in a list $S = \{(R_1, B_1, E_1), (R_2, B_2, E_2), ..., (R_k, B_k, E_k)\}$. Initially $R_1 = R$ with $k = 1$, B_1 and E_1 are computed. There are many possible adaptive strategies that may be used to dynamically refine the list S. A generic *globally* adaptive algorithm has a main loop with the following form:

while $(\sum_{i=1}^{k} E_i > \epsilon$ **and** $k < k_{max})$ **do**

    a) determine j with $E_j = max_{i=1}^{k} E_i$

    b) subdivide R_j into p pieces

    c) compute the integral and error estimates for the new pieces of R_j

    d) update the subregion list and set $k = k + p - 1$

end while

The final output from the algorithm is an estimate $\sum_{i=1}^{k} B_i$ for the integral, and an error estimate $\sum_{i=1}^{k} E_i$.

Software using this type of adaptive algorithm was first developed by van Dooren and de Ridder (1976, Fortran implementation HALF). A key feature of this software was the subdivision step b), where the selected subregion is divided in half along the coordinate axis where the integrand is (locally) most rapidly changing. This clever strategy allows the growth in computational time to be increased in a controlled manner and the adaptation to occur only in those variables which cause most of the variation in the integrand. An improved version of HALF was developed by Genz and Malik (1980, Fortran implementation ADAPT), where a better basic rule and subregion data structure were introduced. A slightly modified version of ADAPT was placed in the NAG library (1981, D01FCF). A vector version of ADAPT was also placed in the NAG library (1984, D01EAF). This allowed a vector of integrals over a common region to be computed simultaneously. More recently, an improved subregion adaptive algorithm for a vector of integrals was developed by Berntsen, Espelid, and Genz (1991a, with Fortran software SCUHRE, 1991b). This algorithm uses a more robust error estimation method and allows the user to select different basic rules. These vector algorithms are particular useful in statistics applications because all of the integrals for the different $f(\boldsymbol{\theta})$ functions can be computed simultaneously.

3 Transformations

3.1 Background

Subregion adaptive integration algorithms require repeated subdivision of the integration region. The globally adaptive algorithm described in the previous section uses a strategy that chooses a priority subregion to be halved along some coordinate axis. When the integration region is infinite, the point selected along the axis where the current subregion is to be cut is not clearly defined. Since any strategy for choosing these cutting points implies some transformation an infinite interval to a finite one, it is convenient to first consider some appropriate transformation from the infinite region to a finite region. Then the adaptive algorithm can be used directly on the transformed integrand over the finite integration region.

Another method that is often used to deal with an infinite integration region is to select cutoff values for the infinite integration limits. In this case the adaptive algorithm can work directly with the original integrand on the resulting truncated integration region. These cutoff values are often selected experimentally; one would like to truncate the region as much as is possible without significantly changing the final result. This approach is sometimes hard to implement, particularly when the integral is a difficult one and it is not known whether errors introduced from a particular choice of cutoff values are significant compared to errors that are present in the current estimated value of the integral. Reducing the magnitude of the cutoffs might make it easier to compute a result which is inaccurate because the region has been truncated too much. But increasing the magnitude of the cutoff values sometimes makes the problem so hard that the algorithm does not terminate in the allotted time. Because of these difficulties with cutoffs, the rest of the discussion in this section focuses on transformations.

The choice of a good transformation, like the choice of good cutoff values, is often problem dependent. But many statistics problems are formulated in ways that suggest obvious choices for transformations. And in many of these problems, a good transformation actually simplifies the integral. This is also the key idea behind many importance sampling techniques, where the aim is to choose the random integrand evaluation points in a way that provides a good approximation to random points that would be chosen from the distribution function for $p(\theta)$, if they were available.

3.2 Simple Transformations

There are a number of simple one variable transformations that have been used for integration problems when one or both of the integration limits are infinite (see Davis and Rabinowitz (1984) for further discussion and examples). Some common choices are listed below, first for the half-infinite interval product hypercube $[0, \infty]^m$, and then for the doubly-infinite interval product hypercube $[-\infty, \infty]^m$. These transformations are presented in the way in which they would be used in an adaptive algorithm. The assumption is that a point z produced by the algorithm from a finite interval ($[0,1]$ and $[-1,1]$ are used here for definiteness) is transformed into a point θ from some infinite interval.

A. Singly infinite hypercube transformations: $[0, \infty]^m \leftarrow [0, 1]^m$

    i) $\theta_i = -\alpha ln(z_i)$, $[0, \infty] \leftarrow [1, 0]$, $d\theta_i = -\alpha dz_i/z_i$,

    ii) $\theta_i = -\alpha ln(1 - z_i)$, $[0, \infty] \leftarrow [0, 1]$, $d\theta_i = \alpha dz_i/(1 - z_i)$,

    iii) $\theta_i = z_i/(1 - z_i)$, $[0, \infty] \leftarrow [0, 1]$, $d\theta_i = dz_i/(1 - z_i)^2$,

    iv) $\theta_i = (z_i/(1 - z_i))^2$, $[0, \infty] \leftarrow [0, 1]$, $d\theta_i = 2z_i dz_i/(1 - z_i)^3$.

B. Doubly infinite hypercube transformations: $[-\infty, \infty]^m \leftarrow [-1, 1]^m$

    i) Obvious generalizations of i), ii), iii) and iv) above, e.g.
        iii)' $\theta_i = z_i/(1 - |z_i|)$, $[-\infty, \infty] \leftarrow [-1, 1]$, $d\theta_i = dz_i/(1 - |z_i|)^2$,

    ii) $\theta_i = sign(z_i)(-\alpha ln(|z_i|))^{\frac{1}{2}}$, $[-\infty, \infty] \leftarrow [-1, 1]$, $d\theta_i = \alpha dz_i/|2\theta_i z_i|$,

    iii) $\theta_i = sign(z_i)(-\alpha ln(1 - |z_i|))^{\frac{1}{2}}$, $[-\infty, \infty] \leftarrow [-1, 1]$, $d\theta_i = \alpha dz_i/|2\theta_i(1 - z_i)|$,

iv) $\theta_i = tan^{-1}(\pi z_i/2)$, $[-\infty, \infty] \leftarrow [-1, 1]$, $d\theta_i = (\pi/2)dz_i/(1 + (\pi z_i/2)^2)$.

These simple transformations are often used when the choice of a better transformation is not obvious. As a check on consistency and efficiency, two or more transformations can be used for different computations of the same integral, and the results compared.

3.3 Prior Transformations for Bayesian Analysis Computations

Bayesian analysis integration problems usually have the form

$$\int_R f(\boldsymbol{\theta})p(\boldsymbol{\theta})d\boldsymbol{\theta} = \int_R f(\boldsymbol{\theta})L(\boldsymbol{\theta})\pi(\boldsymbol{\theta})d\boldsymbol{\theta}.$$

The *likelihood* function $L(\boldsymbol{\theta})$ includes the problem data and the *prior* density function $\pi(\boldsymbol{\theta})$ has been constructed using prior information about the expected statistical behavior of the data. Many of these problems (including the problem that will be discussed in Section 4.2) have a prior that is a product of commonly-occurring one or two variable density functions. In this case an obvious choice for a transformation is simply to use the appropriate cumulative distribution function to transform each of the variables.

One of the most common situations occurs when one of the integration variables θ_i, in the prior, has an associated factor $e^{-((\theta_i - \mu)/\sigma)^2/2}/(\sigma\sqrt{2\pi})$, with integration limits $-\infty$ and ∞. In this case the change of variable $\theta_i = \mu + \sigma\Phi^{-1}(z_i)$, with

$$\Phi(t) = \frac{1}{\sqrt{2\pi}} \int_{-\infty}^t e^{-y^2/2}dy,$$

removes the associated exponential factor from the prior, and the z_i variable integration limits become 0 and 1.

Another common prior factor has the form $\frac{2^{-\nu/2}}{\Gamma(\nu/2)}e^{-\theta_i/2}\theta_i^{\nu/2-1}$ with integration limits 0 and ∞. Here, the change of variable $\theta_i = \Gamma^{-1}(z_i, \nu)$, using the inverse incomplete Gamma function defined by

$$\Gamma(t, \nu) = \frac{2^{-\nu/2}}{\Gamma(\nu/2)} \int_0^t e^{-t/2}t^{\nu/2-1}dt,$$

removes the associated prior factor from the integrand.

Both of these transformations require the availability of reliable and efficient software for computation of the inverse distribution functions. A good general source for statistical computation software is *statlib*. This repository of public domain software can be easily accessed using electronic mail: the one line message "send index", sent to the Internet address "statlib@stat.cmu.edu" will return a detailed summary of the software available in *statlib*, along with instructions about how to obtain particular parts of the library. While software for many of the standard inverse distribution functions is now available, the time for computation of these functions is typically an order of magnitude longer than the time for the computation of standard elementary functions like $sin(x)$ and e^x. This means that the use of these transformations often significantly increases the evaluation time for $p(\theta)$. However, some one variable distributions functions, like the Cauchy distribution have simple inverses that are defined in terms of standard elementary functions (the appropriate transformation for this distribution has already been given as Biv) in the preceding section).

3.4   Modal Approximation Transformations

The use of a transformation of variables can be viewed as a method of preconditioning the integrand so that the transformed integral is easier to compute. From this point of view, a good transformation is one that concentrates the integrand evaluation points in the most important subregions of the integration region. The main problem with the use of only part of the integrand (e.g. the prior function) as the basis for a preconditioning transformation is that the most important part(s) of the integrand might be neglected. However, if the integrand is complicated, finding a good preconditioning transformation for the whole integrand can be difficult. In the rest of this section a fairly general whole integrand transformation method which is particularly appropriate for many statistics integrals will be described.

The general transformation method is based on the assumption (Chen, 1985) that the integrand $p(\boldsymbol{\theta})$ is approximately multivariate normal. This means $p(\boldsymbol{\theta}) \simeq K e^{-\frac{1}{2}(\boldsymbol{\theta}-\boldsymbol{\mu})^t \Sigma^{-1}(\boldsymbol{\theta}-\boldsymbol{\mu})}$ for some constant K. Here, the integration region is assumed to be $[-\infty, \infty]^m$, so that an initial transformation of the original integral might also be necessary for some problems. In most problems of interest, multivariate optimization techniques can be applied to the function $log(p(\boldsymbol{\theta}))$ to obtain the *mode* $\boldsymbol{\mu}$ (the point where the maximum of $log(p(\boldsymbol{\theta})$ is attained) and the modal *covariance* matrix Σ (the inverse of the Hessian matrix at the maximum). This optimization would usually be done numerically. Although numerical optimization for functions of many variables can be difficult and time consuming, the optimization problem here is simplified because of the unimodality assumption and a low accuracy requirement. The time for the approximate solution of this type of numerical optimization problem will usually be insignificant compared to the numerical integration time.

Once Σ and $\boldsymbol{\mu}$ are available, the Cholesky decomposition CC^t of Σ can be found. Then, using the transformation $\boldsymbol{\theta} = \boldsymbol{\mu} + C\mathbf{y}$, (and the result $\boldsymbol{\theta}^t \Sigma^{-1} \boldsymbol{\theta} = \mathbf{y}^t C^t C^{-t} C^{-1} C\mathbf{y} = \mathbf{y}^t \mathbf{y}$), followed by inverse normal transformations $y_i = \Phi^{-1}(z_i)$ on the $\mathbf{y}$ components yields

$$\int_{-\infty}^{\infty} \cdots \int_{-\infty}^{\infty} f(\boldsymbol{\theta}) p(\boldsymbol{\theta}) d\boldsymbol{\theta} = (\sqrt{2\pi})^m |C| \int_0^1 \cdots \int_0^1 f(\boldsymbol{\mu} + C\mathbf{y}(\mathbf{z})) g(\mathbf{y}(\mathbf{z})) d\mathbf{z},$$

where $\mathbf{y}(\mathbf{z}) = (\Phi^{-1}(z_1), ..., \Phi^{-1}(z_m))^t$ and $g(\mathbf{y}) = e^{\frac{1}{2}\mathbf{y}^t \mathbf{y}} p(\boldsymbol{\mu} + C\mathbf{y})$.

This general transformation is often more straightforward to apply than one that would consist of separate (prior based) transformations for each of the integration variables. In addition, as long as the stated assumption of approximate normality is correct, this method should be reasonably efficient because the transformed integrand should be approximately constant.

4 Examples

4.1 The Cumulative Normal Distribution

A problem that arises in many statistics applications is that of computing the cumulative normal distribution function

$$F(\mathbf{a}) = \frac{1}{\sqrt{|\Sigma|(2\pi)^m}} \int_{-\infty}^{a_1} \int_{-\infty}^{a_2} \cdots \int_{-\infty}^{a_m} e^{-\frac{1}{2}\boldsymbol{\theta}^t \Sigma^{-1} \boldsymbol{\theta}} d\boldsymbol{\theta},$$

where Σ is an $m \times m$ symmetric positive definite covariance matrix. There is reliable and efficient software available for the cases $m = 1$ and $m = 2$ (see *statlib* and, for m = 2, the recent paper by Drezner and Wesolowsky, 1990), so assume $m > 2$. If the a_i's are all ∞, the transformation $\boldsymbol{\theta} = C\mathbf{y}$, where CC^t is the Cholesky decomposition of Σ, reduces Φ to the product of one dimensional integrals, all with value one. When some of the a_i's are ∞, an appropriate transformation similarly allows the associated variables to be integrated out and reduces the number of variables in the problem, so assume that all a_i's are finite. An algorithm by Schervish (1984) is implemented for $m < 8$. This algorithm uses a locally adaptive strategy based on Simpson's rule. It also provides a true error bound for F. However, tests which will be briefly reported at the end of this section show that this algorithm can require very long running times for $m > 4$. In the remainder of this section an appropriate set of transformations will be given to put the problem into a form that allows reliable and efficient computation of F for m as large at 10 using subregion adaptive integration.

A sequence of three transformations will be used to transform the original integral into an integral over a unit hyper-cube. This sequence begins with a Cholesky decomposition transformation $\boldsymbol{\theta} = C\mathbf{y}$. Since $\boldsymbol{\theta}^t \Sigma^{-1} \boldsymbol{\theta} = \mathbf{y}^t C^t C^{-t} C^{-1} C \mathbf{y} = \mathbf{y}^t \mathbf{y}$, $d\boldsymbol{\theta} = |C| d\mathbf{y}$ and $|C| = |\Sigma|^{\frac{1}{2}}$,

$$F(\mathbf{a}) = \frac{1}{\sqrt{(2\pi)^m}} \int_{-\infty}^{b_1} e^{-\frac{y_1^2}{2}} \int_{-\infty}^{b_2(y_1)} e^{-\frac{y_2^2}{2}} \cdots \int_{-\infty}^{b_m(y_1,y_2,\ldots,y_{m-1})} e^{-\frac{y_m^2}{2}} d\mathbf{y},$$

with $b_i = (a_i - \sum_{j=1}^{i-1} c_{ij} y_j)/c_{ii}$. If each of the y_i's are now transformed separately using $y_i = \Phi^{-1}(z_i)$, $F(\mathbf{a})$ becomes

$$F(\mathbf{a}) = \int_0^{e_1} \int_0^{e_2(z_1)} \cdots \int_0^{e_m(z_1,z_2,\ldots,z_{m-1})} d\mathbf{z},$$

with $e_i = \Phi((a_i - \sum_{j=1}^{i-1} c_{ij} \Phi^{-1}(z_j))/c_{ii})$. The integral in this form is much simpler than the original integrand. However, the integration region is much more complicated, and cannot be handled directly with subregion adaptive software. This integral is put in constant limit form using $z_i = e_i w_i$, so that

$$F(\mathbf{a}) = e_1 \int_0^1 c_2(w_1) \int_0^1 c_3(w_1, w_2) \cdots \int_0^1 e_m(w_1, w_2, \ldots, w_{m-1}) d\mathbf{w},$$

with $e_i(w_1, w_2, \ldots, w_{i-1}) = \Phi((a_i - \sum_{j=1}^{i-1} c_{ij} \Phi^{-1}(e_j(w_1, w_2, \ldots, w_{j-1}) w_j))/c_{ii})$.

The inner integral over w_m can be removed (this inner integral is in fact done directly using the one variable cumulative normal distribution function Φ), because e_m has no dependence on w_m, so this sequence of transformations has reduced the number of integration variables by one. If efficient software for the bivariate normal distribution function is available, a similar, but slightly modified sequence of transformations, can be used to reduce the number of integration variables by two. In this case the last factor in the integrand will be a bivariate normal distribution function but the complete integrand will be more difficult.

The sequence of transformations described here might appear to have made the final integrand more complicated. However, the overall transformation has forced a priority ordering on the integration variables. The w_1 variable is the most important, because all of the integrand factors e_i for $i > 1$ depend on it. The w_2 variable is the next most important, and so on. This feature of the problem actually makes the problem more suitable for the type subregion adaptive algorithm described in Section 2. This type of adaptive algorithm subdivides along one coordinate axis at a time, and therefore works most efficiently when the integrand has a few priority variables, which define the directions of most variation in the integrand.

In order to test this transformation method, 25 random covariance matrices were generated (using a method described in Marsaglia and Olkin, 1984) for $m = 3 - 6, 8, 10$. The upper limits were chosen using $a_i = r_i, i = 1, 2, ..., m$, where the r_i's were different for each covariance matrix and were uniformly distributed random numbers chosen from the interval [-1,4]. Subregion adaptive integration software (a code similar to the one given by Genz and Malik, 1980) was used to compute $F(\mathbf{a})$ for each test, with requested absolute accuracy set at 0.005. For $m = 3$ and $m = 4$ the Schervish (1984) software was used for comparison. A Digital DECstation 3100 ($\simeq$ 14 mips) was used for the tests. For $m > 4$, however, the Schervish software required widely varying times for each test and would not terminate after 50 hours for one of the tests, so it could not be used for $m > 4$. In Table 4.1.1 the results for these tests are reported. In this table and the one following, A represents a result from the subregion adaptive algorithm, S a result from the Schervish algorithm, and the reported times are in seconds.

Table 4.1.1: Random Covariance Matrix Results

m	Max(A-S)	Ave. S Time	Ave. A Time	Ratio
3	0.001768	0.010	0.007	1.5
4	0.004207	5.672	0.020	287.7
5	-	-	0.073	-
6	-	-	0.256	-
8	-	-	0.822	-
10	-	-	6.623	-

These results indicate that the transformation used here when combined with a good subregion adaptive method should provide a reliable and efficient method for computing $F(\mathbf{a})$ for m as large as 10. As a further check on this approach a second test was done using a series of nine fixed covariance matrices Σ for each m. These matrices had all diagonal entries equal to 1 and all offdiagonal entries equal to ρ, for $\rho = 0.1, 0.2, ..., 0.9$. These tests all used $\mathbf{a} = (1, 1, ..., 1)^t$ and the absolute accuracy requirement was 0.005. In this case the actual values for $F(\mathbf{a})$ can be easily computed by another method (see Tong, 1990). For this test it was feasible to use the Schervish software for m as large as 6. Many of the results were cross-checked using tables from Tong (1990), and all results checked were accurate to 4-5 decimal places. Table 4.1.2 summarizes the results from this test.

Table 4.1.2: Fixed Covariance Matrix Results

m	Max(A-S)	Ave. S Time	Ave. A Time	Ratio
3	0.000371	0.004	0.005	0.8
4	0.000522	0.258	0.018	14.7
5	0.000666	8.318	0.065	127.6
6	0.000546	171.001	0.124	1377.0
8	-	-	1.343	-
10	-	-	9.789	-
12	-	-	28.279	-

4.2 A Bayesian Analysis Example

In this section there is a discussion of an integral that arose in a Bayesian analysis problem. An article by Carlin, Kass, Lerch and Huguenard (1990) considers two cognitive models for predicting error rates in computer-based tasks using the cognitive psychological concept of human working memory. One of the models (the full model) requires the evaluation of six-dimensional integrals. A reduced model involves five-dimensional integrals. Genz and Kass (1991) describe how a modal approximation transformation can be used for the numerical evaluation of the full model integrals. Here, a similar transformation for the reduced model integrals is considered. The required integrals have the form

$$I(f) = \int_{-\infty}^{\infty} \int_{-\infty}^{\infty} \int_{-\infty}^{\infty} \int_{0}^{\infty} \int_{-\infty}^{\infty} f(\boldsymbol{\lambda}) L(\boldsymbol{\lambda}) \pi(\boldsymbol{\lambda}) d\boldsymbol{\lambda}$$

with $\boldsymbol{\lambda} = (\gamma, \mu_\theta^{(1)}, \mu_\theta^{(2)}, \sigma_\theta, \alpha)^t$, and $f = 1$ and $f = \boldsymbol{\lambda}$

The posterior density function $\pi(\boldsymbol{\lambda})$ is given by

$$\pi(\boldsymbol{\lambda}) = \frac{e^{-\frac{1}{2}((\frac{\mu_\theta^{(1)}-a_1}{b_1})^2 + (\frac{\mu_\theta^{(2)}-a_2}{b_2})^2 + (\frac{\gamma-\mu_\gamma}{\sigma_\gamma})^2 + (\frac{2}{d\sigma_\theta^2}) + (\frac{\alpha-\mu_\alpha}{\sigma_\alpha})^2)}}{\sqrt{2\pi}b_1\sqrt{2\pi}b_2\sqrt{2\pi}\sigma_\gamma cd^c\sigma_\theta^{2(c+1)}\sqrt{2\pi}\sigma_\alpha}.$$

This is a product of one variable normal densities, except for the σ_θ variable. All of the parameters $(a_1, b_1, etc.)$ in this prior density function are modeling parameters, chosen in advance as part of the design of the model.

The likelihood function $L(\boldsymbol{\lambda})$ has the form

$$L(\boldsymbol{\lambda}) = \prod_{i=1}^{2} \prod_{j=1}^{10} \int_{-\infty}^{\infty} \frac{e^{-\frac{1}{2}(\frac{\theta_j^{(i)}-\mu_\theta^{(i)}}{\sigma_\theta})^2}}{\sqrt{2\pi}\sigma_\theta} g(\theta_j^{(i)}, \boldsymbol{\lambda}) d\theta_j^{(i)},$$

with

$$g(\theta_j^{(i)}, \lambda) = \prod_{k=1}^{3} \prod_{l=1}^{3} \frac{(e^{\theta_j^{(i)}+\gamma(k-1)+\alpha[l/2]})^{z_{i,j,k,l}}}{(1 + e^{\theta_j^{(i)}+\gamma(k-1)+\alpha[l/2]})^{n_{i,j,k,l}}},$$

where [x] is used to denote the integer part of x and the $z_{i,j,k,l}$ and $n_{i,j,k,l}$ values come from experimental data. Putting everything together, the computation $I(f)$ could actually be considered a twenty-five dimensional numerical integration problem. However, the problem will be treated as a five dimensional integral with a difficult integrand, part of which is a product of twenty one dimensional integrals.

This problem is typical of many problems in Bayesian analysis (see Wolpert, 1991). The prior density function is a product of simple density functions and the likelihood function is a much more complicated function that includes data from some experiment. The complete integrand is unimodal, and only 2-3 decimal digit accuracy is needed. Without a careful choice of transformation this type of integral can be very difficult. As an initial experiment, several different types of numerical integration software (Monte-Carlo, Gauss product and subregion adaptive with a simple transformation of each variable) were used directly on this integral and no accurate results were obtained for any of these methods when using several hours of computation time.

Because of the complicated likelihood function it is natural to first consider transformations for the prior, where appropriately chosen inverse normal transformations can be used on all of the variables except for σ_θ. For this variable $\sigma_\theta = d'/x$, with $d' = (\frac{2}{d})^{\frac{1}{2}}$, followed by an inverse normal transformation gives (in one variable)

$$\int_0^\infty \frac{e^{-\frac{1}{d\sigma_\theta^2}}}{\sigma_\theta^{2c+2}} g(\sigma_\theta) d\sigma_\theta = \frac{(d\pi)^{\frac{1}{2}}}{2^{c+1}} \int_0^1 \Phi^{-1}(\frac{z+1}{2})^{2c} g(\frac{d'}{\Phi^{-1}(\frac{z+1}{2})}) dz.$$

Using these transformations the integral $I(f)$ now becomes

$$I(f) = \frac{(d\pi)^{\frac{1}{2}}}{2^{c+1}} \int_0^1 \cdots \int_0^1 \tilde{f}(\mathbf{z}) \Phi^{-1}(\frac{z_4+1}{2})^{2c} \tilde{L}(\mathbf{z}) d\mathbf{z},$$

where

$$\tilde{L}(\mathbf{z}) = \prod_{i=1}^{2} \prod_{j=1}^{10} \int_0^1 g(\theta_j^{(i)}(s_{i,j}), \lambda(\mathbf{z})) ds_{i,j},$$

$\tilde{f}(\mathbf{z}) = f(\lambda(\mathbf{z}))$ and $\theta_j^{(i)}(s_{i,j}) = \mu_\theta^{(i)} + \sigma_\theta \Phi^{-1}(s_{i,j})$.

The integral calculations all require a five-dimensional outer integral of a function that is a product of twenty one-dimensional inner integrals. The inner integrals can all be done with a subregion adaptive one-dimensional quadrature algorithm. The algorithm that was used is similar to the algorithm used by the QUADPACK (Piessens, deDoncker-Kapenga, and Kahaner, 1983) subroutine QAG, with a 7-15 point Gauss-Kronrod pair chosen for the basic integration rule. The outer integrals were computed with a subregion adaptive m-dimensional algorithm using the SCUHRE (Berntsen, Espelid and Genz, 1991a, 1991b) subroutine for vectors of integrals.

One problem with the inner integrals involves scaling. Some experimentation showed that these inner integrals had values that were typically about 10^{-6}. Because a product of twenty of these could cause underflow, the log of integrand for the outer integrals was computed as a sum of the logs of the inner integrals. The sum was initialized to the value 210, and the integrand value was obtained by exponentiating the final sum. The effect of the initialization was to scale the integral value by e^{210}, but since the numbers of interest all require a division by the integral value when $f = 1$, these scale factors cancel.

In Table 4.2.1 the prior transformation results are given. The SCUHRE adaptive integration software computed the vector of integrals $I(f)$ for $f = 1, \mu_\theta^{(1)}, \mu_\theta^{(2)}, \gamma, log(\sigma_\theta)$ and α, and then scaled the results by $I(1)$ to obtain the required expected values. The $log(\sigma_\theta)$ f was used instead of σ_θ for comparison with the modal transformation results, to be discussed next. Since the major part of the computation is the computation of the inner integral products, the use of the vector integration software reduces the computation time by an approximate factor of $1/6$. In spite of this saving, the prior transformation results required approximately two hours of single precision computation time on a Digital DECstation 3100 workstation. The numbers in the rows labeled "$L(\lambda)$'s" are the numbers of evaluations of $L(\lambda)\pi(\lambda)$. Several columns of results are given for each case, to illustrate the speed of convergence, and allow some estimation of the accuracy of the final results. The computation was continued until there was approximate agreement in the second decimal digit from one iteration to the next.

Table 4.2.1: Prior Transformation Results

| $L(\lambda)$'s | \multicolumn{6}{c}{Expected Values} | | | | | |
	1	$\bar{\gamma}$	$\bar{\mu}_\theta^{(1)}$	$\bar{\mu}_\theta^{(2)}$	$\overline{log\sigma}$	$\bar{\alpha}$
3193	0.362E-10	0.999	-4.241	-4.916	0.226	1.775
6283	0.364E-10	1.021	-4.255	-4.880	0.179	1.793
12669	0.374E-10	1.013	-4.247	-4.860	0.178	1.786
25441	0.377E-10	1.015	-4.253	-4.859	0.179	1.792

Now consider the use of the modal approximation transformation discussed in Section 3.3. This transformation can be used almost directly with the integral I. The only difficulty occurs with the variable σ_β, which has limits 0 and ∞. If the initial transformation $\sigma'_\beta = log(\sigma_\beta)$, is followed by numerical optimization to obtain the mode $\hat{\lambda}$ and modal covariance matrix Σ, with Cholesky factor C, the integral I can then be put into the form

$$I(f) = (\sqrt{2\pi})^5 |C| \int_0^1 \cdots \int_0^1 p(\tilde{\lambda}(z))dz,$$

with

$$p(\tilde{\lambda}_i(z)) = e^{w_4 + \frac{y(z)^t y(z)}{2}} f(\tilde{\lambda}_i(z)) L_i(\tilde{\lambda}(z)) \pi(\tilde{\lambda}(z)),$$

where $\tilde{\lambda}(z) = (w_1, w_2, w_3, e^{w_4}, w_5, w_6)^t$ is defined using $\mathbf{w} = \hat{\lambda} + C\mathbf{y}(z)$ with $\mathbf{y}(z) = (\Phi^{-1}(z_1), \dots, \Phi^{-1}(z_m))^t$.

Results in Table 4.2.2 were obtained using the SCUHRE software. These results required approximately eight minutes of computation time and are clearly more accurate than the

278

Table 4.2.1 results.

Table 4.2.2: Modal Transformation Results

$L(\lambda)$'s	Expected Values					
	1	$\bar{\gamma}$	$\bar{\mu}_\theta^{(1)}$	$\bar{\mu}_\theta^{(2)}$	$\overline{log\sigma}$	$\bar{\alpha}$
309	0.375E-10	1.015	-4.249	-4.852	0.171	1.794
721	0.376E-10	1.015	-4.249	-4.855	0.172	1.794
1545	0.376E-10	1.015	-4.249	-4.858	0.173	1.794

The modal transformation results in the first row of Table 4.2.2 appear to be accurate to 2-3 digits, but the prior transformation results do not have this level of accuracy until the last row. In this example the modal transformation method takes about 1/100 of the time taken by the prior transformation method to achieve a comparable level of accuracy.

4.3 Eigenvalue Integrals

A family of integrals that are used for estimating the eigenvalues of a random covariance matrix has been described by Luzar and Olkin (1991). These integrals $I_{s,m}(g)$ are defined by

$$I_{s,m}(f) = C_{s,m} \int_S e^{-\sum_{i=1}^m \theta_i/2} \prod_{1 \leq j < i \leq m} (\theta_i - \theta_j) \prod_{i=1}^m \theta_i^{(s-m-1)/2} f(\boldsymbol{\theta}) d\boldsymbol{\theta},$$

where S is the infinite simplex defined by $0 \leq \theta_m \leq \theta_{m-1}... \leq \theta_1 \leq \infty$, and $C_{s,m}$ is a known constant defined by $I_{s,m}(1) = 1$. The integral was needed for the vector $\mathbf{f} = \boldsymbol{\theta}$, with $1 < m < s$.

These integrals were computed to 3-4 decimal digit accuracy for $1 < m < 8$ and $s < 16$ using subregion adaptive algorithms. A paper by Genz (1991b) describes how a subregion adaptive algorithm for vectors of integrals over a simplex was used. In that work the transformations $\theta_i = (z_i/(1-z_i))^2$, for $i = 1, 2, ..., m$, were used to transform the infinite simplex S to the finite simplex defined by $0 \leq z_m \leq z_{m-1}... \leq z_1 \leq 1$, and all of the $m(16-m)$ integrals for each particular m value were done in one computation. The $m = 2$ results required about 500 evaluations of the associated 28-vector of integrands. The number of required integrand evaluations for successive m values rapidly increased to where the $m = 7$ results required about 250,000 evaluations of the associated 63-vector of integrands. The computations for all of the results required only about fifteen minutes of Digital DECstation 3100 time. Similar results were obtained when the finite simplex was transformed to a hypercube and a subregion adaptive algorithm for hypercubes was used. The factors $e^{-\theta_i/2}\theta_i^{(s-m-1)/2}$ in the original integrand suggest that the transformations $\theta_i = \Gamma^{-1}(z_i, (m-s+1)/2)$ might be more suitable for this problem. These transformations were also tried, but the results were no better than those obtained using the simpler transformations, and the computations took more time because of the inverse Γ function evaluations. For a comparison, consider the integral $I(\theta_1)$ for $m = 3$, $s = 15$, which was computed using a Monte-Carlo method by Luzar and Olkin (1991). Their computation required 30,000 integrand values to achieve 0.001 accuracy.

5 Concluding Remarks

The examples discussed in this paper suggest that subregion adaptive integration algorithms can be used effectively in some statistics multiple integration problems. Care must be taken in the selection of the transformation that is used to precondition an integral before a subregion adaptive algorithm is used, but statistical theory often provides helpful information to aid in the transformation selection process. The results presented do not support the common assumption that subregion adaptive algorithms are useful only for problems of low dimensionality.

This work was supported in part by grant DMS-9008125 from the US National Science Foundation.

References

Berntsen, J., Espelid, T.O. and Genz, A. (1991a) 'An Adaptive Algorithm for the Approximate Calculation of Multiple Integrals', *ACM Trans. Math. Soft.*, to appear.

Berntsen, J., Espelid, T.O. and Genz, A. (1991b) 'An Adaptive Multiple Integration Routine for a Vector of Integrals', *ACM Trans. Math. Soft.*, to appear.

Carlin, B.P., Kass, R.E., Lerch, F.J. and Huguenard, B.R. (1990) 'Predicting Working Memory Failure: A Subjective Bayesian Approach to Model Selection', Technical Report No. 503, Department of Statistics, Carnegie Mellon University.

Chen, C.F. (1985) 'On Asymptotic Normality of Limiting Density Functions with Bayesian Implications', *J. Royal Statist. Soc.*, **47**, pp. 540-546.

Davis, P. J. and Rabinowitz P. (1984) *Methods of Numerical Integration*, Academic Press, New York.

van Dooren, P. and de Ridder, L. (1976) 'An Adaptive Algorithm for Numerical Integration over an N-Dimensional Rectangular Region', *J. Comp. Appl. Math.* **2**, pp. 207-217.

Drezner, Z. and Wesolowsky, G. O. (1990) 'On the Computation of the Bivariate Normal Integral', *J. Statist. Comput. Simul.* **35**, pp. 101-107.

Gelfand, A. and Smith, A. (1990) 'Sampling-Based Approaches to Calculating Marginal Densities', *J. Amer. Stat. Assoc.* **85**, pp. 398-409.

Genz, A. (1991a) 'Subregion Adaptive Algorithms for Multiple Integrals', in N. Flournoy and R. K. Tsutakawa (eds.) *Statistical Numerical Integration*, Contemporary Mathematics **115**, American Mathematical Society, Providence, Rhode Island, pp. 23-31.

Genz, A. (1991b) 'An Adaptive Numerical Integration Algorithm for Simplices', in N. A. Sherwani, E. de Doncker and J. A. Kapenga, (eds.) *Computing in the 90's, Proceedings of the First Great Lakes Computer Science Conference*, Lecture Notes in Computer Science **507**, Springer-Verlag, New York, 1991, pp. 279-292.

Genz, A. and Kass, R. (1991) 'An Application of Subregion Adaptive Numerical Integration to a Bayesian Inference Problem', Department of Statistics Technical Report No. 523, Carnegie Mellon University, Pittsburgh, Pennsylvania.

Genz, A. and Malik, A. (1980) 'An Adaptive Algorithm for Numerical Integration over an N-Dimensional Rectangular Region', *J. Comp. Appl. Math.* **6**, pp. 295-302.

Geweke, J. (1989) 'Bayesian Inference in Econometric Models Using Monte Carlo Integration', *Econometrica* **57**, pp. 1317-1340.

Luzar, V. and Olkin, I. (1991) 'Comparison of Simulation Methods in the Estimation of Ordered Characteristic Roots of a Random Covariance Matrix', in N. Flournoy and R. K. Tsutakawa (eds.) *Statistical Numerical Integration*, Contemporary Mathematics **115**, American Mathematical Society, Providence, Rhode Island, 1991, pp. 189-202.

Marsaglia, G. and Olkin, I. (1984) 'Generating Correlation Matrices', *SIAM J. Sci. Stat. Comput.* **5**, pp. 470-475.

Metropolis, N., Rosenbluth, A., Rosenbluth, M., Teller, M. and Teller, E. (1953) 'Equation of State Calculations by Fast Computing Machines', *J. Chem. Phys.* **21**, pp. 1087-1092.

NAG, Numerical Algorithms Group Limited, Wilkinson House, Jordan Hill Road, Oxford, OX2 8DR, UK

Naylor, J. C. and Smith, A. F. M. (1982) 'Applications of a Method for the Efficient Computation of Posterior Distributions', *Appl. Stat.*, **31**, pp. 214-225.

Naylor, J. C. and Smith, A. F. M. (1988) 'Econometric Illustrations of Novel Numerical Integration Strategies for Bayesian Inference', *J. Economet.*, **38**, pp. 103-125.

Ogata, Y. (1989) 'A Monte Carlo Method for High Dimensional Integration', *Numer. Math.* **55**, pp. 137-157.

Piessens, R., de Doncker-Kapenga, E., Uberhuber, C. W. and Kahaner, D. K. (1983) *QUADPACK*, Springer-Verlag, Heidelberg.

Schervish, M. (1984) 'Multivariate Normal Probabilities with Error Bound', *Appl. Stat.* **33**, pp. 81-87.

Shaw, J. E. H. (1988) 'Aspects of Numerical Integration and Summarisation', in J. M. Bernado, M. H. Degroot, D. V. Lindley and A. F. M. Smith (eds.) *Bayesian Statistics 3*, Oxford University Press, Oxford, pp. 411-428.

Tong, Y. L. (1990) *The Multivariate Normal Distribution*, Springer-Verlag, New York.

Wolpert, R. L. (1991) 'Monte Carlo Integration in Bayesian Statistical Analysis', in N. Flournoy and R. K. Tsutakawa (eds.) *Statistical Numerical Integration*, Contemporary Mathematics **115**, American Mathematical Society, Providence, Rhode Island, pp. 101-116.

SINC INDEFINITE INTEGRATION AND INITIAL VALUE PROBLEMS

FRANK STENGER, BRIAN KEYES,
MIKE O'REILLY and KEN PARKER
Department of Computer Science
University of Utah
Salt Lake City, UT 84112 USA

ABSTRACT. The Sinc indefinite integral has been discussed by several authors: by Kearfott (1983), by Haber (1991), and by Stenger (1981). The following presentation summarizes some of the results in the monograph (Stenger, 1991), involving both Sinc indefinite integration and the application of this formula to the solution of initial value problems in ordinary differential equations.

Let $\mathcal{R} = (-\infty, \infty)$, let $\mathcal{C} = \{z = x + iy : x \in \mathcal{R}, y \in \mathcal{R}\}$, and let $\mathcal{Z}$ denote the integers. N will always denote a positive integer, and $n = 2N + 1$. Let $\mathcal{D}$ denote a simply connected domain in $\mathcal{C}$, let $\mathbf{C}(\mathcal{D})$ and $\mathbf{Hol}(\mathcal{D})$ denote respectively the set of all functions that are continuous in $\bar{\mathcal{D}}$ and holomorphic in $\mathcal{D}$.

Definition *Let $\mathcal{D}$ be a simply-connected domain having boundary $\partial \mathcal{D}$. Let a and b denote two distinct points of $\partial \mathcal{D}$, let $d > 0$, let $\mathcal{D}_d = \{z \in \mathcal{C} : |\Im z| < d\}$, let ϕ denote a conformal map of $\mathcal{D}$ onto $\mathcal{D}_d$, such that $\phi(a) = -\infty$, and $\phi(b) = \infty$. Let $\psi = \phi^{-1}$ denote the inverse map, and let $\Gamma = \psi(\mathcal{R})$. Given ϕ, ψ and a positive number h, let $z_k = \psi(kh)$, $k \in \mathcal{Z}$. For example if $\Gamma = (0,1)$, we can take $\phi(x) = \log\{x/(1-x)\}$; if $\Gamma = (0,\infty)$, we can take $\phi(x) = \log\{\sinh(x)\}$. We set $X = \mathbf{C}(\mathcal{D}) \cap \mathbf{Hol}(\mathcal{D})$ and for any $f \in X$ we define the norm $\|f\| = \sup_{(x \in \Gamma)} |f(x)|$. Let us also set $\rho = e^{\phi}$, and corresponding to a positive number α let $\mathbf{L}_\alpha(\mathcal{D})$ denote the family of all functions F in $\mathbf{Hol}(\mathcal{D})$ for which there exists a constant C such that $|F(z)| \leq C|\rho(z)|^\alpha/[1 + |\rho(z)|]^{2\alpha}$ for all $z \in \mathcal{D}$. We shall henceforth in this note assume that $0 < \alpha \leq 1$ and $0 < d < \pi$. Set $\mathrm{sinc}(x) = \sin(\pi x)/(\pi x)$, and for $h > 0$ and $k \in \mathcal{Z}$, set $S(k,h) \circ (x) = \mathrm{sinc}([x - kh]/h)$. For $k \in \mathcal{Z}$, set $\delta_k = \int_{-\infty}^k \mathrm{sinc}(x)\,dx$.*

Theorem 1 *Let $F/\phi' \in \mathbf{L}_\alpha(\mathcal{D})$. Let $I^{(-1)}$ denote the square matrix of order n with (k,ℓ)th element equal to $\delta_{k-\ell}$. Set $J_k = \int_a^{z_k} F(t)\,dt$, $F_k = F(z_k)$, $\hat{J} = (J_{-N}, \cdots, J_N)^T$, and $\hat{F} = (F_{-N}, \cdots, F_N)^T$, where the superscript "T" denotes the transpose. Let $D(1/\phi')$ denote the diagonal matrix with entries $1/\phi'(z_{-N}), \cdots, 1/\phi'(z_N)$. Then*

$$\hat{J} = h\, I^{(-1)} D(1/\phi')\hat{F} + \hat{\varepsilon}, \tag{1}$$

where $\hat{\varepsilon} = (\varepsilon_{-N}, \cdots, \varepsilon_N)^T$, and where for $j = -N, \cdots, N$, we have $\varepsilon_j \leq K e^{-(\pi d\alpha N)^{1/2}}$, where K is a constant which is independent of N. Moreover, there exists a constant K_1 which is independent of N, such that

$$\left| \int_a^x F(t)\,dt - \frac{\rho(x)J_N}{[1 + \rho(x)]} - \sum_{k=-N}^N (J_k - \frac{e^{kh}J_N}{[1 + e^{kh}]})S(k,h) \circ \phi(x) \right| \leq K_1\sqrt{N}\,e^{-\sqrt{\pi d\alpha N}}.$$

T. O. Espelid and A. Genz (eds.), Numerical Integration, 281–282.
© 1992 Kluwer Academic Publishers.

282

The model differential equation of this section is the problem

$$\frac{dy}{dx} = F(x, y), \quad x \in \Gamma, \quad y(a) = y_a. \tag{2}$$

This differential equation is often solved by successive approximations, via the scheme

$$y^{(0)}(x) = g(x), y^{(m+1)}(x) = (Sy^{(m)})(x) \equiv \int_a^x F(t, y^{(m)}(t))dt + y_a, \tag{3}$$

where g is a suitable initial approximation, which we assume to be in X.

We shall also assume that there exists a number $r > 0$, such that whenever $y \in B(r; g) \equiv \{f \in X : \|f - g\| < r\}$, then $F(\cdot, y)/\phi' \in \mathbf{L}_\alpha(\mathcal{D})$.

Let us now set $h = [\pi d/(\alpha N)]^{1/2}$, and let us define a sequence of vectors $\{\mathbf{y}^{(m)}\}$, by applying the *indefinite integration formula* of Equation (1) to approximate the integrals in (3). Starting with $\mathbf{y}^{(0)} = (g(z_{-N}), \cdots, g(z_N))^T$, and using $\mathbf{S}(\mathbf{y}) \equiv hI^{(-1)}D(1/\phi')\mathbf{f} + \mathbf{y}_a$, we thus get

$$\mathbf{y}^{(m+1)} = \mathbf{S}(\mathbf{y}^{(m)}), \quad m = 0, 1, 2, \cdots, \tag{4}$$

where $\mathbf{y} = (y_{-N}, \cdots, y_N)^T$, $\mathbf{f} = (F(z_{-N}, y_{-N}), \cdots, F(z_N, y_N))^T$ and $\mathbf{y}_a = (y_a, \cdots, y_a)^T$.

Theorem 2 *If the sequence of functions $\{y^{(m)}\}$ defined as in (3) converges in X to a function $Y \in B(g; r)$, then Y solves the initial value problem (2). In this case the sequence of vectors defined as in (4) also converges, for all N sufficiently large, to a vector $\mathbf{y} = (y_{-N}, \cdots, y_N)^T$, and if we define $Y_N(x)$ by*

$$Y_N(x) = [y_a + \rho(x)y_N]/[1 + \rho(x)] + \sum_{j=-N}^{N} (y_j - [y_a + e^{jh}y_N]/[1 + e^{jh}])S(j, h) \circ \phi(x),$$

then there exists a constant C, independent of N, such that $\|Y - Y_N\| \le C N^{1/2} e^{-(\pi d\alpha N)^{1/2}}$.

It is worthwhile to mention that in practice it is nearly always more efficient to use the *Seidel* iteration, instead of the *Neumann* iteration described in (4). In using the Seidel version, we evaluate the equations of the system $\mathbf{y} = \mathbf{Sy}$ one at a time, starting at the top and working down. We then use the newly computed component of $\mathbf{y}$ on the right-hand side of the next equation immediately after it has been computed.

References

Haber, S. (1991) 'Two Formulas for Numerical Indefinite Integration', to appear in *Math. Comp.*.

Kearfott, R. B. (1983) 'A Sinc Approximation for the Indefinite Integral', *Math. Comp.* **41** pp. 559-572.

Stenger, F. (1981) 'Numerical Methods Based on the Whittaker Cardinal, or Sinc Functions', *SIAM Rev.* **23** pp. 165-224.

Stenger, F. (1991) *Sinc Numerical Methods*, textbook, to appear.

SOFTWARE FOR INTEGRATION OVER TRIANGLES AND GENERAL SIMPLICES.

PATRICK KEAST,
Department of Mathematics, Statistics
and Computing Science,
Dalhousie University,
Halifax, Nova Scotia, Canada, B3H 3J5.

ABSTRACT. Numerical integration over the triangle and n dimensional simplex is generally carried out using rules based on polynomials. The theory of construction of these rules is based on the invariant polynomials over the simplex, and leads to the same kinds of difficulties of non-linear moment equations as arise in the construction of formulas for regions such as the hypercube. This paper will discuss methods of construction of arbitrary degree rules. There will also be a discussion of the more robust methods for automatic integration over triangles, and of a package which may be used for automatic integration of functions which have singularities whose type and location are known.

1. Introduction

Triangulation of domains over which integrals are to be computed is a common approach to the problem of quadrature in non-standard domains. The more general problem of n-dimensional integration can often be made tractable by decomposition of the domain into simplices. Consequently, the construction of simplex integration formulas has an important role in the area of quadrature and cubature. In section 2, this paper will review some of the theory of the construction of formulas for the n-simplex, and will discuss the methods which have appeared in the literature. A technique for arriving at feasible minimal point structures will be described. In section 3, the software available for the n-simplex, $n > 2$, is mentioned. For the triangle there is a wide choice of software, and in section 4 we give a brief survey. In section 5 an extrapolation method tailored to functions with a particular type of singularity is discussed.

2. Basic rules for the simplex, and simplicial symmetries

The n-dimensional standard simplex is defined by:

$$T_n = \{ \, x \mid 0 \le x_i \le 1, \Sigma \, x_i \le 1 \, \}$$

which has the $n+1$ vertices $e_i = (\, 0,0,...,0,1,0,...0)$, $i = 1, . . ., n$, and the origin 0. The integral of $f(x)$ over T_n is denoted by

$$I_n(f) = \int_{T_n} f(x) \ dx.$$

An approximation to $I_n(f)$ is defined by

$$Q_n(f) = \Sigma \, w_i f(a_i),$$

where it is usually assumed that $a_i \in T_n$, and $w_i > 0$. Such a formula is called **PI** (Positive weights, points Inside) in the literature, (see [21]). Requiring $a_i \in T_n$ is probably reasonable, since there may be no assurance that the function even exists outside the simplex,

T. O. Espelid and A. Genz (eds.), Numerical Integration, 283–294.
© 1992 *Kluwer Academic Publishers.*

although if the formula is applied by breaking the simplex up into sub-simplices, i.e. in a *cytolic* manner, then it is likely that only a small proportion of the sub-simplices will be on the boundary of T_n and will consequently have points outside the region. However, in general one may want to avoid external points. On the other hand, $w_i > 0$ is not really necessary. What *is* necessary is that the quantity

$$\rho = \Sigma |w_i| / (\Sigma w_i)$$

be small. This is because ρ is a measure of the *round-off* error in the calculation of the sum. Frequently, too much is made of the presence of a negative weight. For example, in [28] the formula

$$Q_n(f) = (-27/48)f(1/3,1/3) + (25/48)(f(2/15,2/15) + f(2/15,11/15) + f(11/15,2/15))$$

is given for T_2, with the footnote that "this formula is not recommended due to the negative weight and round-off error". In fact, $\rho = 102/48 \cong 2.1$, so that at most one binary bit is lost in each sum. (The same text gives a formula for T_3 for which the absolute value sum of the weights is 2.6, but no recommendation is made.)

The formulae we consider will be designed to integrate exactly all polynomials up to a fixed degree. We need, first, some notation:

DEFINITION: A *monomial of degree d* is given by

$$x_1^{i_1} x_2^{i_2} \dots x_n^{i_n} , \qquad \sum_{j=1}^{n} i_j = d$$

Then:

DEFINITION: A *formula Q_n* for T_n is of *degree d* if $Q_n(f) = I_n(f)$ for all monomials f of degree $\leq d$, and there is at least one monomial of degree $d + 1$ for which $Q_n(f) \neq I_n(f)$.

The dimension of the space of polynomials of degree d, ie the number of distinct monomials of degree d, is denoted by $N(n,d) = (n+d)!/(n!d!)$. Thus, in principle, in order to obtain a formula of degree d for T_n we need simply to solve the following set of *moment equations* for a_i and w_i:

$$Q_n(f) = \Sigma w_i f(a_i),$$

where f runs through all polynomials of degree $\leq d$. However, we can decrease the size of the system of non-linear moment equations by making use of the symmetries of the simplex. One defines a set of polynomials and a summation operator which each possess the same symmetries as the simplex. Then a quadrature formula is defined as a weighted average of summation operators, with the weights and operators being chosen so that all polynomials in the symmetric set of degree less than or equal to some specified degree are integrated exactly. There are various ways of doing this. Beckers and Haegemans [1], use the theory of the extended tetrahedral group. In Keast [16] the set of monomials denoted by $S_{n,d}$ is defined by:

$$x_1^{i_1} x_2^{i_2} ... x_n^{i_n} (1 - x_1 - ... - x_n)^{i_{n+1}}, \quad \sum_{j=1}^{n+1} i_j \le d, i_1 = i_2 \ge i_3 \ge .. \ge i_{n+1} \ge 0$$

where it should be noted that $i_1 = i_2$ is specified. For example, $S_{3,7}$ is:

$$1, \quad xy, \quad xyz, \quad x^2y^2, \quad xyz(1-x-y-z), \quad x^2y^2z, \quad x^3y^3,$$
$$x^2y^2z^2, \quad x^2y^2z(1-x-y-z), \quad x^3y^3z, \quad x^2y^2z^2(1-x-y-z).$$

The summation operators or *basic rules* are defined as follows. Let $x = (x_1, x_2, ..., x_n)$, and let $y = (y_1, y_2, ..., y_n; y_{n+1})$, where $y_i = x_i$, $i = 1, 2, ..., n$, and $y_{n+1} = 1 - x_1 - x_2 - ... - x_n$. Define $\mu = (\mu_1, \mu_2, ..., \mu_{n+1})$ to be a permutation of $(1, 2, ..., n+1)$. Then the set of points given by $x_\mu = (y_{\mu_1}, y_{\mu_2}, ..., y_{\mu_n})$, for all permutations μ, is called E(x), the equivalence class of points under the symmetries of the simplex T_n.

DEFINITION: A *simplicially symmetric basic rule* is one which assigns the same weight to all points in the equivalence class E(a) for any point **a** in T_n.

DEFINITION: Let **a** be a point in T_n. The *basic symmetric rule* $T(a_1, a_2, ..., a_{n+1})$, where $\Sigma a_i = 1$, is defined by

$$T(a_1, a_2, ..., a_n; a_{n+1}) f(x) = \Sigma f(c)$$

where the sum is over all points in $E(a_1, a_2, ..., a_n)$.

Then we have:

THEOREM: A *simplicially symmetric formula*

$$Q_n (f) = \sum_{j=1}^{K} w_j T(\alpha_j) f(x)$$

is of degree d if $Q_n(f) = I_n(f)$ for all $f \in S_{n,d}$, and if $Q_n(f) \ne I_n(f)$ for some f in $S_{n,d+1}$

For a formula of degree d we will have p(d,n+1) moment equations, where p(d,n+1) is the number of unordered partitions of d into n+1 non-negative integers. The forms of basic rules chosen will determine the cost of the rule, and may determine, even before any attempt is made to solve the moment equations, that no possible choice of w_i or α_i can be made to solve the system.

DEFINITION: A basic rule $T(\alpha)$ is of *class* $[m] = [m_1, m_2, ..., m_k]$ if there are k distinct non-zero values of α_i, $i = 1, ..., k$, with α_i being repeated m_i times.

The condition that the α's be non-zero is necessary to ensure that any basic rule lies only in one class. It is not a restriction which reduces in any way the number of available rules.

The *cost* (number of function evaluations) of a simplicially symmetric basic rule of class $[m]$ in n dimensions is given by:

$$\upsilon(\mathbf{m}) = \frac{(n+1)!}{m_1! \, m_2! \, ...m_k!(n+1 - m_1 - m_2 - .... - m_k)!}$$

For example, in the 3-dimensional simplex the only rule classes available, and their costs, are:

Rule:	[4]	[3,1]	[2,2]		[2,1,1]	[1,1,1,1]
Cost:	1	4	6		12	24
Rule	[3]	[2,1]	[1,1,1]	[2]	[1,1]	[1]
Cost:	4	12	24	6	12	4

It can be shown, [16], that the minimum point simplicially symmetric formula of degree 6 is:

$$w_1 T(a_1, a_1, a_1; 1-3a_1) + w_2 T(a_2, a_2, a_2; 1-3a_2) + w_3 T(a_3, a_3, a_3; 1-3a_3) +$$
$$w_4 T(a_4, a_4, b_4; 1-2a_4-b_4).$$

Thus there are three rules of class [3,1] and one of class [2,1,1], giving a cost of 24 points.

DEFINITION: The *rule structure parameters* for a simplicially symmetric formula are integers $K([m])$ giving the number of basic rules of class $[m]$ used in the formula.

Since each basic rule of class $[m_1, m_2, ..., m_k]$ has k-1 free coordinates and one weight, then for each rule of class $[m]$ there are k degrees of freedom. The cost of the formula is

$$\sum_{[\mathbf{m}]} \upsilon(\mathbf{m}) \, K([\mathbf{m}])$$

where the sum is over all classes appearing in the formula. The total number of degrees of freedom is

$$\sum_{[\mathbf{m}]} \sigma(\mathbf{m}) \, K([\mathbf{m}])$$

where $\sigma(\mathbf{m}) = k$, the number of components in $\mathbf{m}$.

Suppose, for example, we wish to derive a formula of degree 7 for T_3. One would, clearly, wish to use as many of the less expensive rules as possible. Associated with each basic rule of class $[m_1, m_2, ..., m_k]$ are k rule parameters, namely the k-1 independent coordinates $\alpha_1, ..., \alpha_{k-1}$, and the weight w. Therefore, in order to integrate the above 11 monomials, we would require 11 rule parameters. If we choose 5 rules of class [3,1] and the single rule of class [4], we have the necessary degrees of freedom, at a cost of $5*4 + 1 = 21$ function evaluations. But there is a hidden difficulty: there are four polynomials of degree ≤ 7 which have non-zero integrals over the simplex T_3, but which give zero identically when operated on by basic rules of class [4] or [3,1]. These are

$$2xyz - 3x^2y^2 - 5xyz(1-x-y-z)$$
$$xyz(1-x-y-z) + 4x^2y^2z - 6x^3y^3 - 26x^2y^2z(1-x-y-z)$$
$$xyz(1-x-y-z) - 9x^2y^2z + 7x^3y^3 + 13x^2y^2z^2$$
$$28x^2y^2z^2(1-x-y-z) + 12x^3y^3z - x^2y^2z^2 - 3x^2y^2z + 3x^3y^3$$

DEFINITION: The *null space*, $N_{n,d}[m]$, of a basic rule of class [m] is the space of polynomials of degree $\le d$ in dimension n which give zero identically for all basic rules $T(\alpha)$ of class [m].

The null space $N_{3,7}[3,1]$ in T_3 contains the above four functions, and therefore if we are to be able to construct a formula of degree 7 in T_3 we cannot do so with rules only of class [3,1] and [4]. It is simple to show, also, that any polynomial integrated to zero by rules of class [3,1] will be integrated to zero by the rules of classes [4],[3,1],[3],[1]. This implies that we must have enough rules of types [2], [1,1], [2,1], [1,1,1], [2,2], [2,1,1], and [1,1,1,1]. This is a condition on the structure parameters, which, if violated, will lead to inconsistent moment equations. Hence, the structure we posed to start with is not possible. A set of *linear consistency conditions* (see Rabinowitz and Mantel [23], and Keast and Lyness [18]) can be obtained, on the numbers of rules $K([m])$ of each class, which are necessary to avoid the problem of inconsistent moment equations. Then one can choose the numbers of rules satisfying these consistency conditions using the least number of points. The problem of solving the moment equations remains.

In Lyness and Keast [18] theorems are proved concerning null spaces and inclusions. Some general theorems on null spaces, and techniques for computing the dimensions, are given in Keast [15] for the hypercube, but the techniques apply to any region with specific symmetries. It was shown in [17] that the consistency conditions and null space dimensions for a formula of degree d in the simplex T_n were the same as for a formula of degree $2d+1$ on the *surface* of the $n+1$ dimensional sphere U_{n+1}. The minimum point structures for U_3 and U_4 are given in [17], and the structures for T_3 are given in [16]. The techniques apply to any region with given symmetries. For example, for the degree 7 formula in T_3, the rule structure parameters are

$K[1], K[2], K[1,1], K[3], K[2,1], K[1,1,1], K[4], K[3,1], K[2,2], K[2,1,1],$
$K[1,1,1,1].$

The consistency conditions are given by:

$K[1], K[2], K[3], K[4] \le 1,$

from the fact that there can only be one rule of each of these classes; also:

$K[1] + K[2] + 2K[1,1] + K[3] + 2K[2,1] + 3K[1,1,1] +$
$K[4] + 2K[3,1] + 2K[2,2] + 3K[2,1,1] + 4K[1,1,1,1] \ge 11,$

since there must be enough free parameters to ensure that the 11 functions in $S_{3,7}$ are integrated exactly;

and, finally, the set of conditions given in Figure 1, which arise from the null space dimensions. In Figure 1, $\mathbf{B}^T$ is the vector $(7,4,3,4,7,3,2,4,1,1,1)$, and $\mathbf{K}$ is the vector of rule structure parameters. The condition mentioned above, on the rule structure parameters for rules *other* than those of classes $[3,1]$, $[4]$, $[3]$, $[1]$, is:

$$K[2] + 2K[1,1] + 2K[2,1] + 3K[1,1,1] + 2K[2,2] + 3K[2,1,1] + 4K[1,1,1,1] \geq 4.$$

This is the 4th constraint, Figure 1. The structure $K[3,1] = 5$, $K[4] = 1$, with all other structure parameters equal to zero, violates this constraint. In addition, there is one polynomial which will be integrated to zero by all rules of classes $[1]$, $[2]$, $[1,1]$, $[3]$, $[2,1]$, $[1,1,1]$, $[4]$, $[2,2]$. This is the single polynomial of degree 7 which lies in the intersection of $N_{3,7}[2,2]$ and $N_{3,7}[1,1,1]$.

There must therefore be at least one rule from the classes $[3,1]$, $[2,1,1]$, $[1,1,1,1]$. This is the meaning of the second last constraint, in Figure 1

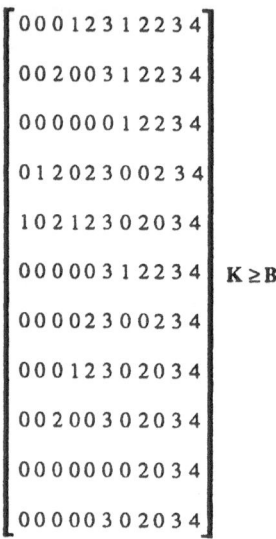

$$\mathbf{K} \geq \mathbf{B}$$

Figure 1.

An integer programming code, such as H02BAF in [24], may be used to find the rule structure parameters which minimize the cost of the rule, ie to minimise the function:

$$F(\mathbf{K}) = 4K[1] + 6K[2] + 12K[1,1] + 4K[3] + 12K[2,1] + 24K[1,1,1] + K[4] + 4K[3,1] + 6K[2,2] + 12K[2,1,1] + 24K[1,1,1,1],$$

subject to the above linear constraints.

3. Available formulae for $n \geq 3$

A comprehensive list of formulae appearing in the literature since the book [26] appeared, has been prepared by Cools and Rabinowitz [6]. In this section we give a brief overview of the most comprehensive *family* of formulae for arbitrary dimensions and odd degree, produced by Grundmann and Möller [11]. For dimension n and degree $d = 2s+1$, these use the basic rules:

$$T(\frac{2i_1+1}{p}, \frac{2i_2+1}{p}, ..., \frac{2i_n+1}{p}; \frac{2i_{n+1}+1}{p})$$

for $p = n+1$, $n+3$, ..., $n+d$, and $i_1 \geq i_2 \geq ... \geq i_{n+1} > 0$, $\Sigma i_j = (p-n-1)/2$. The cost of the formulae is given by $(n+s+1)! / (s!(n+1)!)$. These formulae have the distinct advantage that they are simple to produce and to program. In addition they provide a sequence of embedded formulae, albeit with some negative weights. They use more points than the minimum point simplicially symmetric formulae For example, in [16] it is shown that one minimum point degree 7 rule for T_3 has the structure $K[1] = 1$, $K[3,1] = 3$, $K[2,2] = 2$, which uses 28 points, and that two other structures exist using the same number of points. No rule, however, has been constructed which uses this number of points. The cheapest known symmetric rule uses 31 points. One rule with this cost is given in [16] as $K[2] = 1$, $K[4] = 1$, $K[3,1] = 3$, and $K[2,2].= 1$. The formula in [11] uses 35 points. The formula in [16] has one negative weight, with $\rho = 4$ approximately. There are several formulae using 31 and 33 points reported in [1], but all with one negative weight. However, the values of ρ are not given, and so it is not clear whether these are useful or not. A ρ-value of 4 means that we can expect to lose no more than 4 bits on each sum - not really significant even in 32 bit arithmetic.

The formulae of [11] are included in the NAG library, [24], as D01PAF, with a built in error estimator based on the asymptotic expansions of de Doncker [7]. No other software for the simplex is available in general scientific subroutine libraries.

4. Formulae for the triangle

Triangle formulae of moderate degree from 2 to 11 (excluding 10) are given in Lyness and Jesperson, [21]. The degree 11 rules, for example, use 27 or 28 points, (although the 27 point one has points outside the triangle), while the rule in [11] uses 56 points. The 27 point rule is the optimum rule with triangular symmetry. All the rules given are optimum symmetric formulae, with only the degree 11 rule having points outside the region. The other optimal rules all have positive weights except the degree 7 one. This, however, has a ρ value of only 1.3. In addition, it requires only 13 function values, compared to 56 for the one in [11]. The minimum point symmetric formula of degree 10 would use 22 points, but no formula is known with points inside the region. One formula [5] with points outside is known and there are two 25-point formulae with positive weights and points inside given in [20], and one 25-point rule in [10]. Some higher degree formulae, from degree 12 up to degree 20 are also given in [10]. These are close to being minimum point rules, and include five PI formulae of degrees 12, 13, 14, 17 and 19. The formulae of degrees 15, 16, 18, 20 have points outside the triangle. In [2] a formula of type NO (Negative weights, points Outside) of degree 13 using 36 points is given, together with an independently discovered degree 13, 37 point rule of type PI (similar to one which appeared in [10]). There is also, in [2], a degree 13, 40 point formula which uses the vertices and the mid-points of the sides of the triangle, and which therefore has a *cytolic* point-count of 36 asymptotically.

There are many *automatic* procedures available for the triangle, generally based on elementary polynomial rules, with error estimation and/or extrapolation. These all have the following main components:

     I.   Basic rule approximation.
    II.   Error estimation procedure.
   III. Subdivision strategy.

There are several robust codes available, including the following:

4.1 TRIADA: [12]

Conical product formulae using 36 and 49 points are applied, with the error estimate being the difference between the two rules. At any stage, a collection of subtriangles is given, with error estimates on each. If the total error is less than some user-defined relative or absolute error then the process stops. In this sense the error estimate is *global*. If the error requirement is not satisfied, the triangle with the largest error is subdivided into 4 equal triangles. The data are stored simply as an array, in no particular order. (For a discussion of the problems of robustness in automatic quadrature, and of the choice of data structure to be used in handling the collections of "intervals" - triangles or sub-intervals- see [25]).

4.2 CUBTRI: [19]

Two basic rules are used in the approximation: Radon's degree 5 rule using 7 points, and a degree 8 rule which uses 19 points, including the Radon points. The error estimate and subdivision procedure are essentially the same as in TRIADA. The data are stored in no particular order.
    Both of the above procedures are quite uncomplicated in their approaches to error estimation and subdivision strategy. They work on a single triangle, and, since their acceptance strategies are global, they can handle functions which vary widely over the triangle. Neither uses extrapolation, or makes any attempt to handle a singularity other than by subdivision.

4.3 TRIEX: [9]

This procedure is a carefully engineered piece of software, very much in tune with the ideas in the paper by Shapiro, [25]. The basic rules used are two simplicially symmetric formulae of degree 9 and 11 from [21]. These use, respectively, 19 and 28 points, so that they are the optimum symmetric formulae which are known. The rule structure for the optimum degree 11 formula gives a formula using 27 points, but as mentioned above the only solution to the moment equations results in a formula with points outside the triangle. (As mentioned earlier, it should be noted that, when used adaptively, this would probably result in only a small proportion of the evaluation points, corresponding to subtriangles on the boundary of the main triangle, lying outside the region. But the cost of one extra evaluation point seems a small price for avoiding the requirement of evaluating the integrand outside the specified region.) The basic rule values are also extrapolated. Once composite applications of the basic formulae are known, the ε- algorithm [27] is used to produce higher order approximations.
    As in most similar automatic procedures, the error estimator owes almost as much to heuristics as to theory. The difficulty with error estimators obtained from theoretical arguments is that they are often either too pessimistic, or too easily fooled. For example, extrapolation can actually make things worse if the component of the error being eliminated at any stage is not actually present. From such considerations the idea of *cautious* extrapolation, first used in CADRE, [4] was suggested. In TRIEX, the error estimator is taken as the maximum of three components.
    Let Q_9 be the degree 9 formula, and Q_{11} be the degree 11 formula. For a given triangle, let Δ be the area, and let:

$$M = 1/\Delta Q_9(f);$$
$$c = Q_9(|f - M|);$$
$$X = |Q_9(f) - Q_{11}(f)|;$$
$$Y = \min(1,(20X/c)^{1.5});$$

$$Z = 50Q_9(|f|)eps,$$

where *eps* is machine epsilon. The error estimate is then given by $E = \max(X, Y, Z)$. In [9] it is stated that the 1.5 and the 20 are arrived at experimentally, but no explanation of the 50 is given. (It is in fact an attempt to guard against an unnecessarily strict value of machine epsilon, and since it is not common to seek approximations of the integral to full machine accuracy, this factor does not play a large role in general.) This illustrates the difficulties of obtaining robust error estimators. The only way to produce a "black box" integrator seems to be by experimentation.

The subdivision strategy is to divide triangles into four equivalent sub-triangles. At stage j an approximation to the application of Q_9 and Q_{11} to the 4^j subtriangles is obtained. The triangle to be subdivided is the one with the largest error estimate. The current set of triangles is stored in a linked list, ordered according to error estimates, and no triangle is "accepted" until the estimated over the whole region is small enough, so that the method is *global*.

When singularities are present, TRIEX handles them best when they are on the boundaries of the triangle. This may happen as the triangles are sub-divided, but it is best to do a preliminary subdivision to ensure that no singularities are in the interior of a triangle, if it is possible to do so. This raises the difficulty of how to put a suitable error requirement on each of the subtriangles. In a sense, one is applying the *global* formula in a *local* manner. If the exact integral over the whole triangle is of order 1, while subtriangles have contributions of large order and opposite signs, it may become difficult to decide what error to request on the subtriangles to ensure the required error on the full triangle. To handle this kind of problem a modification of TRIEX has recently been produced:

4.4 TRISET: [8]

This is a generalization of TRIEX, offering a wider choice of basic rules, and allowing the user to specify a set of triangles instead of one. In addition, options are provided to allow extrapolation on either the degree 9 or the degree 11 approximations, or of using a pair of degree 6 and 8 rules from [21]. The latter choices may be more efficient if a high level of extrapolation can be carried out, since they use fewer points (12 and 16 respectively). Finally, giving a triangulated region to TRISET allows for some large grained parallelisation, with each triangle and its subsequent subdivisions being given to a different processor.

4.5 TWODQD: [14]

This (single precision) procedure provides a choice of a relative or absolute error request. An input parameter allows selection of one of two pairs of Lyness and Jesperson formulae - either a degree 8 formula using 16 points, together with a degree 6 formula using 12 points; or the degree 11 formula together with the degree 9 rule used in TRIEX. No extrapolation is attempted. The higher degree rule value is returned as the approximation, and the error estimate is arrived at in essentially the same way as in TRIEX and TRISET. The algorithm accepts a collection of triangles and attempts to satisfy a global error requirement. If the total error is too large, then the triangle with the largest estimate is divided into two triangles along the median to the largest side. The data associated with the set of triangles is stored in two heaps, one for those triangles which have been accepted and one for those yet to be accepted. This data structure allows the user to re-enter TWODQD using the previous computations. Many of the features of TWODQD and TRIEX are therefore combined in TRISET.

4.6 CUTRI: [3]

As with TWODQD, this method allows the user to specify a set of triangles, ie to give a region which has been triangulated by the user. The basic rule used is the degree 13 rule using 37 points given in [2]. No extrapolation is done, and the error estimate is obtained from application of "null rules" of lower degree using the same points as the basic rule. The set of active triangles is kept as a heap, which leads to faster insertion into the data structure, with little extra overhead. The triangle with the largest error estimate is subdivided into four equal parts. The use of multiprocessors is explicitly allowed in the options of the code.

5. Non-Adaptive Extrapolation

In all of the above automatic procedures, the assumption is that the form of any singularity present is not known. In some applications it is known ahead of time what the singularity is, and where it is. In some boundary element applications, for example, it is necessary to integrate functions of the form $r^\alpha f(x,y)$, where $-2 < \alpha < 0$, f is analytic, α is known, and r is the Euclidean distance from a corner of the triangle. If the singularity is of this type then we have the following result.

Let Q be a formula for the standard triangle, and let Q_m be the m^2-copy of Q, applied to the triangle by subdividing it into m^2 equal triangles. Then [22],

$$Q(f) - I(f) = \sum_{j=0} \frac{A_j + C_j \ln(m)}{m^{j+1}} + \sum_{j=1} \frac{B_j}{m^j}$$

If Q is *symmetric*, and of polynomial degree d, then C_{j-1} and B_j are zero for odd values of j, and also for $j = 1,2,...,d$. A *non-adaptive* method can be defined which takes advantage of knowledge of the exact form of the asymptotic error. The terms in the expansion may be eliminated one by one. Fortran subroutines have been written which allow integration in a non-adaptive way over a triangle with extrapolation being carried out on the assumption that the singularity is of the above form. For a sequence $\{m_k\}$ of subdivisions, define $Q_k(f)$ to be the m^2 copy of a basic rule Q on the triangle. Then asymptotically,

$$Q_k(f) = I(f) + a_1 g_1(m_k) + a_2 g_2(m_k) + ...,$$

where $\lim g_{i-1}(m)/g_i(m) = 0$ as m tends to infinity. A general extrapolation process can then be set up. In Hollosi [13], the subroutine ETXRR is described which allows the user to specify one of three low degree basic rules (vertex, centroid or mid-point), and one of three sequences of subdivision: the geometric sequence, $\{m_k\} = \{2^k\}$, the Bauer sequence, given by $\{m_k\} = \{2^k, 3 \cdot 2^k\}$, or the Bulirsch sequence, $m_k = \{3^k, 2 \cdot 3^k\}$. The extrapolation is carried out on the assumption that the form of the asymptotic series is known, and in a *cautious* way, as in [4]. The termination criterion is based on agreement between two pairs of column elements in the extrapolation table.

If an integral is given over a general triangle with the integrand having a singularity of the form r^{α}, an affine transformation can be used to express the integral over the standard 2-dimensional simplex. It may be shown that the same asymptotic error expansion holds, and that we can extrapolate in the same way.

REFERENCES:

[1] M. Beckers and A. Haegemans, The construction of cubature formulae for the tetrahedron, TW 128, 1990, Katholiek Universiteit, Leuven.
[2] J. Berntsen and T.O. Espelid, Degree 13 symmetric formulas for the triangle, Reports in Informatics 44, Department of Informatics, University of Bergen, 1990.
[3] J. Berntsen and T.O. Espelid, DCUTRI: An algorithm for adaptive cubature over a collection of triangles, to appear.
[4] C. de Boor, CADRE: An algorithm for numerical quadrature, Mat. Soft., Academic Press, 1971, (ed. J.R. Rice).
[5] R. Cools and A. Haegemans, Construction of minimal cubature formulae for the square and the triangle using invariant theory, TW 96, 1987, Katholiek Universiteit, Leuven.
[6] R. Cools and P. Rabinowitz, A catalogue of numerical integration formulae for certain symmetric regions. (Manuscript).
[7] Elise de Doncker, New Euler-Maclaurin expansions and their application to quadrature over the s-dimensional simplex, Math. Comp., **33**, 1979, pp. 1003-1018.
[8] E. de Doncker, D. Kahaner, B. Starkenburg, Adaptive integration over a triangulated region, to appear.
[9] E. de Doncker and I. Robinson, Algorithm 612, TRIEX: Integration over a TRIangle using non-linear EXtrapolation, ACM Trans. Math Soft., **10**, 1984, pp. 17-22.
[10] D.A. Dunavant, High degree efficient symmetric gaussian quadrature rules for the triangle, Int. J. Num. Meth. Eng., **21**, 1985, pp. 1129-1148.
[11] A. Grundmann and H.M. Moler, Invariant integration formulas for the n-simplex by combinatorial methods, SIAM J. Numer. Anal., **15**, 1978, pp. 282-290.
[12] A. Haegemans, Algorithm 34, an algorithm for the automatic integration over a triangle, Computing, **19**, 1977, pp. 179-187.
[13] M. Hollosi, Numerical integration of singular functions over a triangle, MSc thesis, University of Toronto, (Technical report 1985CS-3, Computing Science, Dalhousie University, Halifax, Canada).
[14] D.K. Kahaner and O.W. Rechard, TWODQD: an adaptive routine for two-dimensional integration, J. Comp. and Appl. Math., **17**, 1984, pp. 215-234.
[15] P. Keast, On the null spaces of fully symmetric basic rules for quadrature formulas in s-dimensions, Univ of Toronto Technical Report, 134, 1979.
[16] P. Keast, Moderate degree tetrahedral quadrature formulas, Comp. Meth. in Appl. Mech. and Eng., **55**, 1986, pp. 339-348.
[17] P. Keast, Cubature formulas for the surface of the sphere, J Comp. and Appl. Math., **17**, 1987, pp 151-172.
[18] P. Keast and J.N. Lyness, On the structure of fully symmetric multidimensional quadrature formulas, SIAM J Numer. Anal., **16**, 1979, pp. 11-29.
[19] D.P. Laurie, Algorithm 584, CUBTRI: Automatic integration over a triangle, ACM Trans. Math. Soft., **8**, 1982, pp. 210-218.
[20] M.E. Laursen and M. Gellert, Some criteria for numerically integrated matrices, Int. J. Num. Meth. Eng., **12**, 1978, pp. 67-76.
[21] J.N. Lyness and D. Jesperson, Moderate degree symmetric quadrature rules for the triangle, J. Inst. Math. Appl., **15**, 1975, pp. 19-32.
[22] J.N. Lyness and G. Monegato, Quadrature error functional expansions for the simplex when

294

the integrand has singularities at vertices, Math. Comp., **34**, 1980, pp. 213-225.

[23] F. Mantel and P. Rabinowitz, The application of integer programming to the computation of fully symmetric integration formulas in two and three dimensions, SIAM J. Numer. Anal., **14**, 1977, pp. 391-425.

[24] NAG (Numerical Algorithms Group), Scientific Subroutine Library, Mark 13, Oxford, England.

[25] H.D. Shapiro, Increasing robustness in global adaptive quadrature through interval selection, ACM trans. Math. Soft., **10**, 1984, pp. 117-139.

[26] A.H. Stroud, Approximate calculation of multiple integrals, 1972, Prentice Hall.

[27] P.Wynn, .On a device for computing the $e_m(S_n)$ transformation, Math. Tables and Aids to Computation, **10**, 1956, pp. 91-96.

[28] O.C. Zienckiewicz, The finite element method in engineering science, McGraw-Hill, 1971.·

AN ALGORITHM FOR AUTOMATIC INTEGRATION OF CERTAIN SINGULAR FUNCTIONS OVER A TRIANGLE

RICOLINDO CARIÑO*
La Trobe University
Bundoora VIC 3083
Australia

IAN ROBINSON
La Trobe University
Bundoora VIC 3083
Australia

ELISE DE DONCKER
Western Michigan Univ.
Kalamazoo MI 49008
USA

ABSTRACT. We describe an automatic cubature routine which, in addition to being suitable for regular functions, is specifically designed for functions which are singular at a vertex or along an edge of the triangle of integration. The algorithm uses a non-adaptive subdivisional strategy based on the harmonic sequence of mesh ratios and incorporates the Levin u-transformation for extrapolation. Numerical results are presented which indicate favorable comparison with other published algorithms when the integrand function has the prescribed properties.

1. Introduction

Most automatic routines for integration over triangles are adaptive. Such routines use cubature rules with a moderate to high degree of precision and incorporate sophisticated procedures for storage of information relating to subdivisions of the original triangle. Among published routines of this type are TRIADA by Haegemans [9], CUBTRI by Laurie [11], TRIEX by de Doncker and Robinson [5], TWODQD by Kahaner and Rechard [10], and DCUTRI by Berntsen and Espelid [3]. In addition to being adaptive, TRIEX incorporates the epsilon algorithm for extrapolation, while TWODQD and DCUTRI can be used to integrate over a set of triangles.

In this paper, we describe an automatic integration routine which is non-adaptive and is specifically designed for well-behaved integrand functions and for functions which have an algebraic singularity at a vertex or along a side of the triangle of integration. Distinctive of the routine is its combined use of a moderate degree cubature rule, subdivision using the harmonic mesh sequence and the Levin u-transformation (see Levin [12]).

The suitability of the algorithm depends on the asymptotic quadrature error expansion associated with the integrand function (see Cariño, Robinson and de Doncker [4]). In the next section, we review some known asymptotic quadrature error expansions and list the types of integrand functions for which the Levin u-transformation is applicable. In Section 3, we give an outline of the algorithm (implemented as the FORTRAN subroutine TRILEV) and describe its various components. Finally, results of numerical testing are presented in Section 4.

---

*On leave from the University of the Philippines at Los Baños, Laguna, Philippines.

T. O. Espelid and A. Genz (eds.), Numerical Integration, 295–304.

2. Review of the Theory

Without loss of generality, we consider the numerical evaluation of the integral

$$If = \int_0^1 \int_0^{1-x} f(x,y)\, dy\, dx. \tag{1}$$

The exact value If can be approximated by the ν-point cubature rule Q given by

$$Qf = \sum_{s=1}^{\nu} w_s f(u_s, v_s). \tag{2}$$

The m^2-copy version of this rule ($Q^{(m)}f$) is obtained by subdividing the unit triangle into m^2 equal subtriangles using the lines $x = \text{integer}/m$, $y = \text{integer}/m$, and $x+y = \text{integer}/m$, and applying a properly scaled version of Q to each.

We consider three classes of integrand function for which the asymptotic error expansion for $Q^{(m)}f$ is known:

$\mathcal{R}$: f is regular in x and y

$\mathcal{V}$: f has a vertex singularity of the form $f(x,y) = r^\alpha g(x,y)$,
  where $r^2 = x^2 + y^2$, $\alpha > -2$ and g is regular in x and y,

$\mathcal{B}$: f has a boundary singularity of one of the forms
  $f(x,y) = x^\lambda g(x,y)$ or $f(x,y) = y^\lambda g(x,y)$,
  where $\lambda > -1$, and g is regular in x and y.

When $f \in \mathcal{R}$, the expansion for $Q^{(m)}f$ is the well-known Euler-Maclaurin expansion

$$Q^{(m)}f \sim If + \sum_{s \geq 1} \frac{B_s}{m^s}. \tag{3}$$

Further, if Q is symmetric, $B_s = 0$ for odd s, and if Q has degree of precision d, then $B_s = 0$, $s \leq d$.

Lyness and Monegato [16] establish the asymptotic error expansion of $Q^{(m)}f$ when $f \in \mathcal{V}$. In this case $Q^{(m)}f$ behaves like

$$Q^{(m)}f \sim If + \sum_{s \geq 2} \frac{A_{\alpha+s}}{m^{\alpha+s}} + \sum_{s \geq 2} \frac{C_{\alpha+s} \log m}{m^{\alpha+s}} + \sum_{s \geq 1} \frac{B_s}{m^s}. \tag{4}$$

Some of the coefficients may vanish e.g., when Q is symmetric, $B_s = C_s = 0$ for odd s; when Q has degree precision d, then $B_s = C_s = 0$, $s \leq d$; and the C-coefficients vanish if α is non-integer or positive even.

Sidi [18] and Lyness and de Doncker-Kapenga [17] establish the asymptotic error expansion of $Q^{(m)}f$ when $f \in \mathcal{B}$. In this case, $Q^{(m)}f$ behaves like

$$Q^{(m)}f \sim If + \sum_{s \geq 1} \frac{E_{\lambda+s}}{m^{\lambda+s}} + \sum_{s \geq 1} \frac{B_s}{m^s}. \tag{5}$$

If the rule Q is symmetric, then $B_s = 0$ for odd s, and if Q has degree of precision d, then $B_s = 0$, $s \leq d$.

In [4], it is shown that, given a sequence of m^2-copy approximations $Q^{(m)}f$, where $\{1/m\}$ follows the harmonic sequence $\{1, \frac{1}{2}, \frac{1}{3}, \frac{1}{4}, \ldots\}$, the expansions (3), (4) and (5) can be re-expressed in the form

$$Q^{(m)}f \sim If + R_m \sum_{s \geq 0} p_s/m^s + E_d(m), \quad p_0 \neq 0, \tag{6}$$

where $R_m = ma_m$, with $a_1 = Q^{(1)}f$ and $a_m = Q^{(m)}f - Q^{(m-1)}f$, $m \geq 2$. Here, d is the degree of precision of Q and

$$f \in \mathcal{R} \quad \Rightarrow \quad E_d(m) = 0$$

$$f \in \mathcal{V} \quad \Rightarrow \quad E_d(m) = \begin{cases} O(m^{-(d+1)}), & \alpha \text{ non-integer} \\ O(m^{-(d+1)} \log m), & \alpha \text{ integer} \end{cases} \quad \text{and}$$

$$f \in \mathcal{B} \quad \Rightarrow \quad E_d(m) = O(m^{-(d+1)}).$$

Furthermore, the Levin u-transformation successively eliminates terms in the summation in (6). Thus, if $f \in \mathcal{R}$, increasingly accurate estimates of If are obtained until round-off error dominates. When $f \in \mathcal{V} \cup \mathcal{B}$, the accuracy attained is limited by the remainder term $E_d(m)$ (or by round-off).

3. Description of the Algorithm

3.1. OVERVIEW

TRILEV computes an approximation to the integral

$$If = \int_T f(x, y) \, dy \, dx,$$

where T is a triangle with vertices (a_i, b_i), $i = 1, 2, 3$. The user must provide values for the following arguments:

fsum   - name of a subroutine for evaluating $\sum_{i=1}^n f(x_i, y_i)$ (see §3.2 for an explanation of how the points (x_i, y_i) and the value of n are determined);

a, b   - arrays containing vertex abscissae and ordinates of triangle T;

epsa   - absolute accuracy requirement for the computed approximation;

epsr   - relative accuracy requirement for the computed approximation;

maxnfe - a limit on the number of function evaluations to be used;

key   - number of different vertex and/or line singularities of the type $\mathcal{V}$ or $\mathcal{B}$ in the integrand function.

TRILEV returns:

result   - the computed approximation, hopefully satisfying

$$|\text{result} - If| \leq \max(\text{epsa}, \text{epsr} * |If|) \tag{7}$$

errest - an estimate of the absolute error in **result**;
nfe   - the actual number of function evaluations used;
ier   - a termination condition indicator.

The structure of the algorithm is as follows:

subroutine TRILEV **(fsum, a, b, epsa, epsr, maxnfe, key,**
                    result, errest, nfe, ier)
    begin {TRILEV}
        $m = 0$; **ier** = -1;
        [check validity of inputs; set **ier** accordingly];
        while (**ier** = -1) *do*
            $m = m + 1$;
            [compute m^2-copy cubature sum];
            [extrapolate by Levin *u*-transformation];
            [compute error estimates; choose "best" approximation];
            [check for (non-)convergence; set **ier** accordingly]
        endwhile
    end {TRILEV}

Normal termination of the subroutine (**ier**=0) occurs when the error estimate satisfies a condition like (7). Abnormal termination occurs when the number of function evaluations exceeds the user-supplied **maxnfe** (**ier**=1) or when the estimated round-off errors introduced by the extrapolation become dominant (**ier**=2). Other values of **ier** are possible and these are described in the subroutine.

We describe the computation of the cubature sums, extrapolation by the Levin *u*-transformation and the error estimation procedure in the remainder of this section.

3.2.   COMPUTATION OF THE CUBATURE SUMS

The function values used in the cubature sums are computed in the subroutine **fsum** which must be provided in the form:

```
subroutine fsum(n,x,y, sum)
integer n,i
double precision x(n),y(n), sum,fxy
sum = 0
do 10 i=1,n
   [code to compute fxy=f(x(i),y(i)) ]
   sum = sum + fxy
10   continue
end
```

The arrays **x** and **y** in this subroutine contain those abscissae and ordinates of the evaluation points of the m^2-copy rule $Q^{(m)}f$ which have the same weights. For example, if the points of Q which have equal weights are denoted by "•" as in Figure 1a, then the 3^2-copy rule

$Q^{(3)}f$ will have **n**=27 evaluation points shown by "." in Figure 1b. The coordinates of these points are passed to the subroutine **fsum**. This technique reduces the overhead cost associated with what otherwise would be 27 subprogram calls (see Gladwell [8]). At stage m, for this example, the usual strategy of calling a function within a nested loop would require $3m^2$ subprogram instantiations, while using the above method, only one subroutine call is made (at the expense of 2n memory locations). Although the number of function evaluations is unchanged, the overhead is substantially reduced.

Figure 1. Sample point placements of a three-point rule and its 3^2-copy version.

To compute the m^2-copy approximations $Q^{(m)}f$, one of the degree 4, degree 6 or degree 8 Lyness-Jespersen rules [13] is used depending on the value of **key** and on the required accuracy. These rules are symmetric, all of the abscissae lie inside the triangle of integration and the weights are positive. Furthermore, the number of points used by each rule is low with respect to the lower bound on the minimum number possible for rules of the same degree.

For well-behaved integrand functions (**key**=0), (6) indicates that the Levin transformation is capable of eliminating the error in $Q^{(m)}f$ up to round-off level. However, rather than rely on extrapolation in this case, it is generally more cost effective to use a moderately high degree rule; this will normally provide quite accurate approximations and only a few extrapolation steps will be required. The degree 8 rule is used in TRILEV.

If $f \in \mathcal{V} \cup \mathcal{B}$, (6) indicates that the Levin transformation is capable of eliminating the error terms up to the order of the remainder term $E_d(m)$ only. With efficiency in mind, we choose the rule according to the accuracy prescribed by the user, the aim being to render the remainder term less than that of the accuracy requirement. Taking into account the symmetry of the Lyness-Jespersen rules, the remainder term is, respectively, $O(m^{-6})$, $O(m^{-8})$ and $O(m^{-10})$ for the degree 4, degree 6 and degree 8 rules (multiplied by $\log(m)$ if $f \in \mathcal{V}$ and α is an integer). Further, numerical experiments in double precision arithmetic indicate that round-off accumulation generally limits the computational accuracy of the Levin transformation after approximately $m = 10$ subdivisions. Combining these two observations, we arrive at the following rule selection strategy for **key**=1:

> *if* eps $\geq 10^{-5}$ *then*
>   use degree 4 rule
> *elseif* $10^{-7} \leq$ eps $< 10^{-5}$ *then*
>   use degree 6 rule
> *else* {eps $< 10^{-7}$}
>   use degree 8 rule
> *endif*

In the above, **eps** = max(epsr*|**result**|,epsa). If **epsr** is non-zero, the initial value of **result** is obtained by approximating If using the degree 8 rule.

For integrand functions with mixed singularities (**key=2**), the degree 4 rule is chosen since the Levin transformation is only marginally effective in these cases and this choice will minimize the number of function evaluations needed.

3.3. COMPUTATION OF THE LEVIN TRANSFORMATION

Let $f \in \mathcal{R}$ and let $Q^{(m)}f$, $m = 1, 2, \ldots$, be a sequence of m^2-copy rule approximations to If. Then a new approximation T_{kn} to If can be obtained by solving the $k + 1$ linear equations

$$Q^{(m)}f = T_{kn} + R_m \sum_{i=0}^{k-1} p_i/m^i, \ m = n, n+1, \ldots, n+k,$$

where $R_m \neq 0$ for any m.

When $f \in \mathcal{V} \cup \mathcal{B}$, we also assume $k \leq d$, the degree of precision of the rule used. A closed-form solution for T_{kn}, given by Levin [12], is

$$T_{kn} = \frac{\sum_{j=0}^{k}(-1)^j \binom{k}{j}(n+j)^{k-1}(Q^{(n+j)}f/R_{n+j})}{\sum_{j=0}^{k}(-1)^j \binom{k}{j}(n+j)^{k-1}(1/R_{n+j})}.$$

Typically, the approximations T_{kn} can be arranged in a triangular table (see Table 1) which is generated one row at a time, with the entries T_{0m}, $m = n, n+1, n+2, \ldots$ being the cubature sums $Q^{(m)}f$.

Table 1: Levin table

$$
\begin{array}{lllll}
T_{0n} & & & & \\
T_{0,n+1} & T_{1n} & & & \\
T_{0,n+2} & T_{1,n+1} & T_{2n} & & \\
\vdots & \vdots & \vdots & \ddots & \\
T_{0,n+k} & T_{1,n+k-1} & T_{2,n+k-2} & \cdots & T_{kn}
\end{array}
$$

In a FORTRAN implementation by Fessler, Smith and Ford [7], the Levin transformation is calculated as

$$T_{kn} = \frac{N_{kn}}{D_{kn}},$$

where both N_{kn} and D_{kn} are recursively calculated as

$$H_{kn} = H_{k-1,n+1} - c_{kn} H_{k-1,n}.$$

Here, $c_{kn} = n(n+k-1)^{k-2}/(n+k)^k$ and the starting values are $N_{0n} = Q^{(n)}f/(nR_n)$ and $D_{0n} = 1/(nR_n)$.

An equivalent computational scheme is given in [4]:

$$T_{kn} = T_{k-1,n+1} + \frac{c_{kn}D_{k-1,n}}{D_{kn}}(T_{k-1,n+1} - T_{k-1,n}).$$

Calculation of T_{kn} using this formula requires only the storage of T_{kn} and D_{kn}, and round-off is removed to the "correction" term. This is the formula used in the algorithm.

3.4. ERROR ESTIMATION

Two error estimation heuristics are used in TRILEV. The first is based on the difference between successive entries in the last row of the Levin table. We note that the difference $|T_{jn} - T_{j+1,n-1}|$, $j < \min(k,d)$, will be dominated by the error in T_{jn} as $n \to \infty$. However in practice, this difference may be an underestimate of the actual error; for safety, we use

$$|T_{jn} - If| \simeq |T_{jn} - T_{j+1,n-1}| + |T_{jn} - T_{j+2,n-2}|. \tag{8}$$

This error estimate is used when the Levin transformation is known to work well, i.e., when **key**=0 or when **key**=1 and $j \le d$ (degree of rule); otherwise the second heuristic is employed.

Assume the error in T_{jn} is dominated by $Cn^{-\gamma}$, $\gamma > 0$. Then

$$
\begin{aligned}
T_{jn} - T_{j,n-1} &\simeq Cn^{-\gamma}\left[1 - (1 - \tfrac{1}{n})^{-\gamma}\right] \\
&= -C\gamma n^{-\gamma-1} + \cdots,
\end{aligned}
$$

which suggests the use of $n|T_{jn} - T_{j,n-1}|$ as an error estimate for T_{jn}. We add the difference $|T_{jn} - T_{j+1,n-1}|$ (which is of the same order as the error in T_{jn}) to compensate for possible unreliability, giving the error estimate

$$|T_{jn} - If| \simeq n|T_{jn} - T_{j,n-1}| + |T_{jn} - T_{j+1,n-1}|. \tag{9}$$

This is the error estimate used when **key**=2, or when $j > d$, where the remainder term $E_d(n)$ dominates the error in T_{jn}.

The propagation of round-off error is also a common problem in extrapolation when the harmonic sequence is used as the mesh sequence. Since the Levin transformation implicitly solves a linear system, we can rely on the arguments given in [14] to show that the extrapolation condition number C_{kn} can be calculated using precisely the same algorithm as that used to compute T_{kn}, but with initial values $(-1)^m$ replacing T_{0m}. The magnification error resulting from small errors in the cubature approximations (assuming the computations are carried out exactly) is then $\epsilon|C_{kn}|$, where ϵ is set to the relative machine accuracy. If this quantity is greater than the error estimate given by (8) or (9), then round-off is deemed to be dominant. When this occurs, the algorithm is terminated abnormally.

4. Numerical Results

To illustrate the performance of TRILEV for functions in $\mathcal{R}$, $\mathcal{V}$ and $\mathcal{B}$, we use the performance testing technique of Lyness and Kaganove [15] as implemented in the subroutine TRITST by Berntsen [2].

We select test family 7 (oscillatory integrand function) of [2] as representative of class $\mathcal{R}$: the error expansion associated with this family is of the form assumed by the Levin u-transformation. For this family, 500 sample integrands with difficulty parameters $\alpha_1 + \alpha_2 = 30$ and $\beta_1 \in [0,1]$ are used. The effectiveness of TRILEV for this family is exemplified in Figure 2 where the results of running TRILEV, TRIEX and DCUTRI are given.

Although class $\mathcal{V}$ is not directly included in TRITST, we can subdivide the triangle of integration for test family 1 so that the resulting integrands will belong to classes $\mathcal{R}$ or $\mathcal{V}$.

Specifically, the subdivision is obtained by joining the origin with the point $(\beta_1, 1 - \beta_1)$, and joining $(\beta_1, 0)$ and $(\beta_1, 1 - \beta_1)$. The problem is now reduced to integrating two functions belonging to class $\mathcal{V}$ and one function belonging to class $\mathcal{R}$. In Figure 3, we compare the results obtained by TRILEV, TRIEX and DCUTRI for test family 1 after this subdivision. Five hundred (500) sample integrands with $\beta_1 \in [0, 1]$ are used.

To test TRILEV for functions $f \in \mathcal{B}$, we use 500 samples of the test family $f(x, y) = c_1 x^{-1/2} + c_2 y^{-1/2}$, $0 \leq c_1, c_2 \leq 50$, which has algebraic singularities along the x-axis and/or y-axis. A comparison of the performance of TRILEV and TRIEX is given in Figure 4. DCUTRI was not able to return answers with more than an average of 2 correct digits for this family.

The effect of the rule selection strategy in TRILEV is evident in Figures 3 and 4 where the rate of increase in the number of function evaluations required to achieve accuracies $\mathbf{epsr} < 10^{-5}$ (and $\mathbf{epsr} < 10^{-7}$) is greater than for $\mathbf{epsr} \geq 10^{-5}$ (and $\mathbf{epsr} \geq 10^{-7}$). This is a direct result of the need to use higher degree (and therefore more expensive) rules to obtain finer accuracies.

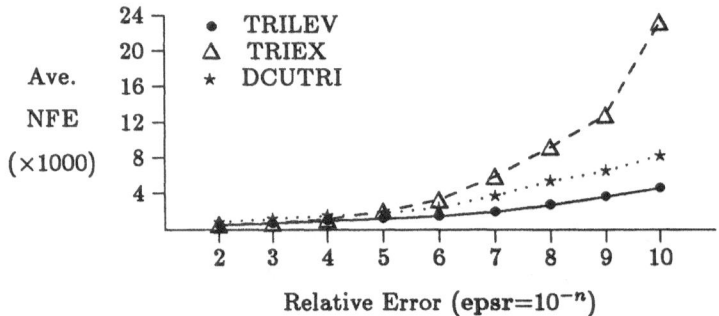

Figure 2. Comparative results of TRILEV, TRIEX and DCUTRI for $\int_0^1 \int_0^{1-x} \cos(2\pi\beta_1 + \alpha_1 x + \alpha_2 y) dy dx$.

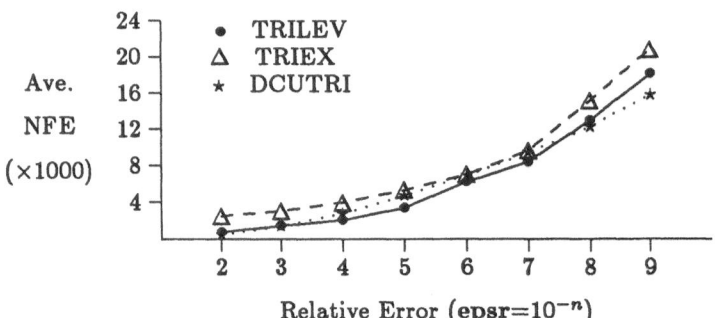

Figure 3. Comparative results of TRILEV, TRIEX and DCUTRI for $\int_0^1 \int_0^{1-x} (|x - \beta_1| + y)^{-0.9} dy dx$ (after subdivision).

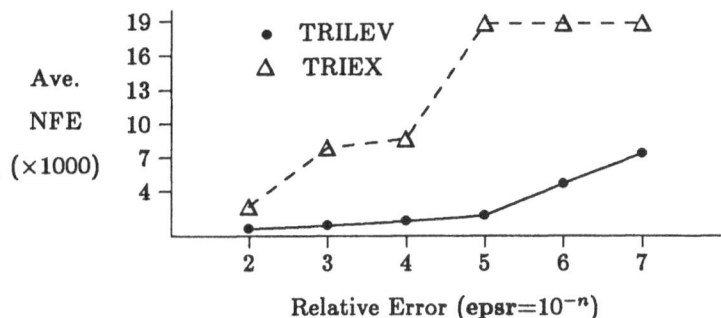

Figure 4. Comparative results of TRILEV and TRIEX for $\int_0^1 \int_0^{1-x}(c_1 x^{-1/2} + c_2 y^{-1/2})dy dx$.

5. Concluding Remarks

Limited numerical testing indicates that TRILEV is capable of efficiently attaining moderate accuracy approximations to integrals over a triangle when the integrand function has certain algebraic vertex or boundary singularities. It can also be adapted for integration over any other region that can be subdivided into triangles so that on each triangle, the integrand function is regular or has an algebraic singularity at a vertex or along a side. In this manner, TRILEV is applicable for a wide range of integration problems, including those where the integrand function has interior point or line singularities. Further, its non-adaptive nature makes it suitable for parallelization, since even for functions with such singularities, a parallelized version of TRILEV would make significant use of available processors.

References

[1] J. Berntsen (1989), "TRITST: A subroutine for evaluating the performance of subroutines for automatic integration over triangles," *Report No. 34*, Dept. of Informatics, University of Bergen, Norway.

[2] J. Berntsen, and T.O. Espelid (1989), "A test of DCUTRI and TWODQD," *Report No. 40*, Dept. of Informatics, Univ. of Bergen, Norway.

[3] J. Berntsen, and T.O. Espelid (1991), "DCUTRI: An algorithm for adaptive cubature over a collection of triangles," *ACM Trans. Math. Softw.* (to appear).

[4] R.L. Cariño, E. de Doncker and I. Robinson (1990), "Approximate integration by the Levin transformation," *Technical Report 9/90*, Dept. of Computer Science and Computer Engineering, La Trobe University, Australia (submitted for publication).

[5] E. de Doncker and I. Robinson (1984), "TRIEX : Integration over a triangle using non-linear extrapolation," *ACM Trans. Math. Softw.*, v. 10, pp. 17-22.

[6] E. de Doncker-Kapenga (1987),"Asymptotic expansions and their applications in numerical integration," *P. Keast and G. Fairweather (eds.), Numerical Integration: Recent Developments, Software, and Applications*, pp. 141-151.

[7] T. Fessler, W.F. Ford and D.A. Smith (1983), "HURRY: An acceleration algorithm for scalar sequences and series," *ACM Trans. Math. Software*, v. 9, pp. 346-354.

[8] I. Gladwell (1987),"Vectorization of one dimensional quadrature codes," *P. Keast and G. Fairweather (eds.), Numerical Integration: Recent Developments, Software, and Applications*, pp. 231-238.

[9] A. Haegemans (1977), "An Algorithm for integration over a triangle," *Computing*, v. 19, pp. 179-187.

[10] D. Kahaner and O. Rechard (1987), "TWODQD: an adaptive routine for two-dimensional integration," *J. Comp. Applied Math.*, v. 17, pp. 215-234.

[11] D. Laurie (1982), "CUBTRI: Automatic Cubature over a Triangle," *ACM Trans. Math. Soft.*, v. 8, pp. 210-218.

[12] D. Levin (1973), "Development of non-linear transformations for improving convergence of sequences," *Intern. J. Computer Math.*, v. B3, pp. 371–388.

[13] J.N. Lyness and D. Jespersen (1975), "Moderate degreee symmetric rules for the triangle," *J. Inst. Maths Applics*, v. 15, pp. 19-32.

[14] J.N. Lyness (1976), "Applications of extrapolation techniques to multidimensional quadrature of some integrand functions with a singularity," *J. Comp. Physics*, v. 20, pp. 346-364.

[15] J.N. Lyness and J.J Kaganove (1977), "A technique for comparing automatic quadrature routines," *Computer Journal*, v. 20, pp. 170-177.

[16] J.N. Lyness and G. Monegato (1980), "Quadrature error expansions when the integrand has singularities at vertices," *Math. Comp.*, v. 34, pp. 213-225.

[17] J.N. Lyness and de Doncker-Kapenga (1987), "On quadrature error expansions - Part 1," *J. Comp. Appl. Math.*, v. 17, pp. 131-149.

[18] A. Sidi (1983), "Euler-Maclaurin expansions for integrals over triangles and squares of functions having algebraic/logarithmic singularities along an edge," *J. Approx. Theory*, v. 39, pp. 39-53.

CUBPACK: PROGRESS REPORT

Ronald COOLS and Ann HAEGEMANS
Department of Computer Science
Katholieke Universiteit Leuven
Celestijnenlaan 200 A
B-3001 Heverlee, Belgium
Email :ronald@cs.kuleuven.ac.be and ann@cs.kuleuven.ac.be

ABSTRACT. In this paper we report on progress in developing CUBPACK and describe some alternative adaptive schemes for numerical multidimensional integration that are being considered for CUBPACK. These strategies are compared with the classical strategy that is implemented in most existing routines. For singular and discontinuous integrands an improvement up to 50% in efficiency is achieved.

1 Introduction

Several routines are now available for adaptive integration in one or more dimensions where the region of integration is a (union of) hyperrectangles or simplices. All these routines are implementations of the same globally adaptive algorithm, which is described in the next section.

Although all routines essentially use the same algorithm, the only piece of code they sometimes share is the function R1MACH/D1MACH that returns some machine constants. Of course not all authors used the same structure to store information. Sometimes an ordered list and sometimes a heap, which hides a partially sorted tree, is used. The choice made influences most lines in all existing codes because the data structure is visible everywhere. It is very time consuming and difficult if you want to change the data structure in an existing routine.

One of the aims of CUBPACK is to offer a collection of routines for automatic n-dimensional ($n \geq 1$) integration of a set of functions over a collection of regions. At this point, only the framework is finished and the program is in an experimental stage, but the code is so constructed that all existing and new algorithms can easily be implemented.

We will first give some general ideas about automatic integration. Then we will report on the data structure we used, on some results with a double adaptive strategy and suggestions for parallel implementation.

T. O. Espelid and A. Genz (eds.), Numerical Integration, 305–315.

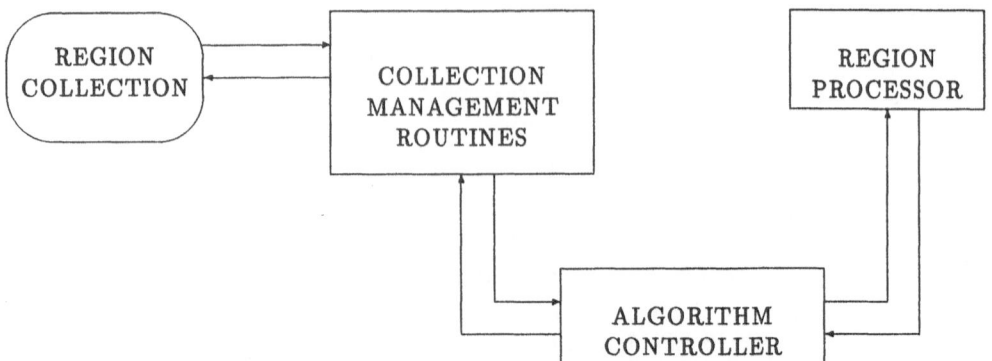

Figure 1: Blockdiagram of global adaptive cubature algorithm.

2 A globally adaptive cubature algorithm

The usual algorithm for adaptive quadrature and cubature can graphically be represented by the block diagram shown in Figure 1. In contrast with Rice [8], we disconnect the link between the region processor and the region collection. A high level description of the classical globally adaptive scheme is given in Algorithm 1.

The region collection, which contains all information about the regions that is needed for future reference, is organized using some data structure. The region collection management routines create and maintain the region collection. The algorithm controller takes region(s) R_k from the collection and passes them to the region processor. In a globally adaptive algorithm the region with the largest error is selected from the collection for further processing. The region processor tries to improve the estimates for the integrals $\hat{Q}_k$ and their error $\hat{E}_k$ over the regions it receives from the algorithm controller. Usually the region processor first divides the region in 2 or 2^N subregions with equal volume (N is the dimension of the region) and then computes new estimates $\hat{Q}_k^{(i)}$ for the integrals over each subregion using a fixed local integral estimator and error estimator. The results from the region processor are returned to the algorithm controller, which passes all relevant information to the region collection management routines. The algorithm controller decides when to terminate the algorithm. This is usually done when the total estimated error $\hat{E}$ becomes smaller than the requested error.

If the integrand is a vector of functions, then $\hat{Q}$, $\hat{Q}_k$, $\hat{E}$ and $\hat{E}_k$ are also vectors. The algorithm controller terminates the algorithm if the largest component of the vector $\hat{E}$ becomes smaller then ϵ.

Initialize the collection of regions with M given regions;

Produce $\hat{Q}_k$ and $\hat{E}_k$ for $k = 1, 2, \ldots, M$;

Put $\hat{Q} = \sum_{k=1}^{M} \hat{Q}_k$ and $\hat{E} = \sum_{k=1}^{M} \hat{E}_k$;

while $\|\hat{E}\|_\infty > \epsilon$ **do**

begin

    Take some region(s) from the collection;

    Process these region(s)

        Divide the region into $s := 2$ or 2^N subregions with equal volume;

        Compute $\hat{Q}_k^{(i)}$ and $\hat{E}_k^{(i)}, i = 1, 2, \ldots, s$ using an 'a priori' chosen integration rule and error estimator;

    Update $\hat{Q}$ an $\hat{E}$

$$\hat{Q} = \hat{Q} + \left(\sum_{i=1}^{s} \hat{Q}_k^{(i)} - \hat{Q}_k\right);$$
$$\hat{E} = \hat{E} + \left(\sum_{i=1}^{s} \hat{E}_k^{(i)} - \hat{E}_k\right);$$

    Put the new regions in the collection

    and put $M = M + s - 1$;

end

Algorithm 1: A globally adaptive quadrature/cubature algorithm with standard Region Processor.

---

3 The data structure

The routines that deal with a certain region only need information about that specific region. They do not need information on other regions. So we have chosen a hidden data structure. The region processor has no access to the data structure. All communication is done via the region collection management routines and controlled by the algorithm controller (see Figure 1).

At the same time we have constructed a flexible data structure, which is independent of the sorting algorithm. Useful information for future use, i.e. history, can be kept, and several independent region collections can be used in one program. This may be useful for parallel computations.

With each region in the collection we associate a record that contains all information we want to save. The record thus contains :

- information that describes the region, i.e. dimension, type of region, number of vertices, vertices, ...;

- the results from the region processor, i.e. approximation for the integrals over the region and error estimates;

- useful information for future use, e.g. the volume of the region, the relation between volume of the original region and this subregion, known function values, difficult spots, degree of last rule used, ...

In the current implementation [3] these records appear in a partially sorted binary tree with the element with the largest sortkey on the top, or in an unsorted pool. The records have a fixed place. Only pointers to these records are modified if a new element is added or if the top is removed. If the top element is removed from the tree, the place where the record was stored can be reused.

At the moment the package contains the following routines :

DSINIT   :   initialize a region collection

DSSPUT   :   add a record to the sorted subcollection

DSUPUT   :   add a record to the unsorted subcollection

DSGET   :   pick the region with the largest error from
               the sorted part of the region collection

DSUSED   :   return the number of used records

DSFREE   :   return the number of free records

DSPINT   :   move all records from the unsorted subcollection
               to the sorted subcollection

The data structure and the algorithm controller are independent of the dimension and the type of the region. The only two region dependent routines are DIVIDE and COMPUTE.INTEGRAL.AND.ERROREST that are called by the region processor.

4   Double adaptivity

In most previously developed algorithms the statement 'Process these subregion(s)' is implemented as described in Algorithm 1. The region processor consists of 2 independent parts that are executed sequentially: first divide, then compute new approximations for each region using an a priory chosen integration rule and error estimator. We would now like to describe alternate subdivision strategies, where the region processor is allowed to decide whether a region should be subdivided or not.

Suppose the region processor has a sequence of integration rules $Q_1, \ldots, Q_n$ and error estimators $E_1, \ldots, E_n$ with Q_i less accurate than Q_{i+1}. With each rule a number $rule.used$ is associated. Then the region processor can follow Algorithm 2. For $n = 1$ Algorithm 2 is the classical Algorithm 1. For $n = \infty$ a nonadaptive integrator is obtained.

The length of the sequence in Algorithm 2 is fixed in advance and the complete sequence is applied to a region before it is subdivided. A more adaptive approach is obtained if the region processor is not forced to use the next rule from the sequence but at each step can choose between taking the next rule or subdividing.

What are reasons to subdivide? One reason is that all available rules were applied to the region. This is again Algorithm 2. Another reason to subdivide is that subdividing

---

If this is the first time the region is passed to the region processor
>   **then**
>> $rule.used := 1;$
>   **else**
>> $rule.used := rule.used + 1;$
endif
If $rule.used \leq n$
>   **then**
>>   Compute new approximations $\hat{Q}$ and $\hat{E}$ using
>> $Q_{rule.used}$ and $E_{rule.used};$
>   **else**
>>   Divide the region into s subregions;
>>   Process these regions;
endif

Algorithm 2: A Region Processor using a sequence of cubature rules.

---

and applying less accurate rules to each subregion may be more cost-effective than using a more accurate rule without subdividing at this stage. One can justify this by considering the behavior of the function as reflected by the sequence of approximations to the integral over the current region. The error estimation procedure of [5] checks if there is asymptotic behavior in the sequence and based on this tries to improve the error estimate of the current region. Let us review this procedure.

Suppose one has a sequence of cubature formulae $\{Q_i : i = 1, \ldots, n\}$. By combining 2 or more successive formulae one obtains a sequence of null rules or error estimators $\{E_i \geq 0 : i = 1, \ldots, n\}$. Let $d_i = degree(E_i)$ be the degree of the null rule E_i and d the degree of the most accurate cubature rule Q_n. The following procedure is used to deal with the difference between E_n, which is an error estimate for Q_{n-1}, and the error in Q_n, which is what is actually needed.
Suppose asymptotic behavior is observed, i.e.

$$r = \max(r_1, r_2, \ldots, r_{n-1}) < c < 1$$

where

$$r_1 = (E_2/E_1)^{\frac{1}{d_2-d_1}}, r_2 = (E_3/E_2)^{\frac{1}{d_3-d_2}}, \ldots, r_{n-1} = (E_n/E_{n-1})^{\frac{1}{d_n-d_{n-1}}}.$$

Then set

$$\hat{E} \approx F.E_n.r^\alpha$$

where $\alpha = degree(Q_n) - degree(E_n)$ and F a positive number. This is a very reliable procedure for error estimation when the constants c and F are well tuned, but this procedure is too pessimistic for double adaptivity. We propose using a higher degree rule instead of

---

If this is the first time the region is passed to the region processor

    then

        apply the first rule of the sequence;

    else

        If there is asymptotic behavior and the sequence can be extended

            then

                Compute new approximations using an extension of

                the sequence;

            else

                Divide the region into s subregions;

                Process these subregions;

            endif

  endif

Algorithm 3: A Region Processor using a double adaptive strategy

---

subdividing if

$$r_{n-1} = E_n/E_{n-1} < c < 1.$$

This leads to Algorithm 3.

5   Numerical results

We compared the three algorithms described in the previous section by computing integrals over triangles. We used the test package TRITST [1].

For Algorithm 1 we simulated DCUTRI [2] which uses a degree 13 rule with 37 points and null rules of degree 7, 5, 3 and 1.

For Algorithm 2 and 3 we used a sequence of 3 cubature rules with their respective null rules:

  Q_1 : degree 9 rule with 19 points [7]

            + orthonormalised null rules of degree 5, 4, 3, 2, 1

  Q_2 : degree 13 rule with 37 points [2]

            + orthonormalised null rules of degree 7, 5, 3, 1

  Q_3 : degree 19 rule with 73 points [4]

            + orthonormalised null rules of degree 10, 9, 8, 7

In the future we plan to construct and use a longer sequence of cubature rules.

In Figure 2 the results of the three strategies are compared for the integration of the function

$$f_1(x, y) = (\mid x - \beta_1 \mid + y)^{d_1}$$

(singularity on the x-axes) over the standard triangle with vertices $(0,0),(1,0),(0,1)$. β_1 is chosen random and d_1 is a difficulty parameter.
The three algorithms behave equally reliably, so we only report on the number of function evaluations (relative to the number of evaluations with Algorithm 1) as a function of the requested relative accuracy. For these test families an improvement up to 50% in efficiency can be achieved using Strategy 3.

In Figure 3 the results are given for different values of the difficulty parameter. For discontinuous functions and C_0 functions we get almost the same results.

For the Gaussian function

$$f_4(x, y) = \exp(-(\alpha_1^2(x - \beta_1)^2 + \alpha_2^2(y - \beta_2)^2))$$

or functions with a peak the results are given in Figure 4 and Figure 5. Here we remark a 20% more work for Algorithm 3. This may be explained by the fact that after subdivision we always start again with the degree 9 rule. Maybe it would be better if at the end of the sequence (where we have used the degree 19 rule) to subdivide and then use the degree 13 rule.

For oscillatory functions

$$f_7(x, y) = \cos(2\pi\beta_1 + \alpha_1 x + \alpha_2 y)$$

the results are given in Figure 6 and 7. here we remark that the degree 13 rule of Algorithm 1 is high enough, at least for weak oscillations.

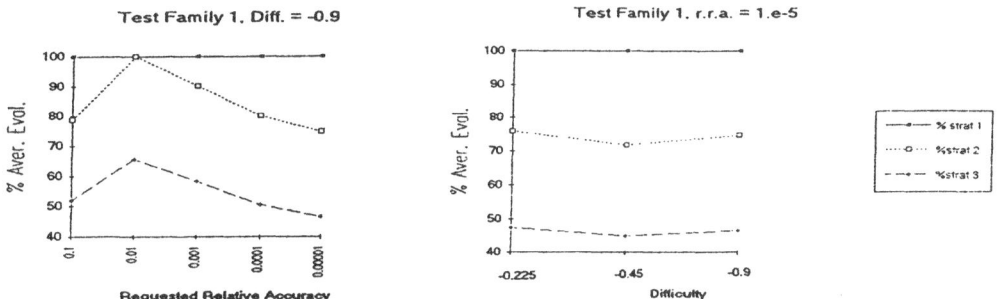

Figure 2 and Figure 3: % of average evaluations for test family 1.

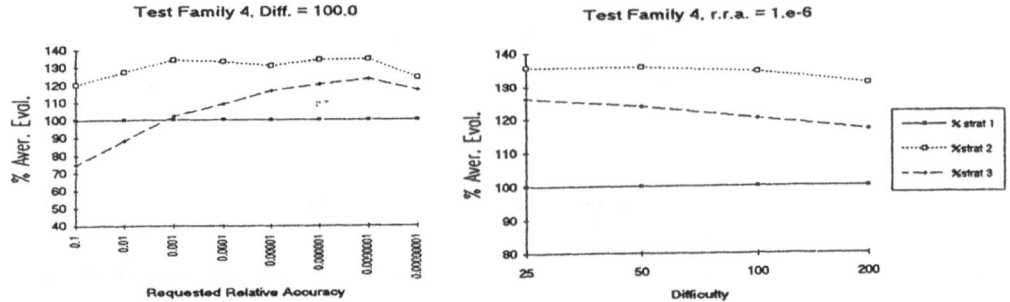

Figure 4 and Figure 5: % of average evaluations for test family 4.

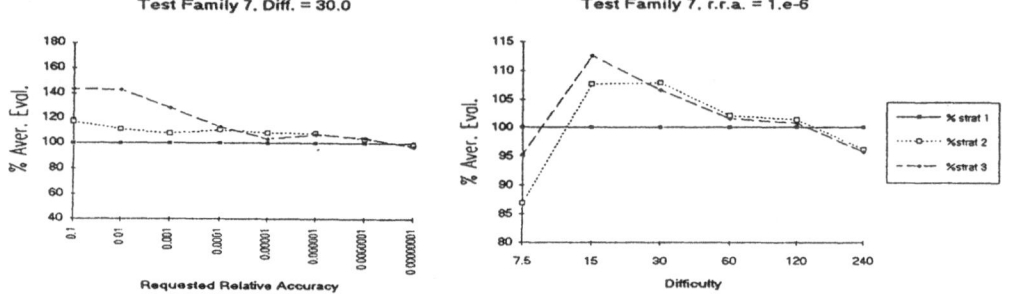

Figure 6 and Figure 7: % of average evaluations for test family 7.

6   One step look-ahead

A final subdivision strategy that we consider involves looking one step ahead. We process the region with and without subdivision and then decide which is the best. This may appear to be waste of time but is it does not need to be.

- In a multiprocessor environment this may be a good way to keep all processors busy.
- One can use a sequence of formulae designed so that the new function values needed for the higher degree formula are the same as those needed by the lower degree formula when the region is subdivided.

So, one obtains 2 new approximations for the integral and the error estimates:

- Compute $\hat{Q}_1$, $\hat{E}_1$ using the higher degree rule Q_1 on the undivided region.
- Compute $\hat{Q}_2^{(i)}$, $\hat{E}_2^{(i)}$, $i = 1, \ldots, s$ using a lower degree rule Q_2 on the s subregions.

This is described in Algorithm 4.

If this is the first time the region is passed to the region processor

    then

        rule.used :=1;

        Compute new approximations $\hat{Q}$ and $\hat{E}$;

        using $Q_{rule.used}$ and $E_{rule.used}$;

    else

        Compute $\hat{Q}_1$ and $\hat{E}_1$ using $Q_{rule.used+1}$ and $E_{rule.used+1}$;

        Compute $\hat{Q}_2$ and $\hat{E}_2$ using $Q_{rule.used}$ and $E_{rule.used}$

        on s subregions;

        rule.used :=rule.used+1;

        If there is no reason to subdivide

            then

                Save $\hat{Q}_1^{(i)}$ and $\hat{E}_1^{(i)}$;

            else

                Divide the region into s subregions;

                Save $\hat{Q}_2^{(i)}$ and $\hat{E}_2^{(i)}$;

        endif

endif

Algorithm 4: Region Processor looking one step ahead

Several criteria can be applied for choosing between $\hat{Q}_1$ and $\hat{Q}_2 = \sum_{i=1}^{s} \hat{Q}_2^{(i)}$. We tested two of them:

- Algorithm 4a : If $\hat{E}_1 < \hat{E}_2 = \sum_{i=1}^{s} \hat{E}_2^{(i)}$ then $\hat{Q}_1$ is preferred, else $\hat{Q}_2$.

- Algorithm 4b : If $\hat{E}_1 < s . \min \left\{ \hat{E}_2^{(i)}, i = 1, 2, \ldots, s \right\}$ then $\hat{Q}_1$ is preferred, else $\hat{Q}_2$.

The second criterion was incorporated in an experimental routine in [6]. In Figure 8, 9 and 10 we give some results for these strategies.

The results for strategy 4a and 4b are comparable. For discontinuous functions, C_0-functions and functions with a singularity on the x-axes (test family 1) with algorithm 4 an improvement in efficiency is achieved for most values of the difficulty parameters.
For other classes of functions algorithm 4 requires about two times as much function evaluations as the classical algorithm. However, using a multiprocessor environment, this proportion can be changed.

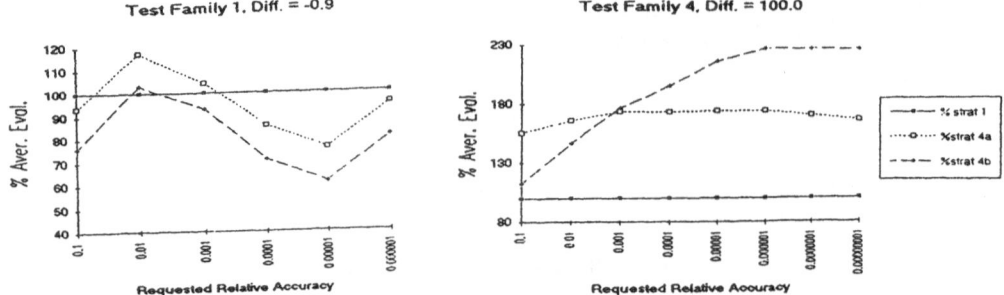

Figure 8 and Figure 9: % of average evaluations for test family 1 and 4.

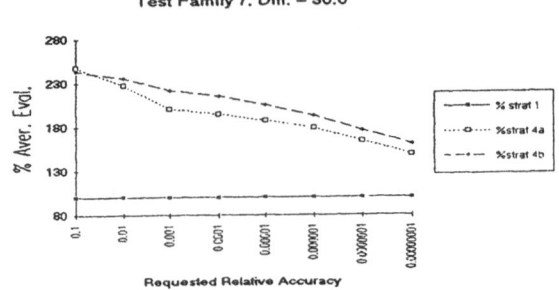

Figure 10: % of average evaluations for test family 7.

7   Conclusions

We tested several alternatives to the classical globally adaptive cubature algorithm. The results of using different degree formulae depending on the behavior of the integrand are promising but more work has to be done. We need longer series of (embedded) cubature formulae of different degrees. Because most of the decisions depend on the error estimator this is the crucial point in an integration routine. The Norwegian error estimator [5] seems to be very reliable but too conservative for certain classes of functions. Also, different division strategies must be tested.

References

[1] J. Berntsen. TRITST: A Subroutine for Evaluating the Performance of Subroutines for Automatic Integration over Triangles. Reports in Informatics 34, Dept. of Informatics, University of Bergen, 1989.

[2] J. Berntsen and T.O. Espelid. An Algorithm for Adaptive Cubature over a Collection of Triangles. Accepted by ACM. Trans. Math. Soft., 1991.

[3] R. Cools. A suite of codes for region collection management in adaptive numerical integration algorithms. Report TW 150, Dept. of Computer Science, K.U. Leuven, 1991.

[4] D. A. Dunavant. High degree efficient symmetrical gaussian quadrature rules for the triangle. *Internat. J. Numer. Methods Engrg.*, 21:1129–1148, 1985.

[5] T.O. Espelid. Integration rules, null rules and error estimation. Report in Informatics 33, Dept. of Informatics, University of Bergen, 1988.

[6] D.K. Kahaner and M.B. Wells. An experimental algorithm for n-dimensional adaptive quadrature. *ACM Trans. Math. Soft.*, 5:86–96, 1979.

[7] J.N. Lyness and D. Jespersen. Moderate degree symmetric quadrature rules for the triangle. *J. Inst. Maths Applics.*, 15:19–32, 1975.

[8] J.R. Rice. Parallel algorithms for adaptive quadrature III. Program correctness. *ACM Trans. Math. Softw.*, 2:1–30, 1976.

PARALLEL CUBATURE ON LOOSELY COUPLED SYSTEMS

ELISE de DONCKER
Western Michigan Univ.
Kalamazoo MI 49008
USA

JOHN KAPENGA
Western Michigan Univ.
Kalamazoo MI 49008
USA

ABSTRACT. We give a survey and results of our work on modifying global adaptive cubature algorithms for use on distributed memory systems. This is an application of a more general task pool management system for MIMD machines, which is under development. On loosely coupled systems, the pool of subregions (tasks) is distributed over the processors. The adaptive nature of the type of algorithm involved causes the grid to be progressively refined in order to meet a set error requirement. In view of a potentially different behavior of the integrand function in subregion sets assigned to different processors, the work loads of the processors may become imbalanced, with a few processors carrying out most of the work. This situation can be alleviated via dynamic load balancing. Results obtained with a simple load balancing strategy show its effects in case of a local integrand problem. An extension of the strategy is proposed, based on a model of the underlying work distribution.

1. Introduction

We introduced a scheme for global adaptive integration on loosely coupled systems in [5]. In the present paper we discuss implementation aspects, give timing results and indicate future extensions.

The next section describes the algorithm and highlights the use of asynchronous communications in an implementation for integration over a set of triangles. Section 3 demonstrates speedups for a problem with a "hot-spot" in mainly one processor. Finally, in section 4, a probabilistic model for the work in a priority queue is given and used to predict the work at a later time. This type of modeling can be employed at the basis of an adaptive dynamic load balancing strategy.

In the remainder of this section, we abstract the algorithm as a task partitioning scheme [8, 5] and deduce its interpretation when mapped to MIMD systems. A meta-algorithm for adaptive integration is represented in Figure 1. In a global adaptive method, the acceptance criterion is based on the size of the total estimated error over the entire domain while the selection of the next region (task) for partitioning is determined according to a priority queue of the local error estimates.

The invocation of partition() in Figure 1 results in one or more subdivision steps, each of which involves the selection of the next region to subdivide, its subdivision, integration over the subregions and their addition to the pool of regions.

As the grid refinement is led by the local cubature error estimates over the individual

317

T. O. Espelid and A. Genz (eds.), Numerical Integration, 317–327.

```
integrate() {
    initialize;
    while(acceptance criterion is not satisfied)
        partition();
}
```

Figure 1: Adaptive integration algorithm

subregions, the pool is managed by a priority queue handler. Within partition(), task pool primitives are needed, such as *get_tasks()* and *put_tasks()* for the selection from the local pool of one or more regions for subdivision, and for the insertion of the new subregions generated by the subdivision. The task pool management primitives are at the heart of a more general parallel task partitioning system which is under development. It is the goal of the latter to provide a layer on top of which the application (in this case cubature) can be ported to various machines and machine types. Note that we achieved an implementation for shared memory MIMD machines by means of a macro interface with the Argonne Labs monitor macros [9, 10, 1]. This was treated in [3, 8, 4].

On a MIMD system, consider the algorithm of Figure 1 mapped to each of a number of designated processes; in general we associate each process with a different processor (node) in the hardware configuration. In a shared memory architecture, the nodes access the same global memory. The task pool is global to the participating processes and is accessed via monitor operations. In loosely coupled systems (multicomputers), where each node has its own local memory, the nodes are connected by a message passing communication network. The pool is distributed over the local memories of the nodes. The abstraction of Figure 1 will be refined to apply to loosely coupled systems in the next section.

2. Parallel approach and implementation

The present implementation is written in C and uses the Reactive Kernel / Cosmic Environment (RK/CE) by Seitz et al. [12] as the node communication system. CE runs on networked Unix hosts. RK is a node operating system for multicomputers; it is the native OS on the Symult 2010 series, while several other parallel machines have RK available through a library interface. The name "Reactive Kernel" relates to the fact that program scheduling is message driven. Apart from the inner kernel and its service routines, the reactive kernel includes the "reactive handler", which dispatches user processes and supports a set of message passing routines callable from C programs [2]. At a future stage, we may move the task pool management primitives we are developing as handlers into the RK kernel.

Although it is possible to spawn (possibly different) programs directly into the nodes, we are using a setup with a node and host program; the node program runs on the nodes, the host program on the host processor. The latter takes care of I/O, spawns the node processes, sends the data to the nodes and receives the result.

```
partition() {
    process_msg_queue();
    if(local_err_est > local_tol)
        for(i=0;i<NS;i++) {
            get_tasks();                    /* select tasks */
            work_tasks();                   /* process tasks */
            put_tasks();                    /* add tasks to pool */
        }
    update();
    load_balance();
}
```

Figure 2: Task Partitioning Step

The algorithm of Figure 1 is run asynchronously by the participating processors of the parallel system, although the control node does not take part in region partitioning in the present version. In the sequel we will distinguish between the control (node 0) and the other (computational) nodes.

The control node is responsible for global initializations, monitoring of global updates, checking termination criteria and communicating global information with the task processors. After receiving the problem data from the host and performing the necessary initializations and memory allocations, the control node waits for messages, checking its message queue by means of the RK/CE non-blocking receive function (recv()), and processing any messages upon arrival according to the type of message received. This is implemented as a switch on the message type value, which we store as the first element in each message structure.

After the initial integration over the supplied subregion set, each computational node sends its integral, error and integrand evaluations contribution to node 0 in a message of type *initype*. All later updates are sent in a message of type *deltatype*. When node 0 receives the last *initype* message, it calculates the global requested absolute accuracy, $\max\{\epsilon_a, |result|\epsilon_r\}$ where ϵ_a and ϵ_r are pre-specified. It checks on the current absolute error estimate and integrand evaluation count and sets a termination flag accordingly, which is broadcast to the participating nodes (using xmsend()), together with the obtained absolute error requirement. Upon receiving later updates of *result*, a new tolerance message is broadcast if the computed tolerance is significantly different from the previous one posted, or a termination criterion is met. At termination, node 0 sends the *result*, estimated error, integrand evaluation count and an error flag to the host.

The partitioning step of Figure 1 can be represented by Figure 2. In the latter, references to task partitioning primitives are printed in italics. The computational nodes execute a message polling loop similar to that of the control, as illustrated by Figure 3. It ends when all messages presently in the queue are processed. Allowing the control to do task partitioning would involve including its message processing into the loop of Figure 3, which is a fairly easy change.

320

```
/*
 * check the message queue
 */
      while(msgptr = (unsigned long *) xrecv()) { /* msg is present */

        switch(*msgptr) {

          case stoptype:                          /* termination msg */
            ps = (struct msgtyp7 *) msgptr;
            pt_term = 1;
            xfree((char *) msgptr);
            break;                                /* out of switch */

          case toltype:               /* tolerance update from node 0 */
            tolflag = 1;
            pt = (struct msgtyp3 *) msgptr;
            if(pt->term)
              pt_term = pt->term;
            else
              local_tol = pt->tol/(double)icnt1;
            xfree((char *) msgptr);
            break;                                /* out of switch */

          case loadtype:                /* load update from neighbor */
            loadflag = 1;
            pload = (struct msgtyp5 *) msgptr;
            for(j=0;j<idim;j++) {            /* check which neighbor */
              nb = j;
              if(pload->nid == myneighborsvalue[j]) break;
                                                  /* out of for(j */
            }
            myneighborsload[nb] = pload->load;
            loadupdate[nb] = 1;
            xfree((char *) msgptr);
            break;                                /* out of switch */

          case worktype:                    /* work from neighbor */
            if(iw++) {                    /* next node in work chain */
              pwork->next = (struct msgtyp6 *) msgptr;
              pwork = pwork->next;
            }
            else {                     /* first node of work list */
              worklist = (struct msgtyp6 *) msgptr;
              pwork = worklist;
            }
            break;                                /* out of switch */

        }                                         /* end of switch */

      }                                     /* end of while(msgptr */
```

Figure 3: Message polling loop

All nodes have a current value of the global tolerated absolute error available. In Figure 2, if the sum of the absolute error estimates over the local pool exceeds the local tolerated error, calculated as the global tolerance divided by the number of computational processors, the computational node will do a number of subdivisions of subregions from its local pool, until either the local tolerance is achieved, or the maximum number of subdivisions or an error condition is reached. In our application to integration over a set of triangles, the actual subdivision of each triangle and the integration over its 4 similar subtriangles in work_tasks() is carried out as in the algorithm Triex [7, 6]. The task pool primitives get_tasks() and put_tasks() in Figure 2 implement removal of items from and addition to the linked list priority queue in each local pool. The incurred integral and error updates are buffered. In the update() primitive, the accumulated buffer is sent to the control node if it has become significant relative to the local tolerance. It is also sent in case of a termination condition or if the integrand evaluation count associated with the buffer is excessive. Note that only updates (differences between current and previous values) are sent to the control. The actual integrand and error values over the local subregion sets are not communicated, since subregions may move to other processors as a result of load balancing; so we do not need to keep track of where they are physically located, to account for their contributions.

The current dynamic load balancing system is a modification of that suggested in [5]. Within the load_balance() primitive, the processor calculates its load. This is the number of subregions with an error estimate that exceeds the maximum of the local tolerance and a fraction of the local estimated error. The load is compared to the load values received from the task partitioning neighbors. In the case of an overload, a number of subregions constituting the excess load are sent to the neighbors with low loads and which were not sent work since their last load update was received. We refer to [5] for details on how the overload is calculated and split up among the neighbors. A processor sends its new load value to its neighbors only if it has changed significantly since the previous value posted. On the receiving end, a processor will retrieve load values and new subregions upon checking of its message queue. The sets of regions received within the same message polling loop are linked together and the list is later merged with the local pool. Note that these operations are done asynchronously with region partitioning in other nodes.

A processor's set of task partitioning neighbors can but does not necessarily depend on its physical neighbors in the architecture. Since much of our development work was done on an iPSC/2, we have been using the processor's physical neighbors in the hypercube topology as its task partitioning neighbors (node 0 so far excluded).

Alternative definitions for the load of a task processor may use the actual sizes of error present in the pool; or the error distribution may be used to predict future loads. A model incorporating the latter is discussed in section 4.

3.   Test results

The benefits offered by concurrent processing are clear for a problem that can be split up over a number of processors, in such a way that the processors keep busy with useful work in completing the individual parts. Multivariate quadrature yields such problems in cases where the integration region can be partitioned among the processors initially and

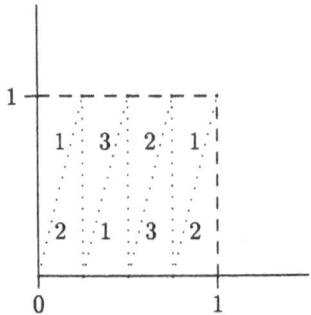

Figure 4: Sample triangles and proc. assignment for 3 comp. proc.

the integrand is behaved similarly in the constituting subregions. All processors do roughly the same amount of work, leading to an almost perfect speedup - apart from the overhead in distributing the data initially and collecting the results at the end.

It is the purpose of this section to show that global adaptive integration also lends itself to parallelization when the integrand is non-uniformly behaved. For problems where processor work loads tend to evolve very differently, the work is spread around adaptively via load balancing.

We chose the "artificial" problem for which we observed and listed the load balancing actions obtained with a specific setting of the parameters in [5]. The integrand function is $\frac{1}{\sqrt{x}}$, to be integrated over the 8 triangles of Figure 4. We arrange the assignment of the original triangles to the processors in such a way that processor 1 receives the triangle with an edge along $x = 0$. This is the bulk of the integrand problem; although processor 2 has a triangle with a point singularity at $x = 0$, the latter is far less severe. We let the routine

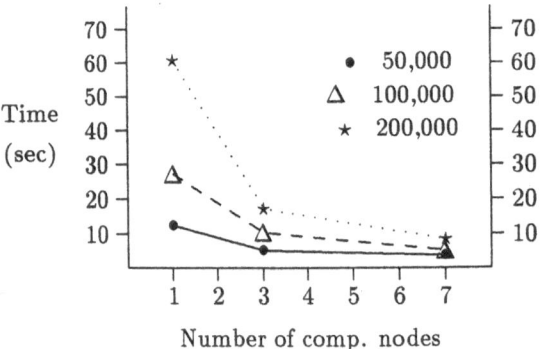

Figure 5: iPSC/2 host times

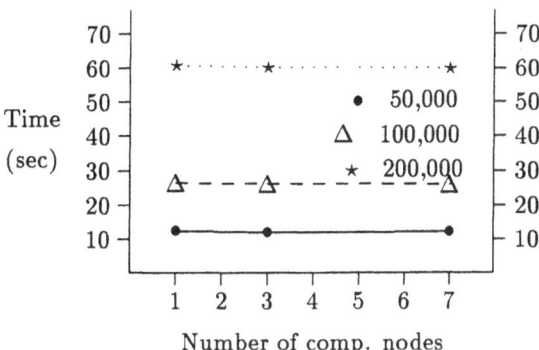

Figure 6: iPSC/2 host times without load balancing

terminate after about 50,000, 100,000 and 200,000 integrand evaluations. Figure 5 plots execution times as obtained with the current load balancing strategy, on an 8-node iPSC/2 (hypercube), using 1, 3 and 7 computational nodes. The timings are performed in the host program, starting before the problem info is broadcast to the nodes, but after the node processes are spawned, and ending after the results are received from node 0. As such we simulate the timing of integration processes which exist in the nodes. The parameter NS (the maximum number of tasks per partitioning step in Figure 2) was set to 5.

Figure 6 shows timings obtained when the problem is run without dynamic load balancing enabled. Evidently these depict a serial behavior; since the work is generated and remains in mainly one processor, it has to be carried out by this processor sequentially.

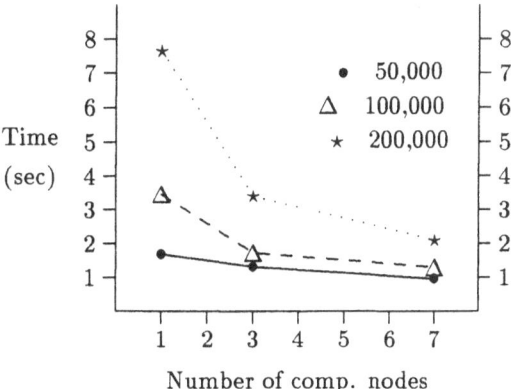

Figure 7: iPSC/860 host times

For comparison, Figure 7 gives the corresponding timings (with dynamic load balancing)

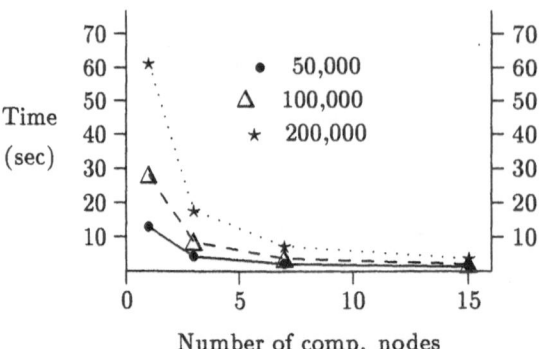

Figure 8: Symult host times

on the iPSC/860. Note that especially in view of the greater computational performance of the i860 (vs. the 386) nodes, it appears that this problem requiring even 200,000 evaluations of a non-time-consuming integrand is too "small". More benefit can be expected indeed when function evaluations are expensive and for multivariate integration in more than two dimensions, where the rules involve more points. Note also that the host times shown include the time needed to distribute the data to all the processors, which is significant.

Timing results are shown in Figure 8 for a problem with an oscillating integrand behavior, $\cos(7\pi x)\sin(20y)$, where load balancing is not needed. These were obtained on the Symult 2010 at the Caltech Concurrent Supercomputing Facility. The domain is as in Figure 4 but initially split up into 64 triangles. Similar results were obtained for the singular problem on the Symult.

4. Task Pool Modeling

In this section, a probabilistic model for the work in a priority queue is given and used to predict the work in the queue at a later time. This type of modeling can be used as a basis for an adaptive dynamic load balancing scheme, to be used as an alternative to the current simple one described above.

Consider a priority queue with real priority values scaled to lie in $[0, 1]$, low values having the highest priority. Let $r_0(p)$ be the distribution of the work in the queue; that is the work between p_1 and p_2 is

$$w(p_1, p_2) = \int_{p_1}^{p_2} r_0(p)\, dp\,. \tag{1}$$

Further, let $\alpha(p)$ be the distribution of new work generated from the old work at $p = 0$ and assume this property is shift invariant. Then the new work generated from the original work is given by

$$r_1(p) = \int_0^p r_0(q)\alpha(p - q)\, dq, \tag{2}$$

or as an operator,

$$r_1 = \Lambda(r_0) . \tag{3}$$

Recursively, the new work generated by the previous new work is,

$$r_{i+1} = \Lambda r_i = \Lambda^{(i)} r_0 . \tag{4}$$

Formally, the total derived work that will be in the queue at some time is then,

$$r = (\sum_{i=0}^{\infty} \Lambda^{(i)}) r_0 . \tag{5}$$

We studied conditions under which this converges on $[0,1)$ and will report on these in a subsequent paper.

The work that can be done on the queue in time T_0 is all that up to priority p_0, where

$$\int_0^{p_0} r(p) \, dp = c_0 T_0 \tag{6}$$

and c_0 is a constant specifying the processor's performance in tasks per second. Note that, because of the discrete nature of tasks, this last equation is not strictly true. However, it should often be a realistic enough approximation.

The overall expected importance of the work remaining to be done at time T_0 is thus

$$w(T_0) = \int_{p_0}^1 r(p) e(p) \, dp , \tag{7}$$

where $e(p)$ provides the actual importance of the work with priority p. The choice of e is quite important and can depend on global information, such as how close to termination the procedure is.

This model can be easily extended to the case where there are some tasks classified as singular (the behavior of which is not governed by $\alpha(p)$), initially $s_0(p)$. In that case there are two additional new work distribution functions, which specify how new work is generated from singular work. These are $\beta(p)$ which models new singular work and $\gamma(p)$ which models new regular work. In this case, equation (2) becomes

$$\begin{bmatrix} r_1(p) \\ s_1(p) \end{bmatrix} = \begin{bmatrix} \int_0^p r_0(q)\alpha(p-q) + s_0(q)\gamma(p-q) \, dq \\ \int_0^p s_0(q)\beta(p-q) \, dq \end{bmatrix} , \tag{8}$$

and equations similar to those of (3) to (7) can be given.

Assuming point mass structure for the new error distributions and maintaining counts, r_i and s_i , which indicate the number of regular and singular tasks in the pool, with priority p_j such that $\epsilon^i \le p_j < \epsilon^{i+1}$, allows the computations above to be approximated in a simple single low overhead loop (the integrals become sums).

The temptation to allow more than two classes exists, and there is little extra cost in doing so. However, we believe that there will usually be two dominant types in terms of the work predictions, even when multiple types of singularities exist. Thus, if the model and procedures for adapting it are somewhat robust, a single singularity class should often suffice.

Clearly, an applicable model and the procedures for adapting it, can be used to predict the load at a future time. Only predictions of the local task pool need to be calculated and compared with predictions of neighbors. In many applications, this prediction can be used to balance the queues in a manner which overcomes any communication delay. This requires selecting the future time T_0, which can be made longer for big overhead or slow communications links and shorter for low overhead fast links. It also allows harmless imbalances (which do not effect performance) to go unaltered, thus increasing performance.

There is a need for the underlying load balancing system to monitor the work load as the computation develops, relative to its predictions, and adapt its behavior, if need be.

The model above can be used when processing rates on nodes differ and inter-node communications are not constant. Thus a heterogeneous system can be balanced.

5. Concluding remarks

Extensions to load balancing based on a model for the error distribution will require more analysis and testing of models. Regarding the task pool data structure, we expect that use of the bucket list structure introduced in [8] will improve the efficiency of the operations involved in unlinking work for transfer to another node and in merging it with the receiving pool. We also plan to re-write the code in order to remove all arrays for storing subregion information, which are still present in the program, inherited from a much earlier version where the actual integration was done in Fortran routines called from the C node program. Using linked structures and dynamic memory allocation is more elegant, clearer and will help further development.

For the sake of treating singularities, the method can be extended naturally to the more sophisticated algorithms incorporating an extrapolation strategy in the adaptive process. In the case of such staged computation [11, 7, 6, 4], both the active task pool and the passive task pool can be balanced during a stage. This allows balance to be maintained after the merge of pools at the beginning of a new stage, without inter-node balancing communication at this point.

Applying the approach directly to adaptive integration in more than two dimensions is straightforward and appropriate, since more computation will be done by the processors individually. A challenge, however, comes about in guarding against roundoff error now that we are ready to allow millions of integrand evaluations. That will be an important step in our effort to produce a "portable" parallel library for multivariate numerical integration.

Acknowledgement. The authors thank Paul Messina for access to the Caltech Concurrent Supercomputing Facility machines. We also thank system manager Heidi Lorenz-Wirzba and RK manager Christopher Lee for valuable support.

References

[1] J. BOYLE, R. BUTLER, T. DISZ, B. GLICKFIELD, E. LUSK, R. OVERBEEK, J. PATTERSON, AND R. STEVENS, *Portable Programs for Parallel Processors*, Holt, Rinehart

and Winston, 1987.

[2] SYMULT SYSTEMS, *Series 2010 System C Programmer's Guide*, Symult Systems Corporation, 1989.

[3] E. DE DONCKER AND J. KAPENGA, *Parallelization of adaptive integration methods*, in Numerical Integration; Recent Developments, Software and Applications, P. Keast and G. Fairweather, eds., NATO ASI Series, Reidel, 1987, pp. 207–218.

[4] ———, *A portable parallel algorithm for multivariate numerical integration and its performance analysis*, in Proceedings of the Third SIAM Conference on Parallel Processing for Scientific Computing, Dec. 1987, Los Angeles, 1988, pp. 109–113.

[5] ———, *Parallel systems and adaptive integration*, in Proceedings of the 1989 Joint Summer Research Conference on Statistical Multiple Integration, Contemporary Mathematics, 1990, pp. 33–51.

[6] E. DE DONCKER AND I. ROBINSON, *Algorithm 612, TRIEX: Integration over a triangle using nonlinear extrapolation*, ACM Transactions on Mathematical Software, 10 (1984), pp. 17–22.

[7] E. DE DONCKER AND I. ROBINSON, *An algorithm for automatic integration over a triangle using nonlinear extrapolation*, ACM Transactions on Mathematical Software, 10 (1984), pp. 1–16.

[8] J. KAPENGA AND E. DE DONCKER, *A parallelization of adaptive task partitioning algorithms*, Parallel Computing, 7 (1988), pp. 211–225.

[9] E. LUSK AND R. OVERBEEK, *Implementation of monitors with macros: A programming aid for the Hep and other parallel processors*, tech. rep., Argonne National Laboratories, 1983. MCS ANL-83-97.

[10] ———, *Use of monitors in FORTRAN: A tutorial on the barrier, self-scheduling do-loop and askfor monitors*, tech. rep., Argonne National Laboratory, 1984. Report MCS ANL-84-51.

[11] R. PIESSENS, E. DE DONCKER, C. ÜBERHUBER, AND D. KAHANER, *QUADPACK, A Subroutine Package for Automatic Integration*, Springer Series in Computational Mathematics, Springer-Verlag, 1983.

[12] C. SEITZ, J. SEIZOVIC, AND W.-K. SU, *The C programmer's abbreviated guide to multicomputer programming*, tech. rep., California Institute of Technology, Department of Computer Science, Jan. 1988, Rev.1 - April 1989. Caltech-CS-TR-88-1.

TRANSFORMATION OF INTEGRANDS FOR LATTICE RULES

MARC BECKERS & ANN HAEGEMANS
Dept. of Computer Science
Katholieke Universiteit Leuven
Celestijnenlaan 200A
B-3001 Heverlee, Belgium
Email: marcb@cs.kuleuven.ac.be

ABSTRACT. In recent years a great amount of work has been done searching for what are called good lattice rules. Lattice rules are used for the numerical integration of smooth functions over the unit s-dimensional cube I^s. The integrand has to be periodic with period 1 with respect to each coordinate separately. So, it is important to look for good methods of periodizing functions. In the past, there were already proposed a few methods. We show that IMT-transformations are good methods of periodizing certain families of functions when lattice rules are used.

1 Introduction

We want to compute

$$I[f] = \int_{I^s} f(\vec{x})d\vec{x} \tag{1}$$

with

- I^s the unit s-dimensional cube $[0,1)^s$

- f periodic with period 1 with respect to each coordinate separately, or equivalently $f(\vec{x} + \vec{j}) = f(\vec{x})$, $\vec{j} \in \mathbb{Z}^s$, $\vec{x} \in \mathbb{R}^s$.

Definition 1.1
A 'lattice rule' for approximating $I[f]$ is a rule of the form

$$Q[f] = \frac{1}{N} \sum_{j=0}^{N-1} f(\vec{x}_j) \tag{2}$$

where $\vec{x}_0, \ldots, \vec{x}_{N-1}$ are all the points of a 'multiple integration lattice' L that lie in I^s. The number of points, N, is called the 'modulus' of the lattice rule.

We still have to define the notion of *multiple integration lattice*.

329

T. O. Espelid and A. Genz (eds.), Numerical Integration, 329–340.

Definition 1.2
A lattice is an infinite set L of points in $\mathbb{R}^s$ with the following three properties :

1. *$\vec{x}, \vec{y} \in L \Rightarrow \vec{x} \pm \vec{y} \in L$;*

2. *L contains s linearly independent points;*

3. *There exists a sphere centered at the origin that contains no points of the lattice other than the origin itself.*

Of particular interest to us are lattices that have the same periodicity property as f.

Definition 1.3
A lattice L is a multiple integration lattice if it contains $\mathbb{Z}^s$ as a sub-lattice.

Very popular lattice rules are the rectangle rule (or the product-trapezoidal rule) and the number-theoretic rules of Korobov [7].

Now we have defined lattices and lattice rules, but still we do not know which lattice rules are good approximations of the integral of a function. Therefore, we have to analyze the integration error of a lattice rule. First we define the dual of a lattice, which plays an important role in the error representation.

Definition 1.4
The dual of the lattice L is

$$L^\perp = \left\{ \vec{r} \in \mathbb{R}^s : \vec{r} \cdot \vec{x} \in \mathbb{Z}, \forall \vec{x} \in L \right\}.$$

The dual of a lattice is again a lattice. The dual of a multiple integration lattice is a subset of $\mathbb{Z}^s$. Or more general, if a lattice L contains a sub-lattice L_a, then it follows immediately from the definition of the dual lattice that $L^\perp$ is a sub-lattice of $L_a^\perp$. The dual of $L^\perp$ is clearly L.

Now assume that f can be expanded into an absolutely convergent multiple Fourier series:

$$f(\vec{x}) = \sum_{\vec{r} \in \mathbb{Z}^s} a(\vec{r}) e^{2\pi i \vec{r} \cdot \vec{x}}, \qquad \vec{x} \in I^s \tag{3}$$

with

$$a(\vec{r}) = \int_{I^s} e^{-2\pi i \vec{r} \cdot \vec{x}} f(\vec{x}) d\vec{x}. \tag{4}$$

Theorem 1.1
Let L be a multiple integration lattice with points $\vec{x}_0, \ldots, \vec{x}_{N-1}$ in I^s. Then the corresponding lattice rule Q has an error

$$Q[f] - I[f] = \sum_{\vec{r} \in L^\perp} {}' a(\vec{r}), \tag{5}$$

where the prime indicates that the $\vec{r} = \vec{0}$ term is to be omitted from the sum.

This theorem has been proved by Sloan and Kachoyan [16].

The construction criteria for obtaining what are called *good lattice rules* may be justified in terms of choosing Q so that those terms in which the function dependent part, $a(\vec{r})$, is expected to be most significant are removed from the dual lattice. Mostly one has used two construction criteria:

- the Zaremba index [21],

- the error of *the worst function* [7, 15].

If we want to use these lattice rules in practice, it is important to use more general testing methods for the lattice rules and to look for methods of periodizing functions.

2 Periodizing functions

In [5, 21] the most used methods of periodizing functions are described. There are different ways of transforming a given integrand into a periodic one while preserving regularity properties and the value of the integral, though this has to be done at the expense of further calculations. We have to take care of the following aspects:

- The transformation should not give rise to a lot of extra calculations, because the purpose is to approximate the integral with as few calculations as possible.

- The regularity of the integrand plays an important role in the theoretical error analysis of lattice rules. Thus we benefit from a transformation which preserves regularity properties.

The regularity of a function can be expressed as follows:
$\alpha \geq 1$ is an integer and all the partial derivatives

$$\frac{\partial^{\alpha_1 + \cdots + \alpha_s} f}{\partial x_1^{\alpha_1} \cdots \partial x_s^{\alpha_s}} \text{ with } 0 \leq \alpha_j \leq \alpha \text{ and } \alpha_j \in I\!\!N \text{ for } 1 \leq j \leq s$$

exist and are of bounded variation on I^s in the sense of *Hardy and Krause* [13]. If we have such a function f, we want to replace it by a function ϕ which has the following properties:

1. all the partial derivatives

$$\frac{\partial^{\alpha_1 + \cdots + \alpha_s} \phi}{\partial x_1^{\alpha_1} \cdots \partial x_s^{\alpha_s}} \text{ with } 0 \leq \alpha_j \leq \alpha \text{ and } \alpha, \alpha_j \in I\!\!N \text{ for } 1 \leq j \leq s$$

   are of bounded variation on I^s in the sense of *Hardy and Krause* ;

2.

$$\left. \frac{\partial^{\alpha_1 + \cdots + \alpha_s} \phi}{\partial x_1^{\alpha_1} \cdots \partial x_s^{\alpha_s}} \right|_{x_i=0} = \left. \frac{\partial^{\alpha_1 + \cdots + \alpha_s} \phi}{\partial x_1^{\alpha_1} \cdots \partial x_s^{\alpha_s}} \right|_{x_i=1} \text{ with } 0 \leq \alpha_j \leq \alpha - 1 \text{ for } 1 \leq j, i \leq s ;$$

3.

$$\int_{I^s} \phi(\vec{x})d\vec{x} = \int_{I^s} f(\vec{x})d\vec{x} \ .$$

The first two properties ensure that the Fourier coefficients decay with prescribed speed [21]. The third property ensures that the integral preserves its value. The most used methods of periodizing are the following:

The method of symmetric points : (the easiest transformation)
The function $f(\vec{x})$ is replaced by the average of f over 2^s points:

$$\phi(x_1, \ldots, x_s) = 2^{-s} \sum_{n_1=0}^{1} \cdots \sum_{n_s=0}^{1} f\left(n_1 + (-1)^{n_1} x_1, \ldots, n_s + (-1)^{n_s} x_s\right) \ .$$

This method has two disadvantages: the number of evaluations increases strongly with the dimension and the properties of ϕ are fulfilled only with $\alpha = 1$. Due to the nature of lattices, we can approximately halve the number of evaluations.

Changes of variables [7]: (a polynomial transformation)

$$\phi(\vec{x}) = f\left(\psi_n(x_1), \ldots, \psi_n(x_s)\right) \psi'_n(x_1) \cdots \psi'_n(x_s)$$

with $\psi_n(u) = (2n-1)C_{2n-2}^{n-1} \int_0^u (t(1-t))^{n-1}dt$ and $n \geq 2$. (6)

The properties of ϕ are fulfilled with $\alpha = n - 1$. For example one gets

$$\begin{aligned}
\psi_2(u) &= 3u^2 - 2u^3 \\
\psi_3(u) &= 10u^3 - 15u^4 + 6u^5 \\
\psi_4(u) &= 35u^4 - 84u^5 + 70u^6 - 20u^7.
\end{aligned}$$

The maxima of the derivatives $\psi_n^{(n)}$ increase with n :

- $|\psi_2''(0)| = |\psi_2''(1)| = 6$,
- $|\psi_3'''(0)| = |\psi_4'''(1)| = 60$,
- $|\psi_4''''(0)| = |\psi_4''''(1)| = 840$.

These values play an important role in the error of a lattice rule.

The method with Bernoulli polynomials [21]: From the viewpoint of the regularity condition this transformation is better than the two previous transformations, but the expression for ϕ is very complicated. $\phi(\vec{x})$ is defined as follows:

Definition 2.1
Let $\phi_0(\vec{x}) = f(\vec{x})$ and

$$\phi_j(\vec{x}) = \phi_{j-1}(\vec{x}) + \sum_{\alpha_j=0}^{\alpha} \sum_{n_j=0}^{1} (-1)^{n_j} P_{\alpha_j}(x_j) \left(\left. \frac{\partial^{\alpha_j} \phi_{j-1}(\vec{x})}{\partial x_j^{\alpha_j}} \right|_{x_j=n_j} \right) \quad \text{for } j = 1, \ldots, s$$

$$and \ with \ P_{\alpha_j}(x) = \frac{1}{(\alpha_j + 1)!} B_{\alpha_j+1}(x) \ ,$$

then is $\phi(\vec{x}) = \phi_s(\vec{x})$.

$B_\alpha(x)$ is the Bernoulli polynomial of degree α. The number of terms in ϕ is very high and one needs a lot of partial derivatives of the original function. The main problem is to program the transformation. The easiest way is to use a language with symbolic manipulation capabilities to calculate the partial derivatives, but even for not so difficult functions it is a huge task. In a classical numerical algorithm one has to use finite difference approximations to the derivatives. This means that one has to do a lot of extra calculations and one has to be very careful for the accuracy of the calculations.

3 IMT-transformations

We investigated other transformations which are successfully used in one-dimensional numerical integration. The transformations have proved to be reliable, accurate and efficient even in dealing with integrands exhibiting end point singularities. The transformed function $\phi(x)$ has the property that it vanishes together with all its derivatives at the end points of the integration interval, i.e.

$$\phi^{(\alpha)}(0) = \phi^{(\alpha)}(1) = 0 \ , \quad \alpha = 0, 1, 2, \ldots$$

So $\phi(x)$ fulfills the three required properties.

IMT-transformation : A well-known and successful transformation in one-dimensional numerical integration is the IMT-transformation. This transformation, proposed by Iri, Moriguti and Takasawa [6, 19], is based upon the idea of transforming the independent variable in such a way that all the derivatives of the new integrand vanish at both end points of the integration interval.

Definition 3.1
The transformed function of f with the IMT-transformation is

$$\phi(\vec{x}) = f(\varphi(x_1), \ldots, \varphi(x_s)) \, \varphi'(x_1) \cdots \varphi'(x_s)$$

with

1.

$$\varphi(t) = \frac{1}{\gamma} \int_0^t \psi(u) du$$

2.

$$\gamma = \int_0^1 \psi(u) du$$

3.

$$\psi(u) = \exp\left(-\frac{c}{u(1-u)}\right), \quad c > 0.$$

In one-dimensional integration this transformation is combined with the trapezoidal rule. Since lattice rules are a direct extension of the one-dimensional trapezoidal rule [15], we think that the combination of lattice rules and IMT-transformation can give good results. However, the transformation requires a lot of calculations; $\varphi(t)$ is approximated using a truncated Chebyshev series expansion [2]. Another problem is that some abscissae are very close to the boundary of the integration region, so that one has to be careful in defining the integrand there to prevent underflow or overflow. Tuning the parameter c is very important for the efficiency of the transformation.

TANH-transformation : A possible alternative which requires not so much calculations because we do not have to approximate $\varphi(t)$:

Definition 3.2
The transformed function of f with the TANH-transformation is

$$\phi(\vec{x}) = f\left(\varphi(x_1), \ldots, \varphi(x_s)\right) \varphi'(x_1) \cdots \varphi'(x_s)$$

with

$$\varphi(t) = \frac{1}{2} \tanh\left(-\frac{c}{2}\left(\frac{1}{t} - \frac{1}{1-t}\right)\right) + \frac{1}{2}, \quad c > 0.$$

Sag and Szekeres [14] proposed this transformation with $c = 1$ and observed that such a transformation immediately could be extended to higher dimensions. The NAG routine D01FDF [1] for the numerical integration of high-dimensional integrals is based on this transformation.

IMT-type double exponential transformation : This transformation suggested by Mori [11] gives faster decay of the integrand at the endpoints of the interval than the IMT-transformation.

Definition 3.3
The transformed function of f with the IMT-type double exponential transformation is

$$\phi(\vec{x}) = f\left(\varphi(x_1), \ldots, \varphi(x_s)\right) \varphi'(x_1) \cdots \varphi'(x_s)$$

with

$$\varphi(t) = \frac{1}{2} \tanh\left(a \sinh\left(b\left(\frac{1}{1-t} - \frac{1}{t}\right)\right)\right) + \frac{1}{2}, \quad a, b > 0.$$

As for the two previous transformations the tuning of the parameters a and b affects the efficiency of the transformation. In the one-dimensional case this transformation may be regarded as a formula obtained by the repeated application of the TANH-transformation [12].

The use of this transformation is encouraged by the results of Sugihara [18]. He has introduced a method, which he called *the method of good matrices*, for the numerical integration over the entire space $I\!\!R^s$. This method can be regarded as an extension of the lattice rules. Sloan and Osborn [17] proposed also a lattice method for integrating over $I\!\!R^s$. To approximate $I[f]$, with

$$I[f] = \int_{I\!\!R^s} f(\vec{x}) \, d\vec{x} \,,$$

both use an equal-weight rule instead of a lattice rule (2) :

$$Q[f] = D_L \sum_{\vec{x} \in L} f(\vec{x}) \tag{7}$$

D_L is the determinant of the lattice [16, 17].

Sugihara has used these equal-weight rules (7) with a double exponential transformation [20] to approximate integrals over I^s. He made a very interesting comparison between the lattice rules combined with a polynomial transformation (6) and the equal-weight rules combined with a double exponential transformation. He also examined the robustness against end point singularities of integrands, which is a well-known feature of double exponential formulae. He concluded that his method combined with the double exponential transformation is highly efficient and, especially, robust against the end point singularities of the integrands.

4   The testing method

To test the influence of the transformations on different functions, we use the package of Genz for testing multidimensional integration routines [3]. Genz found his inspiration in the performance profile method developed by Lyness [8] and by Lyness and Kaganove [9, 10] for one-dimensional integration routines.

One tries to classify the functions in *integrand* or *problem families*. A family is characterized by a particular attribute, or in mathematical terms, each member of a family is specified by two types of parameters:

unaffective parameters They should in principle not affect the difficulty of an integration problem. For example, a parameter which affect the position of a peak of the integrand.

affective parameters They affect really the difficulty of an integration problem. For example, a parameter which affect the height of a peak of the integrand.

Genz [3] recognized six integrand families: oscillatory functions, functions with a product peak, functions with a corner peak, Gaussian functions, C0-functions, discontinuous functions.

We test the transformations also on a battery of functions with all sorts of singularities at the boundary of the integration region (point singularities, line singularities, ...). The IMT-transformations are capable to transform such functions in *well-behaved* integrands for lattice rules.

5 Results

We have limited the tests to functions in three variables and to lattice rules of rank 1 found in the appendix of [5].

When we looked for the optimal value of n of the polynomial transformations, we have made the following observations:

- For *smooth functions* like the family of oscillatory functions and of functions with a corner peak in the package of Genz, the optimal n is 4 or 5.

- For functions with a *difficulty* in the integration region a low value of n (2 or 3) seems better (the family of functions with a product peak, of C0-functions and of discontinuous functions).

- For functions with a singularity at the boundary of the region a higher value of n (6 or 7) seems better.

- If one uses lattice rules with a rather low modulus ($N \leq 1000$) it is better to use a polynomial transformation with a lower value of n than just mentioned.

With the help of REDUCE [4], a symbolic manipulation package, we constructed the transformed integrand $\phi(\vec{x})$ of the method with Bernoulli polynomials with $\alpha = 1$ for a few *smooth functions*. For certain functions this transformation give much better results, but in many cases we could not construct the transformed integrand.

Figure 1 and figure 2 illustrate the foregoing.

Tuning the IMT-transformations gives the following results:

IMT-transformation : For all families of functions in the package of Genz the optimal value of the parameter c seems between 0.2 and 0.5, except for the family of functions with a corner peak, where a value of c around 1.5 seems better. For functions with a singularity at the boundary region one has to choose a higher value ($2 \leq c \leq 4$).

TANH-transformation : The same observation, but not so pronounced, as for the IMT-transformation is valid here. For functions with a singularity a higher value of the parameter c is better ($1 \leq c \leq 2$) than for the families in the package of Genz ($0.5 \leq c \leq 1.0$). There is again an exception for the family of functions with a corner peak: $c \approx 1.5$.

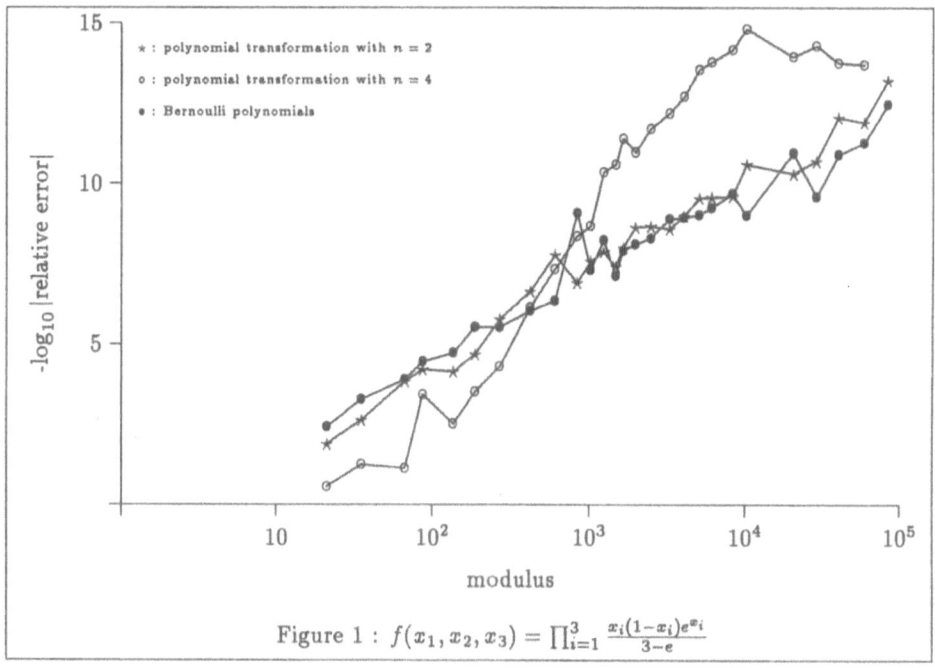

Figure 1 : $f(x_1, x_2, x_3) = \prod_{i=1}^{3} \frac{x_i(1-x_i)e^{x_i}}{3-e}$

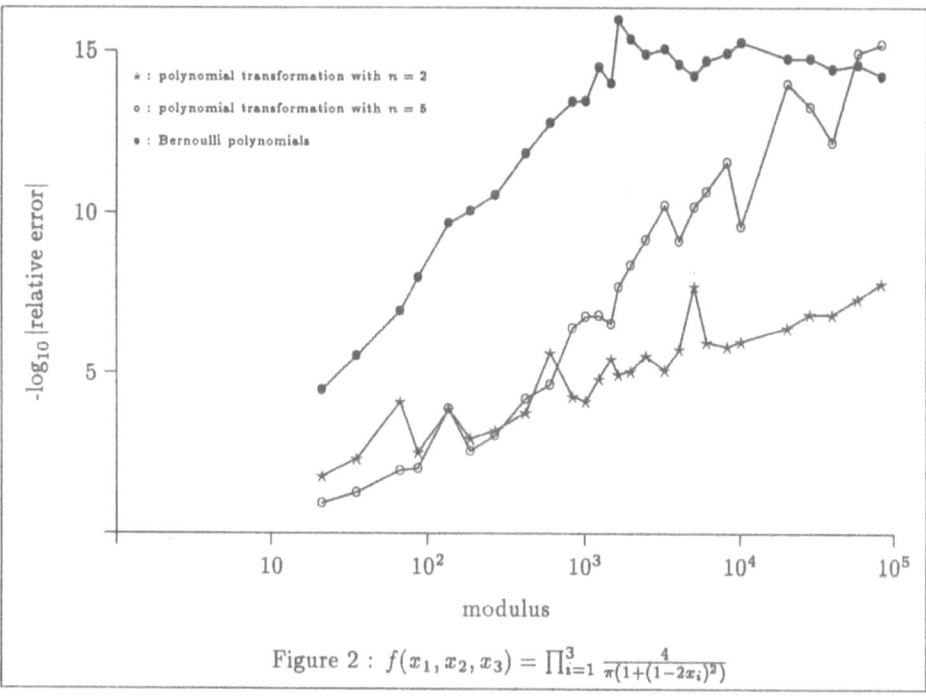

Figure 2 : $f(x_1, x_2, x_3) = \prod_{i=1}^{3} \frac{4}{\pi(1+(1-2x_i)^2)}$

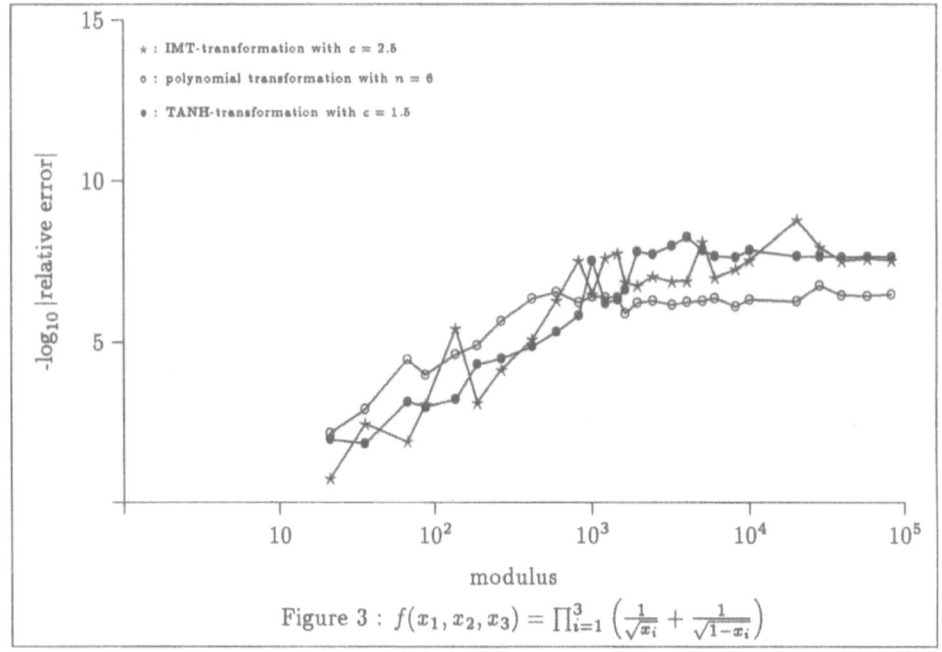

Figure 3 : $f(x_1, x_2, x_3) = \prod_{i=1}^{3} \left(\frac{1}{\sqrt{x_i}} + \frac{1}{\sqrt{1-x_i}} \right)$

IMT-type double exponential transformation : The tuning is more difficult because we have to determine two parameters. The optimal value of the parameter a seems between $\frac{\pi}{4}$ and $\frac{\pi}{2}$ and of the parameter b between $\frac{\pi}{8}$ and $\frac{\pi}{4}$. It seems better to choose a higher value of the parameter a than of the parameter b.

For each family of functions in the package of Genz, the TANH-transformation seems the right transformation of the IMT-transformations, because the results are better than or as good as the results of the other two transformations and it requires the fewest calculations of the three. For functions with a singularity at the boundary of the region it is not so obvious. In some cases the IMT-transformation is much better than the other two, but in other cases there is not a great difference between the IMT-transformation and the TANH-transformation.

Finally we compare the polynomial transformations with the IMT-transformations. For each family in the package of Genz, the results of the best polynomial transformation are slightly better than or as good as the results of the best IMT-transformation. Therefore, and because a polynomial transformation is much easier, we think that for *well-behaved functions* a polynomial transformation is preferable to one of the IMT-transformations. For some functions using a polynomial transformation and ignoring the singularity at the boundary gives comparable results to the IMT-transformation, but for other functions the IMT-transformation is much better.

Figure 3 and figure 4 give two examples of functions with a singularity at the boundary.

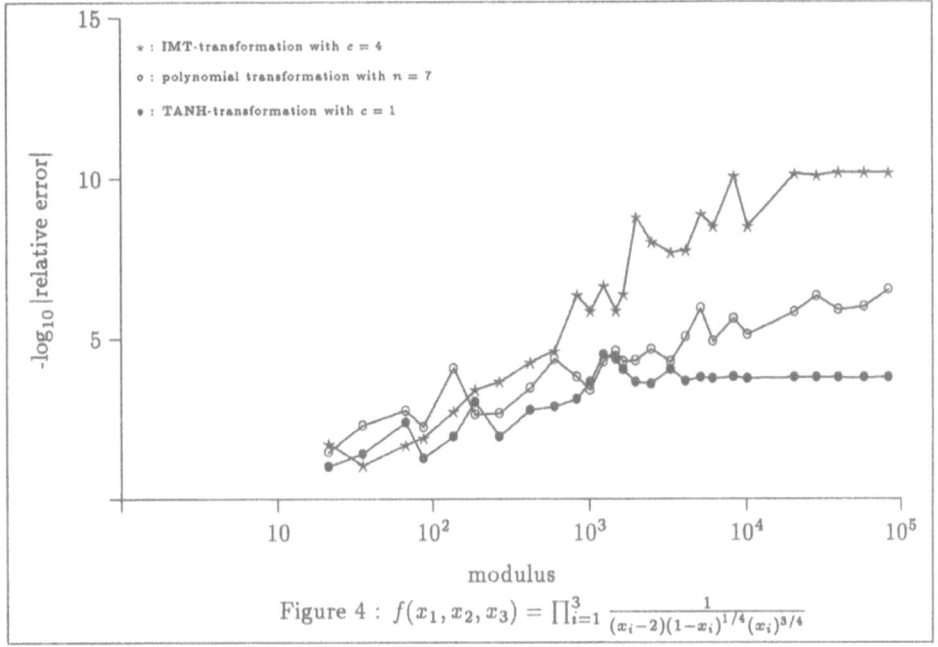

Figure 4 : $f(x_1, x_2, x_3) = \prod_{i=1}^{3} \frac{1}{(x_i - 2)(1 - x_i)^{1/4}(x_i)^{3/4}}$

6 Conclusion

When one wants to use lattice rules for the numerical integration of functions with a singularity at the boundary of I^s, the IMT-transformation to periodize the integrand can give better results than a polynomial transformation. We have only used rules of rank 1. Lattice rules of rank s correspond even better to the one-dimensional trapezoidal rule. Therefore *good* rules of rank s with the IMT-transformation may be more accurate than rules of rank 1 with this transformation.

References

[1] NAG FORTRAN LIBRARY MANUAL, MARK 13, 1989. The Numerical Algorithm Group, Oxford, United Kingdom.

[2] E. De Doncker and R. Piessens. Algorithm 32 - automatic computation of integrals with singular integrand, over a finite or an infinite range. *Computing*, 17:265–279, 1976.

[3] A.C. Genz. Testing multidimensional integration routines. In *Proceedings of INRIA Conference on Software Tools, Methods and Languages*. North Holland, Amsterdam, 1984.

[4] A.C. Hearn. *REDUCE User's Guide, Version 3.3*. The RAND Corporation, 1988.

[5] L.K. Hua and Y. Wang. *Applications of Number Theory to Numerical Analysis*. Springer–Verlag, 1981.

[6] M. Iri, S. Moriguti, and Y. Takasawa. On a certain quadrature formula. *Kokyuroku, Res. Inst. Sci. Kyoto Univ.*, 91:82–118, 1970. (in Japanese).

[7] N.M. Korobov. *Number-Theoretic Methods of Approximate Analysis*. Fitzmatgiz, 1963. (Russian).

[8] J.N. Lyness. Performance profiles and software evaluation. Technical Memorandum 343, Argonne National Laboratory Applied Mathematics Division, 1979.

[9] J.N. Lyness and J.J. Kaganove. Comments on the nature of automatic quadrature routines. *ACM Trans. Math. Software*, 2:65–81, 1976.

[10] J.N. Lyness and J.J. Kaganove. A technique for comparing automatic quadrature routines. *Computer J.*, 20:170–177, 1977.

[11] M. Mori. An IMT-type double exponential formula for numerical integration. *Publ. Res. Inst. Math. Sci. Kyoto Univ.*, 14:713–729, 1978.

[12] K. Murota and M. Iri. Parameter tuning and repeated application of the imt-type transformation in numerical quadrature. *Numer. Math.*, 38:347–363, 1982.

[13] H. Niederreiter. Quasi-Monte Carlo methods and pseudo-random numbers. *Bulletin of the American Mathematical Society*, 84:957–1041, 1978.

[14] T.W. Sag and G. Szekeres. Numerical evaluation of high-dimensional integrals. *Math. Comp.*, 18:245–253, 1964.

[15] I.H. Sloan. Lattice methods for multiple integration. *J. Comput. Appl. Math.*, 12 & 13:131–143, 1985.

[16] I.H. Sloan and P.J. Kachoyan. Lattice mathods for multiple integration: theory, error analysis and examples. *SIAM J. Numer. Anal.*, 14:117–128, 1987.

[17] I.H. Sloan and T.R. Osborn. Multiple integration over bounded and unbounded regions. *J. Comput. Appl. Math.*, 17:181–196, 1987.

[18] M. Sugihara. Method of good matrices for multi-dimensional numerical integrations – An extension of the method of good lattice points. *J. Comput. Appl. Math.*, 17:197–213, 1987.

[19] H. Takahasi and M. Mori. Error estimation in the numerical integration of analytic functions. *Rep. Comput. Centre, Univ. Tokyo*, 3:41–108, 1970.

[20] H. Takahasi and M. Mori. Double exponential formulas for numerical integration. *Publ. of Research institute for Math. Science of Kyoto University*, 9:721–741, 1974.

[21] S.K. Zaremba. La méthode des "bons treillis" pour le calcul des intégrales multiples. In S.K. Zaremba, editor, *Applications of Number Theory to Numerical Analysis*, pages 39–116. Academic Press, 1972.

DQAINT: AN ALGORITHM FOR ADAPTIVE QUADRATURE OVER A COLLECTION OF FINITE INTERVALS

TERJE O. ESPELID
Department of Informatics
University of Bergen
5020 Bergen
Norway
Email: terje@eik.ii.uib.no

ABSTRACT. In this paper we describe a new one-dimensional general purpose algorithm for adaptive quadrature. This algorithm differs from the well known, and high quality, general purpose code DQAG (QUADPACK) [6] in a number of aspects:

1. Three different basic rules are offered: Gauss, Lobatto or Gauss-Kronrod as integration rules over each local interval, while DQAG offers Gauss-Kronrod rules (with a choice of the number of evaluation points). This choice of rules is based on the experience, [1], that Gauss rules are at least as effective as Gauss-Kronrod rules in this kind of software and that closed rules (Lobatto) are better for discontinuous problems.

2. The local error estimation is based on a procedure developed in [1], and this procedure is principally the same for all three basic rules. A key element in this error estimation procedure is the *null rules*, first introduced in [5]. A rule is a null rule, $N[f]$, iff it has at least one nonzero weight and the sum of the weights are zero

$$N[f] = \sum_{i=0}^{n} u_i f(x_i), \quad \sum_{i=0}^{n} u_i = 0.$$

A null rule is furthermore said to have degree d if it integrates to zero all polynomials of degree $\leq d$ and fails to do so with $f(x) = x^{d+1}$. The basic rule $Q[f] = \sum_{i=0}^{n} w_i f(x_i)$ has degree of precision at least n while the null rules have degree of precision at most $n - 1$. In the code we use symmetric rules and a sequence of null rules of decreasing degrees $n - 1$, $n - 2$, ..., which are either symmetric or anti-symmetric. We have chosen $n = 20$ and used a sequence of 8 (this number is based on heuristics) null rules of degrees 19, 18, ... , 12. The null rules are combined into pairs to give 4 independent error estimates in order to create a reliable final local error estimate. Tests in [1] show that this error estimate is as reliable as the one implemented in QAG. The fact that the lowest degree null rule is as high as 12 (in [1] 8 *symmetric* null rules were used: lowest degree 5) makes the code more economic in regions where the function is smooth, without ruining the reliability.

3. As the subdivision strategy a non-uniform 3-division has been implemented based on ideas and experience reported in [4]. In this procedure one attempts to focus on the difficulty in order to reduce the effort in the adaptive algorithm. Tests of the new code confirm the results reported in [4] that this technique does reduce the number of function evaluations when adaptability is important.

T. O. Espelid and A. Genz (eds.), Numerical Integration, 341–342.

4. The initial integration problem may consist of a collection of intervals. This way we allow the user to give information to the code about good subdivision points, but are still treating this as one quadrature problem. Thus the user gives only one error request for the whole problem.

5. Simultaneous integration of the elements in a vector function is also allowed. This idea is not new (e. g. it has been implemented in [2, 3]) however no one-dimensional code has, to the author's knowledge, such an option. We found it necessary to include some modifications of the global strategy because of this feature. This option should be used with care since simultaneous integration of several functions, with different behavior in the integration region, may increase the overall computation time. However, in many problems there may be a substantial gain in simultaneous integration.

6. Finally, the code is designed to take advantage of shared memory parallel computers.

A FORTRAN 77 implementation of the algorithm, DQAINT, has been tested both with respect to efficiency and reliability. We noticed only small differences with respect to reliability for all codes that we tested: DQAINT (Gauss, Lobatto, Gauss-Kronrod) and DQAG (21 point Gauss-Kronrod). Lobatto turned out to be better on discontinuous problems. Economy: the number of function values are in general (all test problems) lower for DQAINT (all options) compared to DQAG. A reduction of the average number of function evaluations up to 50 % was achieved for several of the problem classes. This reduction stems both from the error estimation and the non-uniform 3-division. The overhead is greater for DQAINT than for DQAG: thus for inexpensive functions DQAG is a natural choice while for expensive functions one should choose DQAINT.

References

[1] J. Berntsen and T.O. Espelid. Error Estimation in Automatic Quadrature Routines. *ACM Trans. Math. Software*, 17(2):233–252, 1991.

[2] J. Berntsen, T.O. Espelid, and A. Genz. An Adaptive Algorithm for the Approximate Calculation of Multiple Integrals. *ACM Trans. Math. Software*, 17(4), 1991.

[3] J. Berntsen, T.O. Espelid, and A. Genz. An Adaptive Multidimensional Integration Routine for a Vector of Integrals. *ACM Trans. Math. Software*, 17(4), 1991.

[4] J. Berntsen, T.O. Espelid, and T. Sørevik. On the Subdivision Strategy in Adaptive Quadrature Algorithms. *Journal of Comp. and Appl. Math.*, 35:119–132, 1991.

[5] J.N. Lyness. Symmetric integration rules for hypercubes III. Construction of integration rules using null rules. *Math. Comp.*, 19:625–637, 1965.

[6] R. Piessens, E. de Doncker-Kapenga, C.W. Uberhuber, and D.K. Kahaner. *QUAD-PACK, A Subroutine Package for Automatic Integration*. Series in Computational Math., 1. Springer-Verlag, 1983.

A Note On Variable Knot, Variable Order Composite Quadrature For Integrands With Power Singularities

Christoph Schwab
Department of Mathematics & Statistics
University of Maryland Baltimore County
Baltimore, Maryland 21228
schwab@umbc1.umbc.edu

ABSTRACT. For integrands of the form $x^\alpha g(x), \alpha > 0$ and $g(x)$ analytic, we show that a class of composite quadrature formulas with variable grid, variable orders and positive weights yields exponential convergence in the number of integrand evaluations.

1. Introduction

Many of the adaptive numerical integrators presently available work with (a sequence of) integration formulas and suitable domain subdivisions to compensate for a locally nonsmooth behaviour of the integrand function. Although they generate typically a grid that is refined towards the singular points of the integrand, on each subdomain the same (usually high degree) formula is employed. On the other hand, it is known that in many areas of Numerical Analysis variable order, variable grid strategies perform better than their uniform order counterparts on problems with piecewise smooth solutions (e.g. Runge-Kutta-Fehlberg methods for initial value problems of ordinary differential equations and the $h - p$ finite element method).

In this Note we give error estimates (in terms of the number of integrand evaluations) for a class of variable order, variable grid composite quadrature formulas with positive weights for a one dimensional model integral where the integrand exhibits an algebraic singularity. We show that a composite quadrature formula with low degree and small subdomains near the singular point of the integrand and higher degree away from it will give exponential convergence.

This work was partially supported under grant AFOSR 89-0252.

343

T. O. Espelid and A. Genz (eds.), Numerical Integration, 343–347.

Let us first introduce some notation. We shall be interested in approximating

$$(1.1) \qquad I[f] = \int_0^1 f(x)\,dx$$

by a composite quadrature formula $Q^{\Delta,\underline{m}}$ with *grid* Δ and *order-vector* $\underline{m}$, defined as follows:

$$(1.2) \qquad \Delta = \{x_i | 0 = x_0 < x_1 < \ldots < x_{n(\Delta)} = 1\} \quad,$$
$$(1.3) \qquad \underline{m} = \{m_i \mid m_i \in N_0, i = 1, \ldots, n(\Delta)\}.$$

Then $Q^{\Delta,\underline{m}}$ is built up in the usual way from open elementary quadrature formulas (properly scaled) of exact orders m_i on the subdomains $I_i^\Delta := (x_{i-1}, x_i)$. We denote

$$(1,4) \qquad N := \sum_{i=1}^{n(\Delta)} (m_i + 1),$$

and define the space of discontinuous, piecewise polynomial functions of degrees m_i on [0,1] by

$$(1.5) \qquad S^{\underline{m}}(\Delta) = \{s(x) \mid s(x)|_{I_i^\Delta} \in \Pi_{m_i}, i = 1, ..., n(\Delta)\}.$$

2. A Model Problem and Estimate of the Quadrature Error

Consider (1.1) with
$$(2.1) \qquad f(x) = x^\alpha g(x) \quad, \quad \alpha > 0$$
where the function $g(x)$ is analytic in the disk

$$(2.2) \qquad D_\delta = \{\dot{z} \mid z - 1 \mid \le \cosh(\delta)\}$$

for some $\delta > 0$.

We are interested in the *quadrature error*

$$E^{\Delta,\underline{m}}[f] = I[f] - Q^{\Delta,\underline{m}}[f].$$

The following classical result relates the quadrature error to the best-approximation error of $f(x)$ by functions in $S^{\underline{m}}(\Delta)$ (see also [B]):

Lemma 1 *Let $f \in C^0([0,1])$. Then*

$$(2.3) \qquad |E^{\Delta,\underline{m}}[f]| \le 2 \inf_{s \in S^{\underline{m}}(\Delta)} \| f - s \|$$

where $\| \circ \|$ denotes the sup-norm on $[0,1]$.

Proof: Observe that
$$E^{\Delta,\underline{m}}[s] = 0 \qquad \forall s \in S^{\underline{m}}(\Delta).$$

The estimate (2.3) is hence a consequence of the Lebesgue Inequality

$$|E^{\Delta,\underline{m}}[f]| \leq \| E^{\Delta,\underline{m}} \| \inf_{s \in S^{\underline{m}}(\Delta)} \| f - s \|$$

and
$$\| E^{\Delta,\underline{m}} \| = 2,$$

since

$$
\begin{aligned}
|E^{\Delta,\underline{m}}[f]| &\leq \left| \int_0^1 f(x)dx \right| + |\Sigma w_i f(s_i)| \\
&\leq (1 + \Sigma |w_i|) \| f \| = 2 \| f \|,
\end{aligned}
$$

provided the weights of $Q^{\Delta,\underline{m}}$ are positive. $\qquad\square$

Now the approximation result [S, Theorem 3] and Lemma 1 allow to prove

Theorem 1 *Let $f(x) = x^\alpha g(x)$ with $\alpha > 0$ and $g(x)$ as in (2.1) be given. Then we have the estimate*
(2.4) $$|E^{\Delta,\underline{m}}[f]| \leq C e^{-2\ln(1+\sqrt{2})\sqrt{\alpha N}}$$

where N is as in (1.4), provided that
1. the grid Δ in (1.2) is geometrical with factor $\rho = (\sqrt{2} - 1)^2 \approx 0.17...$, i.e.

$$x_{i-1} = \rho x_i \qquad i = 2, \ldots, n(\Delta),$$

2. the order vector $\underline{m}$ in (1.3) is linear with slope $2\sqrt{\alpha}$, i.e.

$$m_i = 2i\sqrt{\alpha} \qquad i = 1, \ldots, n(\Delta).$$

Here C depends on α in (2.1) and δ in (2.2), but is independent of N. Based on this result we can also show exponential convergence for integrands with algebraic singularities in (0,1). We have the

Corollary 1 *Let $0 < s < 1$ and $f(x) = |x - s|^\alpha g(x), \alpha > 0$, with $g(x)$ analytic in*

$$\{z \mid |z - 1| \leq (1 - s)\cosh\delta\} \cup \{z \mid |z| \leq s\cosh\delta\}.$$

Then for the composite quadrature formula $Q^{\Delta,\underline{m}}$ we still have the estimate (2.4), provided that
1. For all $N, s \in \Delta$ (i.e. s is a grid-point),
2. The two grids $\Delta \cap [0, s]$ and $\Delta \cap [s, 1]$ are each geometrically graded towards s with the grading factor $\rho = (\sqrt{2} - 1)^2$ and the corresponding degree vectors are linearly increasing with slope $2\sqrt{\alpha}$ and increasing distance to s.

3. Remarks

1. We point out that Theorem 1 holds for any family of elementary quadrature rules on [-1, 1] used to build up $Q^{\Delta,\underline{m}}$; they only need to have positive weights.

2. Regardless of the particular elementary quadrature rules chosen, N is proportional to the number of integrand evaluations in (2.4), i.e. we have exponential convergence in terms of the number of function evaluations.

3. Of course a scaled Gauss-Jacobi quadrature (with weight $x^\alpha(1-x)^\beta$ and $\beta = 0$) on (0,1) would yield exponential convergence for $f(x) = x^\alpha g(x)$. However, if the singularity is inside [0,1], such formulas are not readily available. According to Corollary 1, however, the composite formula $Q^{\Delta,\underline{m}}$ works independently of the location of the singularity. Moreover, $Q^{\Delta,\underline{m}}$ is a rule of the type typically constructed by adaptive integrators.

4. Present adaptive integrators in one (and higher) dimensions tend to focus on the refinement of the grid Δ while employing high degree formulas uniformly on all I_i^Δ. We emphasize that for f as in (2.1) this will lead asymptotically at best to *algebraic* convergence

$$|E^{\Delta,\underline{m}}[f]| \le C N^{-\gamma},$$

with some $\gamma = \gamma(\Delta, \alpha) > 0$.

On the other hand, if $f(x)$ is analytic in [0, 1] and admits an analytic extension to a neighborhood $U \subset C$ of [0, 1], it is well known that the Gauss-Legendre formulas G_m converge exponentially as the degree $m \to \infty$. Again, keeping the degree fixed and subdividing [0,1] will not achieve this rate of convergence.

5. Although N (respectively the number of integrand evaluations) represents some sort of work measure, an adaptive integrator which allows for formulas like $Q^{\Delta,\underline{m}}$ involves also a significant portion of data management work not accounted for by N.

6. By far the most important issue in the adaptive construction of $Q^{\Delta,\underline{m}}$ is the *reliable and accurate estimation of the local quadrature error* on I_i^Δ, i.e. the development of so-called "error-indicators". They may be summed up to yield the "error estimator" for the global quadrature error and are the basis for deciding if I_i^Δ should be divided or m_i should be raised. Recently progess has been made in this direction, see e.g. [BE].

7. The estimate (2.4) and the condition $\alpha > 0$ in Theorem 1 can probably be sharpened to $e^{-2\ln(1+\sqrt{2})\sqrt{(\alpha+1)N}}$ for $\alpha > -1$ and this bound is, most likely, almost optimal. This and the question of the optimal grid selection if $\underline{m}$ is constant will be addressed, also for higher dimensional domains, in a forthcoming paper.

4. References

[BE] Berntsen, J. and Espelid, T.O. (1990) "Error Estimation in Automatic Quadrature Routines", to appear.

[B] Brass, H. (1991) "Error Bounds Based On Approximation Theory", these Proceedings.

[S] Scherer, K. (1981) "On Optimal Global Error Bounds Obtained By Scaled Local Error Estimates", Numerische Mathematik, 36, 151-176.

Computation of Oscillatory Infinite Integrals
by Extrapolation Methods

Avram Sidi
Computer Science Department
Technion - Israel Institute of Technology
Haifa 32000, Israel

In a recent work [2] a nonlinear extrapolation method, the D- transformation, was proposed, and this method proved to be very useful in accelerating the convergence of infinite integrals $\int_a^\infty f(t)dt$, $a \geq 0$, of different kinds. The D-transformation was analyzed for its convergence properties in [3] within the framework of the generalized Richardson extrapolation process. Two modifications of the D-transformation for oscillatory infinite integrals were proposed in [5], which were denoted the $\bar{D}-$ and $\tilde{D}-$transformations. Another modification, the W-transformation, useful for "very oscillatory" infinite integrals, was given in [6], and this modification was extended in [8] to divergent oscillatory infinite integrals that are defined in the sense of (Abel) summability. The W-transformation was modified significantly and made very user-friendly in [9], where a detailed convergence analysis for it is also given. For additional convergence results see [4] and [11]. The advantage of these modifications over the D-transformation is that they can achieve a given level of accuracy with considerably less computing than the D-transformation.

All the methods of extrapolation mentioned in the first paragraph are based on the assumption that the integral $\int_x^\infty f(t)dt$ has a well structured asymtotic expansion as $x \to \infty$. For example, in [2] it is shown that for a very large family of functions $f(t)$ (denoted $B^{(m)}$) that are integrable at infinity, asymptotic expansions of the form

$$\int_a^\infty f(t)dt - \int_a^x f(t)dt = \int_x^\infty f(t)dt \sim \sum_{k=0}^{m-1} x^{\rho_k} f^{(k)}(x) \sum_{i=0}^\infty \bar{\beta}_{ki} x^{-i} \quad \text{as } x \to \infty, \quad (1)$$

exist, where $\rho_k \leq k + 1$ are some integers that depend on $f(x)$, and the $\bar{\beta}_{ki}$ are coefficients that are independent of x.

Once the existence of such an asymptotic expansion has been established, the integral $\int_a^\infty f(t)dt$ can be approximated by the D-transformation of [2] as follows: Pick a sequence of points $a < x_0 < x_1 < x_2 < ...$, such that $\lim_{i \to \infty} x_i = \infty$, and compute the finite integrals $\int_a^{x_i} f(t)dt$, $i = 0, 1, 2, ...$, by employing appropriate numerical quadrature formulas. For given nonnegative integers j and $n_k, k = 0, 1, ..., m - 1$, define the approximation $D_n^{(m,j)}$

349

T. O. Espelid and A. Genz (eds.), Numerical Integration, 349–351.
© 1992 Kluwer Academic Publishers.

with $n \equiv (n_0, n_1, ..., n_{m-1})$ to be the solution to the linear system of equations

$$D_n^{(m,j)} - \int_a^{x_\ell} f(t)dt = \sum_{k=0}^{m-1} x_\ell^{\rho_k} f^{(k)}(x_\ell) \sum_{i=0}^{n_k} \beta_{ki} x_\ell^{-i}, \quad j \le \ell \le j + N, \tag{2}$$

where $N = \sum_{k=0}^{m-1}(n_k + 1)$ and β_{ki} are the remaining N unknowns. The D-transformation can be made more user-friendly by replacing ρ_k in (2) by $k+1$, $0 \le k \le m-1$, as suggested in the review paper [10], without affecting its accuracy very much. We note that the solution of (2) for $D_n^{(m,j)}$ can be achieved recursively and very efficiently by using the W-algorithm of [7] for $m = 1$, and the $W^{(m)}$-algorithm of [1] for all other values of m.

The limiting process in which the n_k are all fixed and $j \to \infty$ is called *Process I*, whereas that in which j is held fixed and the $n_k \to \infty$ simultaneously is called *Process II*. Both numerical experiments and the theoretical results of [3-6], [8], and [9] suggest that *Process II* has more powerful convergence properties than *Process I*.

Now by picking the x_i in a suitable manner we can attain a given level of accuracy in $D_n^{(m,j)}$, the approximation to $\int_a^\infty f(t)dt$, with less computational effort than required for arbitrary x_i. For instance, the $\bar{D}$-transformation of [5] achieves this for oscillatory integrals by picking the x_i to be consecutive zeros of some of $f^{(k)}(x), 0 \le k \le m-1$. This eliminates a large number of the unknowns β_{ki}, thus reducing the number of equations, and hence the number of finite integrals $\int_a^{x_i} f(t)dt$ substantially. The philosophy behind the $\bar{D}$-transformation of [5] and the W-and modified W-transformations of [6],[8], and [9] is very similar except that the functions $x^{\rho_k} f^{(k)}(x)$ in (2) are now replaced by other easily computable functions $v_k(x)$ for which

$$\int_x^\infty f(t)dt \sim \sum_{k=1}^m v_k(x) \sum_{i=0}^\infty \beta'_{ki} x^{-i} \quad \text{as } x \to \infty. \tag{3}$$

(The $v_k(x)$ are not necessarily unique.) Consequently, the W- or $W^{(m)}$-algorithms can again be used for implementing the $\bar{D}-, \tilde{D}-$ and W-transformations. Of these methods, the $\tilde{D}$- transformation, which, in fact, is a collection of methods, and the modified W-transformation, seem to be the most user-friendly and numerically effective at the same time.

References

[1] W.F. Ford and A. Sidi, An algorithm for a generalization of the Richardson extrapolation process, SIAM J. Numer. Anal., 24 (1987), pp. 1212-1232.

[2] D. Levin and A. Sidi, Two new classes of non-linear transformations for accelerating the convergence of infinite integrals and series, Appl. Math. Comp., 9 (1981), pp. 175-215.

[3] A. Sidi, Some properties of a generalization of the Richardson extrapolation process, J. Inst. Math. Applics., 24 (1979), pp. 327-346.

[4] A. Sidi, Analysis of convergence of the T-transformation for power series, Math. Comp., 35 (1980), pp. 851-874.

[5] A. Sidi, Extrapolation methods for oscillatory infinite integrals, J. Inst. Math. Applics., 26 (1980), pp. 1-20.

[6] A. Sidi, The numerical evaluation of very oscillatory infinite integrals by extrapolation, Math. Comp., 38 (1982), pp. 517-529.

[7] A. Sidi, An algorithm for a special case of a generalization of the Richardson extrapolation process, Numer. Math., 38 (1982), pp. 299-307.

[8] A. Sidi, Extrapolation methods for divergent oscillatory infinite integrals that are defined in the sense of summability, J. Comp. Appl. Math., 17 (1987), pp. 105-114.

[9] A. Sidi, A user-friendly extrapolation method for oscillatory infinite integrals, Math. Comp., 51 (1988), pp. 249-266.

[10] A. Sidi, Generalizations of Richardson extrapolation with applications to numerical integration, in Numerical Integration III, (1988) (eds. H. Brass and G. Hämmerlin), pp. 237-250.

[11] A. Sidi, On rates of acceleration of extrapolation methods for oscillatory infinite integrals, BIT, 30 (1990), pp. 347-357.

FINAL PROGRAM

Monday June 17

Morning Session - Chairman : Tor Sørevik (Norway)

0900 - 0915   Opening Remarks
                 Director of ARW, Terje O. Espelid
                 Rector of University of Bergen, Ole Didrik Lærum

0915 - 1005   Ian Sloan (Australia)
                 Numerical Integration in High Dimensions -
                 the Lattice Rule Approach

1040 - 1110   Harald Niederreiter (Austria)
                 Existence Theorems for Efficient Lattice Rules

1110 -1140   Marc Beckers (Belgium)
                 Transformation of Integrands for Lattice Rules

1140 - 1200   Discussion

Afternoon Session - Chairman : James N. Lyness (U.S.A.)

1400 - 1450   Günther Hämmerlin (Germany)
                 Developments in Solving Integral Equations Numerically

1450 -1520   Christoph Schwab (U.S.A.)
                 Numerical Integration of Singular and Hypersingular
                 Integrals in Boundary Element Methods

1550 - 1620   David B. Hunter (U.K.)
                 The Numerical Evaluation of Definite Integrals Affected
                 by Singularities Near the Interval of Integration

1620 - 1650   Frank Stenger (U.S.A.)
                 The Sinc Indefinite Integration and Initial Value Problems

Tuesday June 18

Morning Session - Chairman : Elise De Doncker (U.S.A.)

0900 - 0950   Walter Gautschi (U.S.A.)
               Remainder Estimates for Analytic Functions

1030 - 1100   Ricolindo Cariño (Australia)
               An Algebraic Study of the Levin Transformation in
               Numerical Integration

1100 - 1130   Philip J. Davis (U.S.A.)
               Gautschi Summation and the Spiral of Theodorus

1130 - 1200   Avram Sidi (Israel)
               Computation of Oscillatory Infinite Integrals by Extrapolation Methods

Afternoon Session - Chairman : Jarle Berntsen (Norway)

1400 - 1450   Patrick Keast (Canada)
               Software for Integration over Triangles and General Simplices

1450 - 1520   Ian Robinson (Australia)
               An Algorithm for Automatic Integration of
               Certain Singular Functions over a Triangle

1550 - 1620   Terje O. Espelid (Norway)
               DQAINT: An Algorithm for Adaptive Quadrature
               over a Collection of Finite Intervals

1620 - 1650   Harry V. Smith (U.K.)
               The Numerical Evaluation of Integrals of Univalent Functions

Wednesday June 19

Morning Session - Chairman : Günther Hämmerlin (Germany)

0900 - 0950   James N. Lyness (U.S.A.)
               On Handling Singularities in Finite Elements

1030 - 1100   Ken Hayami (Japan)
               A Robust Numerical Integration Method for 3-D Boundary Element
               Analysis and its Error Analysis using Complex Function Theory

1100 - 1130   Bernard Bialecki (U.S.A.)
              SINC Quadratures for Cauchy Principal Value Integrals

1130 - 1200   Discussion

Afternoon Session - Chairman : Patrick Keast (Canada)

1400 - 1450   Jarle Berntsen (Norway)
              On the Numerical Calculation of Multidimensional Integrals
              Appearing in the Theory of Underwater Acoustics

1450 - 1520   Ann Haegemans (Belgium)
              CUBPACK : Progress Report

1550 - 1640   Alan Genz (U.S.A.)
              Statistics Applications of Subregion Adaptive
              Multiple Numerical Integration

1640 - 1710   Discussion

Thursday June 20

Morning Session - Chairman : Ian Sloan (Australia)

0900 - 0950   Ronald Cools (Belgium)
              A Survey of Methods for Constructing Cubature Formulae

1030 - 1100   Hans J. Schmid (Germany)
              On the Number of Nodes of Odd Degree Cubature Formulae
              for Integrals with Jacobi Weight on a Simplex

1100 - 1130   Karin Gatermann (Germany)
              Linear Representations of Finite Groups and the Idealtheoretical
              Construction of G-Invariant Cubature Formulas

1130 - 1200   Nikolaos I. Ioakimidis (Greece)
              Application of Computer Algebra Software to The Derivation of
              Numerical Integration Rules for Singular and Hypersingular Integrals

Afternoon Session - Chairman : Alan Genz (U.S.A.)

1400 - 1450   Elise De Doncker (U.S.A.)
              Parallel Quadrature on Loosely Coupled Systems

1450 - 1520   Ignatios Vakalis (U.S.A.)
              A Parallel Global Adaptive Algorithm for Integration over
              the N-Dimensional Simplex on Shared Memory Machines

1550 - 1620   John Kapenga (U.S.A.)
              Automatic Integration using DIME

Friday June 21

Morning Session - Chairman : Helmut Brass (Germany)

0900 - 0950   Philip Rabinowitz (Israel)
              Interpolatory Product Integration in the Presence
              of Singularities: L_p Theory

1030 - 1100   Takemitsu Hasegawa (Japan)
              Application of a Modified FFT to Product Type Integration

1100 - 1130   Klaus-Jürgen Förster (Germany)
              On Quadrature Formulae near Gaussian Quadrature

1130 - 1200   P. Gonzales-Vera (Spain)
              On Estimations of Stieltjes-Markov
              Functions and Best Approximating Polynomials

Afternoon Session - Chairman : Walter Gautschi (U.S.A.)

1400 - 1450   Helmut Brass (Germany)
              Error Bounds Based on Approximation Theory

1520 - 1550   Knut Petras (Germany)
              One-Sided L_1-Approximation and Bounds for Peano Kernels

1550 - 1630   Discussion and Closing Remarks

LIST OF PARTICIPANTS

Marc Beckers
Department of Computer Science
University of Leuven
Celestijnenlaan 200 A
B-3001 Leuven, Belgium

Jarle Berntsen
Havforskningsinstituttet
Postboks 1870
5024 Bergen-Nordnes, Norway

Bernhard Bialecki
Department of Mathematics
University of Kentucky
Lexington, Kentucky 40506-0027, U.S.A.

Helmut Brass
Institut für Angewandte Mathematik
Technische Universität Braunschweig
Pockelstrasse 14
3300 Braunschweig, Germany

Ricolindo Cariño
Department of Computer Science
La Trobe University
Bundoora, Victoria 3083, Australia

Ronald Cools
Department of Computer Science
University of Leuven
Celestijnenlaan 200 A
B-3001 Leuven, Belgium

Philip J. Davis
Applied Mathematics Division
Brown University
Providence, Rhode Island 02912, U.S.A.

Mishi Derakhshan
NAG Ltd
Wilkinson House
Jordan Hill Road
Oxford OX2 8DR, U.K.

Elise De Doncker
Computer Science Department
Western Michigan University
Kalamazoo, Michigan 49008, U.S.A.

Terje O. Espelid
Department of Informatics
University of Bergen
Høyteknologisenteret
5020 Bergen, Norway

Klaus-Jürgen Förster
Institut für Mathematik
Universität Hildesheim
Marienburger Platz 22
D-3200 Hildesheim, Germany

Karin Gatermann
Konrad-Zuse-Zentrum Berlin
Heilbronner Str. 10
D-1000 Berlin 31, Germany

Walter Gautschi
Computer Science Department
Purdue University
West Lafayette, Indiana 47907-1398, U.S.A.

Alan Genz
School of EE and Computer Science
Washington State University
Pullman, Washington, 99164-2752, U.S.A.

358

P. Gonzales-Vera
Universidad de la Laguna
Facultad de Matematicas
38271 La Laguna, Tenerife
Canary Island, Spain

Ann Haegemans
University of Leuven
Applied Mathematics Division
Celestynenlaan 200
B-3001 Leuven, Belgium

Günther Hämmerlin
Mathematisches Institut
Universität München
Theresienstr. 39
D-8000 München 2, Germany

Takemitsu Hasegawa
Fukui University
Fukui 910, Japan

Ken Hayami
C & C Information Technology
Research Laboratories
NEC Corporation
4-1-1, Miyazaki, Miyamae
Kawasaki 213, Japan

David Hunter
Department of Mathematics
University of Bradford
Bradford BD7 1DP, England

Nikolaos I. Ioakimidis
P.O. Box 1120
GR-261.10 Patras, Greece

John Kapenga
Computer Science Department
Western Michigan University
Kalamazoo, Michigan 49008, U.S.A.

Patrick Keast
Department of Mathematics
Statistics and Computing Science
Dalhousie University
Halifax
Nova Scotia B3H 3J5, Canada

James N. Lyness
Mathematics and Computer Science Division
Argonne National Laboratory
9700 South Cass Avenue
Argonne, Illinois 60439-4844, U.S.A.

Harald Niederreiter
Institute for Information Processing
Austrien Academy of Sciences
Sonnenfelsgasse 19
A-1010 Wien, Austria

Sotirios Notaris
University of Missouri
Department of Mathematics
Mathematical Sciences Building
Columbia, MO 65221, U.S.A.

Knut Petras
Institut für Angewandte Matematik
Technische Universität Braunschweig
Pockelstrasse 14
3300 Braunschweig, Germany

Philip Rabinowitz
Department of Applied Mathematics
Weizmann Institute of Science
Rehovot, Israel

Ian Robinson
Department of Computer Science
La Trobe University
Bundoora, Victoria, 3083 Australia.

Hans J. Schmid
Mathematisches Institut
Universität Erlangen-Nürnberg
Bismarckstrasse 1 1/2
D-8520 Erlangen, Germany

Christoph Schwab
Department of Mathematics
University of Maryland
Baltimore County
Baltimore, Maryland 21228, U.S.A.

Avram Sidi
Technion Computer Science Department
Technion-City
Haifa 32000, Israel

Ian Sloan
School of Mathematics
University of New South Wales
Kensington, N.S.W. 2033, Australia

Harry V. Smith
22 Hodgson Avenue
Leeds LS17 8PQ, England

Frank Stenger
Department of Computer Science
University of Utah
Salt Lake City, Utah 84112, U.S.A.

Tor Sørevik
Department of Informatics
University of Bergen
Høyteknologisenteret
5020 Bergen, Norway

Christoph Überhuber
Wiener Hauptstrasse 8-10/115
A-1040 Wien, Austria

Ignatios Vakalis
Computer Science Department
Western Michigan University
Kalamazoo, Michigan 49008, U.S.A.

LIST OF CONTRIBUTORS

An asterisk (*) denotes a joint author who did not attend the Workshop.

Marc Beckers
Department of Computer Science
University of Leuven
Celestijnenlaan 200 A
B-3001 Leuven, Belgium

*H. Berens
Mathematisches Institut
Bismarckstrasse 1 1/2
D-8520 Erlangen, Germany

Jarle Berntsen
Havforskningsinstituttet
Postboks 1870
5024 Bergen-Nordnes, Norway

Bernhard Bialecki
Department of Mathematics
University of Kentucky
Lexington, Kentucky 40506-0027, U.S.A.

Helmut Brass
Institut für Angewandte Mathematik
Technische Universität Braunschweig
Pockelstrasse 14
3300 Braunschweig, Germany

Ricolindo Cariño
Department of Computer Science
La Trobe University
Bundoora, Victoria 3083, Australia

Ronald Cools
Department of Computer Science
University of Leuven
Celestijnenlaan 200 A
B-3001 Leuven, Belgium

Elise De Doncker
Computer Science Department
Western Michigan University
Kalamazoo, Michigan 49008, U.S.A.

Terje O. Espelid
Department of Informatics
University of Bergen
Høyteknologisenteret
5020 Bergen, Norway

Klaus-Jürgen Förster
Institut für Mathematik
Universität Hildesheim
Marienburger Platz 22
D-3200 Hildesheim, Germany

Karin Gatermann
Konrad-Zuse-Zentrum Berlin
Heilbronner Str. 10
D-1000 Berlin 31, Germany

Walter Gautschi
Computer Science Department
Purdue University
West Lafayette, Indiana 47907-1398, U.S.A.

Alan Genz
School of EE and Computer Science
Washington State University
Pullman, Washington, 99164-2752, U.S.A.

Ann Haegemans
University of Leuven
Applied Mathematics Division
Celestynenlaan 200
B-3001 Leuven, Belgium

Günther Hämmerlin
Mathematisches Institut
Universität München
Theresienstr. 39
D-8000 München 2, Germany

Ken Hayami
C & C Information Technology
Research Laboratories
NEC Corporation
4-1-1, Miyazaki, Miyamae
Kawasaki 213, Japan

David Hunter
Department of Mathematics
University of Bradford
Bradford BD7 1DP, England

Nikolaos I. Ioakimidis
P.O. Box 1120
GR-261.10 Patras, Greece

John Kapenga
Computer Science Department
Western Michigan University
Kalamazoo, Michigan 49008, U.S.A.

Patrick Keast
Department of Mathematics
Statistics and Computing Science
Dalhousie University
Halifax
Nova Scotia B3H 3J5, Canada

*Brian Keyes
Department of Computer Science
University of Utah
Salt Lake City, Utah 84112, U.S.A.

James N. Lyness
Mathematics and Computer Science Division
Argonne National Laboratory
9700 South Cass Avenue
Argonne, Illinois 60439-4844, U.S.A.

Harald Niederreiter
Institute for Information Processing
Austrien Academy of Sciences
Sonnenfelsgasse 19
A-1010 Wien, Austria

*Mike O'Reilly
Department of Computer Science
University of Utah
Salt Lake City, Utah 84112, U.S.A.

*Ken Parker
Department of Computer Science
University of Utah
Salt Lake City, Utah 84112, U.S.A.

Knut Petras
Institut für Angewandte Matematik
Technische Universität Braunschweig
Pockelstrasse 14
3300 Braunschweig, Germany

Philip Rabinowitz
Department of Applied Mathematics
Weizmann Institute of Science
Rehovot, Israel

Ian Robinson
Department of Computer Science
La Trobe University
Bundoora, Victoria, 3083 Australia.

Hans J. Schmid
Mathematisches Institut
Universität Erlangen-Nürnberg
Bismarckstrasse 1 1/2
D-8520 Erlangen, Germany

Christoph Schwab
Department of Mathematics
University of Maryland
Baltimore County
Baltimore, Maryland 21228, U.S.A.

Avram Sidi
Technion Computer Science Department
Technion-City
Haifa 32000, Israel

Ian Sloan
School of Mathematics
University of New South Wales
Kensington, N.S.W. 2033, Australia

*William E. Smith
Department of Applied Mathematics
Univ. of New South Wales
Kensington, NSW 2033, Australia

Frank Stenger
Department of Computer Science
University of Utah
Salt Lake City, Utah 84112, U.S.A.

*Wolfgang L. Wendland
Mathematisches Institut A
Universität Stuttgart
Pfaffenwaldring 57
D-7000 Stuttgart 80, Germany

INDEX

Applications
  artificial intelligence 130
  Bayesian analysis 271-2,275-78
  contact problems 123
  crack problems 123
  elasticity problems 123
  exponential integral 128
  multivariate normal distribution 273-275
  statistics integrals 267-279
  underwater acoustics 249
Approximation theory 147-162
  one-sided 165

Bateman approximation 187,197
Bernoulli polynomials 63,173
Boolean sum 193
Boundary element method 203,219,235,292

Cauchy's formula 134
Collocation methods 189
Consistency conditions 3-19,287
Convergence of quadrature rules 46
  mean 100
  exponential 343,346
Convergence order of solutions 194
Cytolic 284
Cubature - see integration, multiple

Data structure 305-307
Degree of approximation 166
Dynamic load balancing 317,326

Error 194,240
  asymptotics 134
  asymptotic expansions 176,211,289-293,
    296,349-350
  bounds 147-162
  constants 46
  distribution model 317,326
  estimate 133-142,235,254,289-293,
    306-314,341
    Norwegian 253-260,315,341

Epsilon algorithm 290
Euler-Maclaurin formula 221
Extrapolation 176,211,220-232,283,
  289-293,299,349-350
  cautious 290
  generalized Richardson 349
  sequence
    Bauer 292
    Bulirsh 292
    harmonic 295

Finite elements 219
Formula - see integration rule
Fourier coefficients 154
Function
  analytic 133
  associated 137
  Christoffel 45
  gamma 130
  Hilbert 13-15
  hypergeometric 130
  holomorphic 153
  homogeneous 220
  Lebesgue 107
  meromorphic 135
  periodic 329

H-bases 13-14,27-31

Ideal theory 3-21,31-35
Initial value problems 281
Integral equation
  Cauchy singular 89
  Fredholm 187
  Hammerstein 187,190
Integration
  adaptive 249-262,267-279,305-314,341,346
    double adaptivity 305,308
    global 269,306,317
    strategy 290-291
  automatic 249,283,289-293,295,305
  contour 133

Integration - continued
  indefinite 281
  infinite 350-351
  interpolatory product 93
  multiple 249-262,267-279,329
  parallel 317
  region
    hypercube 249,283
    square 25
    simplex 25,37,249,283
    triangle 25,249,283
  rule
    basic 283-285
    Clenshaw-Curtis 148
    composite 343
    embedded 289
    extremal 169
    Filippi 148
    inside 283
    interpolatory 2-15,134,161
    Gaussian 45,81,93,134,150,167,181,220
      Chebyshev 99,121-128,140-142,188
      Kronrod 18,20,341
      Legendre 15,100,121-128,239,341
      Lobatto 93,142,172,341
      Radau 93,142,167
      Radon 3,34,290
    lattice 55,71,329-338
      classification 72
      figure of merit 73
      invariants 60,72
      rank 55,72
    multiple 1-20,25,37,309-310
    nonpositive 50,289
    null 254,341
    outside 289
    polynomial 289
    positive 37,48-53,165,283
    product 55,94
    structure parameters 286-288
    triangle
    weighted 207

Integration - continued
  parallel 312
  Romberg 220
  surface 203

Invariant theory 4-19,31-35,283
  G-invariant 25-31

Jacobi weight function 37-42,94,139-142
  generalized 94
  integrals over the simplex 37

Kernel
  degenerate 187-197
  Green's 187,195
  non-smooth
  Peano 52,53,152,166
  remainder functional 135-142
  spline-blended approximation 196

Lattice 58,71,329-330
  good 55,331
  reciprocal 59
Linear constraints 288
Linear representations 25-34
Loosely coupled systems 317
L_p theory 93

Mangler-type integral 121,128
MIMD computer 292,317
Minimum point structure 287
Molien series 4
Moment
  equations 283-285
  modified 98
Monotone integrable 98
Monomial 284

Null space 287
Numerical instability region 129

Orthogonal polynomials 3-20,26-32,40,138

Oscillatory integral 250-251,349-350
  divergent 349

Parallel computer 342
Plemelj's formula 82-88,128,136
Pseudodifferential operator 203

Quadrature - see integration

Reliability 250,253,342
Remainder estimates 133-142
Residue theorem 86,135
Riemann integrable 98

Sinc 81-91,281
Singular integrals
  Cauchy 111,117,127-128
  nearly 235
  hypersingular 121-129
Singularities 111,225,283,290-293,295
  algebraic 176
  branch point 111,242
  Cauchy principal value 81-91,136
  endpoint 99,332-338
  interior 105,153
  logarithmic 176,250
Software 249-256,283
  parallel 291-292
Splines 191-193
  tensor-product 192

Subdivision strategy 253,289-293,311-313
Surface integrals 203
Symbolic computations 122-127,177
  Macsyma 177
  Maple 122
  Mathematica 121-123
  Reduce 26,129
Symmetry 25,284
Syzygie 25,28

Task partitioning 317
Testing 251-62,301,334-338
Tetrahedral group 285
Transformation 267-279,329-340
  affine 220-222
  angular 238
  D 249-250
  Duffy 208,227
  IMT 332-338
  iterated 180
  Levin 175-184,295
  log 239
  PART 235-237
  periodising 329-338
  radial 238
  substitution kernel 196
  tanh 83,333-338
  W 249-250

Variational principle 189